Global Climate Change

The Pragmatic Guide To Moving the Needle

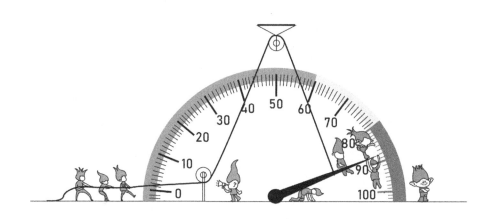

Ivo Welch and Bradford Cornell

Copyright Ivo Welch, Brad Cornell, 2022.

This book was typeset in LuaLaTeX.
Primary Font: Charter, 11pt.
Highlight Fonts: LMSans, Helvetica.
Primary Source of Cartoons: cartoonstock.com.
Primary Graphics Generator: R Language.
Front Cover Design: Midul Hasan and Ivo Welch.

This book is accompanied by a free online version at https://climate-change.world, which has hyperlinked references (merely underlined in the printed version) and fully colorized graphs.

Contents

The Physical Problem   1

1   Humanity   3
    1   Capitalism and Population 4
    2   The Population Explosion 5
    3   Regional Population Variation 8
    4   Rich and Poor: OECD and non-OECD 9
    5   Population Taboos 11
    6   What Now? 12
    7   We Are Most of The Problem 14

2   Energy   17
    1   How To Measure Power and Energy 18
    2   Where Does All Our Energy Go? 26
    3   Rich and Poor Today 28
    4   The Future 32
    5   Clean and Dirty Energy 38
    6   Important Details and Clarifications 43
    7   The Situation Today 45

3   Greenhouse Gases   49
    1   Measuring Human Emissions 50
    2   Earth's Natural Carbon Cycle 58
    3   Accumulating Human Emissions 61
    4   The Balance Sheet 65
    5   Growth in Human CO_2 Emissions 68
    6   Energy and Emissions 72
    7   The Kaya Components 77
    8   Reducing the World's Emissions 86

4		**Climate Science**	**91**
	1	Climate Versus Weather	92
	2	The Global Thermostat	98
	3	The Temperature Record	101
	4	Greenhouse Gases and Temperature	113
	5	Scientific Agreement and Disagreement	116
	6	Were the Models Wrong in the Past?	124

5		**A Warmer Future**	**131**
	1	The Expected Warming Path	133
	2	The Expected Warming Harm	138
	3	Inevitable Change and Associated Harm	145
	4	Temperature Change Summary	148
	5	The Well-Known Unknowns	150
	6	The Less-Known Unknowns	151
	7	Appropriate Perspectives	157
	8	Planetary Roulette, Anyone?	161

The Social Problem 165

6		**A Crash Course in Economics**	**167**
	1	Human Self-Interest and Free Riding	168
	2	The Tragedy of the Commons	171
	3	When Should Governments Intervene?	174
	4	Practical Problems of Pollution Taxes	183
	5	Margins, Costs, and Scale	187
	6	The Economics of Innovation	192
	7	Moving the Needle Now	195

7		**Modeling The World Economic Impact**	**197**
	1	An Economic Sketch of Earth	198
	2	What Goes In and What Comes Out?	200
	3	What is the Scientific Consensus?	205
	4	CO_2 Taxes and Consumption Losses	207
	5	What are the Best Model Parameters?	210
	6	What Else Should Be in the Model?	216
	7	What Have We Learned from IAMs?	218

8	**The Wrong Questions**	**223**
1	Problems, Choices, and Outcomes	224
2	Can We Go It Alone?	226

9	**Unrealistic Approaches**	**235**
1	Why It's So Difficult	237
2	Why A Global CO_2 Tax is Unrealistc	239
3	Why Climate Treaties are Unrealistc	243
4	Why Corporate Solutions are Unrealistic	253
5	Why Divestment Makes No Sense	255
6	Why Individual Solutions Are Doomed	257
7	Is Climate Change About Social Justice?	265

10	**Realistic Approaches**	**273**
1	Basic Requirements for Success	274
2	Enact Local Fossil-Fuel Taxes!	276
3	Promote Technological Change!	284
4	Recommendable Activism	287

The Technology Problem   291

11	**Leaving Fossil Fuels**	**293**
1	Ongoing Growth	294
2	Fossil Fuel Advantages	294
3	Hydrogen	296
4	Nuclear Power	298
5	Batteries	304
6	How To Read Technology Forecasts	305
7	The Politics of Defending Fossil Fuels	307
8	The War on Climate-Change	312

12	**Electricity**		**317**
	1	Why Electricity?	317
	2	Not All Electricity Is the Same	319
	3	Basic Electricity Provision	320
	4	Base, Intermittent, Dispatch Power	325
	5	Technologies For Generation	327
	6	Tech for Storing Electric Energy	336
	7	Transmitting Electricity	347
	8	Earth's Economic Energy Problem	350
	9	The Business Perspective	354
	10	The Role of The Market	356
	11	Current Power Plans and Forecasts	359
	12	Reliability	362
13	**Beyond Electrification**		**369**
	1	Hydrogen	370
	2	Industrial Heat	375
	3	Agriculture	377
	4	More Methane Problems	384
	5	Construction and Efficiency	389
14	**Remediation and Geoengineering**		**393**
	1	The Social Cost of CO_2 (Yet Again)	395
	2	CO_2 Removal	397
	3	Solar Radiation Management	405

The Transitioning Problem   411

15	**Making It Happen**		**413**
	1	What Can Countries Do?	415
	2	What Can Individuals Do?	436
	App. A	Some Exciting Green Tech	444
	App. B	Recommendations By Others	445

Appendix   451

Crib Sheet — Summary of Facts   453

Thanks

Many people have helped us with advice on subjects that we have struggled with. This does not necessarily mean that they endorsed our final perspectives. We want to thank especially the following:

David Archer, University of Chicago
John Cochrane, Stanford University
Andrew Dessler, Texas A&M University
Maynard Holt, Tudor, Pickering, Holt & Co.
Peter Keller, Berkeley Research Group
Steven E. Koonin, New York University
Ronald K. Linde, California Institute of Technology
Bjørn Lomborg, Kopenhagen Institute
Michael E. Mann, Pennsylvania State University
William D. Nordhaus, Yale University
Edward A. Parson, UCLA
Paul Sztorc, formerly Yale University
Nicholas Stern, London School of Economics
Daniel Swain, UCLA
Jan Veizer, University of Ottawa
Griffin M. and Julian S., UCLA
Kelly Yang, UCLA

Special Callout

It may be a bit unusual, but our editor Mary Clare McEwing did such a superb job on helping us with the first edition of this book that she deserves a special callout. Thanks, Mary Clare — we could not have written this book without you!

Authors' Preface

Civilization may well be facing the biggest crises in its ten-thousand year history: global warming, and the related problems of pollution and mass extinction of species — all related to one factor: large increases in the human population.

With a myriad of books available on climate change, why do we need another? Notably, it is because our book has a stronger emphasis on the social sciences and especially the free-riding problems. Although our book also explains the earth science, atmospheric science, and engineering challenges, it focuses a lot more on the critical political and economic obstacles confronting any serious attempts to solve these crises. And just as some technical engineering is impossible today (like nuclear fusion power), so it is with some social engineering.

Perhaps the best way to illustrate our perspective is with a simple analogy. Advocating that the world should stop burning fossil fuels and emitting greenhouse gases is like advocating that the world should stop spending on its armed forces. Although we agree that widespread disarmament would be wonderful from a collective point of view, it's simply not going to happen. This is because countries have primarily their own interests at heart, not the world's.

It is the same with fighting climate change. It is not worth it for each of the world's 195 countries to make large sacrifices that benefit the other 194 countries more than themselves.[1] Worse, the benefits and harms of climate change are unevenly distributed across countries, and expensive sacrifices over many decades will not yield noticeable benefits for many decades.

One may retort that there are politicians who have been signing treaties that are against their own self-interests. However, we argue this perception is illusory. If they — or more likely their successors — were to ever attempt to execute and enforce measures that would result in meaningful reductions in living standards for the sake of the planet, they would be quickly replaced by politicians who would instead offer higher local living standards sooner.

Many scientific and even most economic approaches to climate change have been distracted by arguments over globally optimal responses. Yet

---
[1] Contrary to popular belief and thus not contrary to our main thesis, the EPA had calculated that the 1987 Montreal Protocol did not impose costs on the United States. See Chapter 9.

whether it is globally optimal to stop arming or to stop emitting greenhouse gases seems largely irrelevant to us. We believe that most of the important countries (locked in their own geopolitical struggles) will do neither.

Please do not shoot the messenger. We would love to see good new international agreements, a global carbon tax, and fewer military expenditures. We just don't believe they is a realistic perspective. Other serious researchers are more optimistic in this respect than we are. We hope that they will be proven right, but we are not counting on it.

Climate change is also not a problem that the U.S. and Europe can solve alone. The U.S. accounts for about 12 percent of global emissions. Add Europe, and it is 20 percent. China alone already emits more than both together. India and the rest of Asia are ramping up aggressively. Even if the U.S. and Europe had an iron will — and they do not — and halved their emissions — an unachievable goal for many decades unless a technological breakthrough occurs — that would still leave the world with more than 90% of today's emissions. (See Figure 3.10 on page 69.) Indeed, world emissions would exceed today's emissions even if the U.S. and Europe were to vanish from the earth within one generation.

In total, the emissions of all OECD countries together will account for less than one quarter of world emissions by 2050. Thus, the central problem today can no longer be solved by the one billion people in rich countries alone (let alone the 0.1 billion wealthier 10% therein). Emissions are a problem that needs to be tackled even more by the seven (soon to be eight) billion people in poorer countries. These people also aspire to reach modest standards of living and who won't tolerate any delay for the sake of the planet.

As social scientists, we need to think about what environmentalism can accomplish when faced with the tough realities of the world. Environmentalists should be aware of what they can and cannot achieve. Reality denial is as unproductive as climate-change denial.

Today's Activism

In our view, much climate-change activism today seems performative — akin to a "personal wellness approach to climate change" — more feel-good than effective. The evidence is in the air, and it suggests that most climate-change activism to date has been wasted energy. Earth would probably be in roughly the same spot now if climate-change activism had never occurred in the first place.

Naïve environmentalists have been committing two sins that, sadly, don't cancel out each other — they have been thinking too big, and they have been thinking too small.

Thinking too big, too many analysts (including many economists) have focused on advocating for coordinated global action. We also would love to see this happen, but as we just explained, debating how to achieve this goal seems to us like arguing about how many angels can dance on the head of a pin. The fact is that there is no global government that could enforce global coordinated action. Treaties are no substitutes, either. In the real world, treaties only have a reasonable chance of success if they are in the signers' self-interests or at least not greatly against them. Treaties cannot succeed when they demand large sacrifices today and will show tangible results only after decades. Most of humanity is too poor to afford large payments for the world's common good, and the remainder would never be willing to pay for all of humanity in order to avoid a climate disaster in fifty years.[2] Whether you or others personally believe the world's people should or should not voluntarily take on large sacrifices doesn't matter. It simply won't happen.

Thinking too small, many environmentalists have advocated slogans such as "every little bit counts," "we have to start somewhere," "we must do our fair share," "we must set an example," or "we must reduce our personal carbon footprints." These approaches will not be able to meaningfully change the worldwide CO_2 concentration in the atmosphere, either. When solutions do not have a dynamic that will make them scale to billions of people, they will never move the needle in any meaningful way. If the goal is to truly bring down the CO_2 concentration in the atmosphere, the equivalents of New Year Resolutions simply won't make a difference. It is important to have the appropriate perspective: not only does your own carbon footprint not matter,

---

[2]For perspective, you can think of the cost as approximately the equivalent of 1 to 3 months' rent per year; we explain this yardstick later in the book.

but even the entire United States' carbon footprint is no longer what matters most. Eliminating the emissions of all 0.3 billion Americans is of much less importance now than reducing the emissions of the other 7.5 billion people.

Environmentalists also tend to have another collective shortcoming. They need to be not just *against* but also *for* some big policies, even painful ones. The requirement of zero environmental harm on every dimension cannot possibly change the world for the better. It only empowers the *status quo* when many environmentalists are against fossil fuels, nuclear power, hydroelectric dams, geothermal plants, lithium mining, solar cells, windmills, new electric transmission lines, and tree felling at the same time. What compromises is environmentalism really ready for? Activists who want to change the world should be able to articulate alternatives, be clear as to what difference the results will make, why the alternatives can be net cost-effective, and explain who will *realistically* pay for them. Environmentalist alternatives must also assure that poor and rich economies alike have reliable energy — or the affected people will not accept them. Our book limits its recommendations to alternatives that we and many experts consider to be viable and effective — even though they are also imperfect.

We are not against feel-good activism — unless misguided activism saps the energy from a smarter environmentalism that can make the world better. Rearranging the deck chairs on the Titanic is fine, but only if it does not distract from readying the rescue boats.

Viable Approaches

Despite the enormous magnitude of the problem, we remain cautiously optimistic. Yes, change will be difficult in an era of pandemics, global poverty, wealth inequality,[3] economic stagnation, high taxes and tax evasion, political, cultural and religious strife, and many oppressive and corrupt governments. These are all serious problems — but as bad as they are today, many of them are not as bad today as they were in the past. Yet the world also faces novel challenges: our much larger human population, a bad head-start on environmental degradation, and more pervasive misinformation and effective propaganda.

---

[3] Fair contributions are not just a problem across rich and poor countries, and across rich and poor within the same country. Depending on the country and cutoff, the rich in poor countries are often much richer than the poor in rich countries.

Our book argues that the only viable solutions to the world's climate problem will be based on human ingenuity. And fortunately the rapidly declining prices of cleaner technologies are now giving civilization the ability to solve many of its environmental problems. Our most important suggestion for countries and individuals is therefore to work on accelerating research and development. Engineering progress can nudge the big ship that is Earth — with all of humanity that it contains — in the right direction, even if the ship turns more slowly than activists would wish. Prodding, pushing, and nudging — smart environmentalism can move the needle now. And part of what we want to happen is action now, rather than debating what we should do in decades hence. We wrote this book partly to explain how.

Because we think it is irrelevant, we do not need to get involved in the debate about exactly how bad the world's situation is. We can take it as given that fossil fuel pollution today is already bad enough to warrant appropriate reductions. From a social perspective, humanity today is burning way too much fossil fuel (resulting in millions of deaths every year), and it will almost surely burn more fossil fuels *very soon*. We can thus shift away from some of the most heated arguments among politicians and climate scientists over how aggressively and quickly humanity must wean itself off fossil fuels to stem global warming — 10 years, 30 years, or 100 years, with carbon taxes of $50 or $500 per ton of CO_2. Although we would like to see such CO_2 taxes, too, we think that this debate is as divisive as it is irrelevant in practice. We don't think high CO_2 taxes are going to be enacted in the countries housing 7-8 billion people today.

Instead, we think everyone should be focusing on how to induce decision-makers — including in poorer countries — to move more aggressively towards cleaner solutions. What actions can countries, organizations, and individuals take to *move the needle* now? Fortunately, there are many actions that are cheap and locally advantageous enough to be already worthwhile, many that are possible to accelerate at modest cost, and many more that will become feasible soon. We will cover many promising approaches in Chapter 15.[4]

Of course, many readers will think that we are going too far, and many others will think that we are not going far enough. We expect only a small minority of readers to like everything we write. At some points, we will

---

[4] In particular, if the energy storage cost problem (explained in Chapter 12) can be solved, it will be lights out for most uses of fossil fuels.

offend many climate-change activists; at other points, many climate-change skeptics. We consider no cow sacred. However, we hope that all our readers will appreciate the honesty of our presentation and analysis, whether they will end up agreeing with us or not.

The Educational Component

Our book has a strong educational component. It is designed to be suitable not only as a general interest book for the concerned citizens of the world but also as a course textbook. It is written for interested students at any level — from our teenage children to our political leaders (though sometimes we are not sure which is which). Our readers should come away with a clear understanding of the problems and tradeoffs associated with energy provision, emissions, and climate change. We have tried to keep the book brief and to the point, self-contained, and easy to understand. (The key facts on which our book relies are summarized in our six-page Appendix, itself enough a reason to buy this book.) Our book's target audience is anyone who enjoys reading a newspaper like the *New York Times* or the *Wall Street Journal*.

One problem in learning about energy, emissions, climate change, economics, and technology is not a lack of information. Instead, there is too much information (often biased and false) all over the Internet. You can probably find everything that we discuss here somewhere else. An important goal of our book is vetting and distilling the most important facts.

Our book's first task is thus organizing information into a form that does not miss the forest for the trees. Although many concerned people have read about global warming, most do not fully understand it. (Admittedly, though interested, we did not, either, until we researched more to write this book.) Most people have glimpsed only parts of the elephant. And this elephant is big. It includes earth-science aspects (energy, emissions, planetary changes); social-science aspects (public goods, cost economics, social costs of carbon, coordination); engineering aspects (the viability of potential technologies, electricity, storage, agriculture, geoengineering); and an understanding of feasible large-scale potential solutions. For example, batteries work well for cell phones — but could they really power the daily needs of eight billion people? Where are we right now on the "clean-energy revolution"? And so on.

We want to present information in a fair and unbiased way. Both climate activists and climate skeptics sometimes share a penchant to suppress <u>inconvenient truths</u>. Many environmentalists do not like contrary arguments, because they are afraid that their airing could reduce the alarm and poison the determination of the faithful. Many climate-change skeptics are in a worse predicament, because they have to disavow much of the scientific evidence.[5] Moreover, the vitriol and accusations of ignorance and bad faith on both sides when there are legitimate uncertainties, unresolved issues, and dilemmas is, at times, stunning. Hired trolls further fan the flames.

Our goal is to air all reasonable and important arguments — even if we disagree with them — and to do so with appropriate respect. (And for non-reasonable arguments, this can mean zero respect.) We do not want to discourage our readers from forming their own points of view and disagreeing with us. However, we do want our readers to understand why answers may not be as obvious as they might have thought. In many cases — though not all — honest, smart people can and do come to different conclusions.

Our book does not cover one important area of climate change — adaptation. This is not because this area is unimportant (on the contrary, it is *very* important!), but because adaptation is usually in the self-interest of the affected parties. The focus of our book is on how global greenhouse gas emissions could be reduced.

About the Authors

We are not earth scientists; we are not engineers; we are social scientists. We are not even among those economists who have dedicated their lives to studying energy provision and climate change. We think that our background offers both advantages and disadvantages. On the positive side, being economists allows us to see more forest than trees and not get lost in too many specific details. As first-time authors on these subjects (though not others in our area of economics), we also have no horse in any particular race. We can present what we believe to be fair and objective perspectives on the evidence. On the negative side, our detail knowledge is not as deep as that of our fellow scientists in their specific fields. We are not producers, but rather intelligent consumers of the research we present. We have fact-checked

---

[5]The journalist Katie Worth has written an interesting account of how climate change has become <u>a divisive and partisan issue</u> full of distortions and misinformation.

the information we are presenting, but we always welcome corrections and clarifications.

We are not claiming originality for the many ideas in this book. Our impression is that most good ideas have occurred many times to many people, often independently. Thus, our book should not be viewed as an original research treatise. Its originality is in the exposition of the information in our own particular way.

◊

There is one bias that we do have. We are not simply uninvolved scientists objectively analyzing an interesting intellectual puzzle. We view ourselves as cool-headed environmentalists, advocating choices that are technically viable, economically affordable, and politically feasible. We want to tackle environmental problems sooner rather than later. We want to *move the needle* now in the most sensible ways possible.

> *Grant me the serenity to accept the things I cannot change, the courage to change the things I can, and the wisdom to know the difference.*
> — Serenity Prayer, Reinhold Niebuhr

Conflict of Interest Disclosure: Neither of us has ever been supported by grants or otherwise from lobbies, either environmentalist or fossil-fuel. We are not shills for anyone.

PS: The book contains cartoons (mostly from cartoonstock.com) and jokes. Even if the subject matter is deadly serious, jokes are exactly that — not to be taken seriously much less indicative of our own views.

---
quote

> Ridicule is man's most potent weapon. It is almost impossible to counteract ridicule.
> — Saul Alinsky, 1971.

Part

The Physical Problem

Chapter 1

Humanity

There are only four ways to reduce the emissions of greenhouse gases:

1. We can reduce the number of people on the planet.
2. We can reduce how much energy each of us consumes by reducing our economic activities — for example, by working, producing, and consuming less.
3. We can improve the energy efficiency of our economic activities, reducing how much energy each of our activities consumes — for example, by insulating our buildings better.
4. We can improve the emissions efficiency of our energy use — for example, by switching from coal to solar cells.

That's it. The list is exhaustive. There are no other alternatives. All more specific proposals have to work through one or more of these four categories.[1]

We will start our book with the first of these four levers — population.

---

[1] This decomposition is essentially due to the Japanese economist Yoichi Kaya. We will return to each of the components repeatedly in the first part of our book.

1 Capitalism and Population

Many environmentalists believe that the emissions problem must be solved via #2, a reduction of our economic activities. They believe that capitalism is unsustainable and greed lies at the core of our planet's global climate problem. It is the desire of the wealthy to consume more that is responsible for endangering our planet. And it is why fossil fuels kill millions every year. (The latter part of the statement is correct. Fossil fuels indeed kill millions of us with their small particle emissions.)

Many of their proposed solutions want to tamper with civilization, society, and modern industry for the sake of a greater good. Unfortunately, they misunderstand the problem (and capitalism). This is because the most important cause of the problem is not capitalism.

Instead, the most important cause has been and will continue to be *us*. That is, the world's problem is mostly #1, not #2 from the list above — you and us and 8 billion other human beings. And the large human population is mostly not a rich-country problem. The richer countries of the OECD house only a small minority of us. It is the poorer countries outside the OECD that house the majority of us.

To the extent that environmentalists are correct that increasing per-capita consumption causes more environmental degradation, the problem has been and will continue to be primarily about poorer people. They desire reasonable living standards. They want to be able to commute to jobs that allow them to send their kids to school, have basic lighting and sanitation, and boil water when they want to.

On a planetary scale, human emissions are no longer primarily a problem of providing luxury goods for the wealthy few in rich countries, housing 1–2 billion people now. They are primarily a problem of delivering healthier living standards to the 6–7 billion people in poor countries.

Our list of possible levers also included two efficiency improvements, specifically energy efficiency (#3) and emission efficiency (#4). These items are already seeing a lot of progress. The great hope of humanity is the second of these — clean energy — and we will cover clean technologies in great detail later in the book.

But we are getting ahead of ourselves. Let's start with the most important factor, population. It's a big topic.

2 The Population Explosion

Figure 1.1. Population Growth

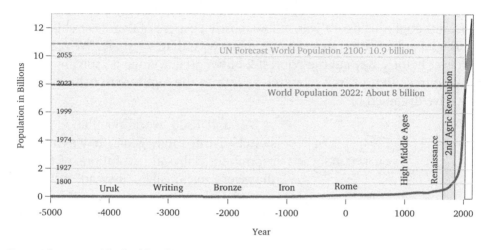

Note: The part with the blue line over gray background is history. The part with the red line over white background is the forecast to 2100. A reasonable forecast for 2100 is about 10–11 billion people. The graph is deliberately not drawn on a logarithmic scale. This is because the planet has a fixed scale.

Source: United Nations and worldometers.info.

Let us start by looking at the history of our planet's population. Figure 1.1 shows that even as recently as 5,000 BC — almost 200,000 years after the rise of _homo sapiens_ — the human population was still tiny, approximately 5 million people. Around the time of the Han, Mayan, Parthian, and Roman empires, the planet hosted about 150–200 million people. By the end of the Middle Ages, it stood around 400 million. Humanity's impact on the global environment was still largely negligible — though humanity was not as innocuous as often imagined. We probably did manage to wipe out some species — not only the large edible kind, like Woolly Mammoths and Ground Sloths, but also other human species like Neanderthals and Denisovans; and humans converted forests and peatlands into fields and pastures.

The slow growth of humanity — often interrupted by wars and diseases — continued largely unchanged until around the time of the U.S. and French revolutions. At that point, Earth hosted about 1 billion people.

1. Humanity

You should not feel nostalgic about these bygone days. It was not a pastoral landscape of peace and harmony. The vast majority of humans lived short lives at or below subsistence level. Hunger was prevalent. Pandemics were common. Technology was rudimentary. Transportation was cumbersome. Communication took days. Heating was minimal. Cooling was (mostly) unavailable. As late as 1900, global average life expectancy was still only 31 years, although this figure was so low mostly because of deaths at child-birth. Today, global life expectancy is more than twice that, 72 years. We are richer than ever before. Beyond the poorest 2 billion, the average person alive today has resources that even a monarch as late as the eighteenth century could only have dreamed of.

Around 1800, world population growth switched into high gear. It was not the prominent political revolutions of the era that were responsible. Instead, it was the second agricultural revolution (mostly crop rotation but also newer tools and breeds), the hygiene revolution, the industrial revolution, and the medical revolution (probably in that order). Human population accelerated. But the human impact was still not a meaningful environmental problem for at least another century. By 1900, the world population was still "only" 1.6 billion — easily supportable by our planet and its natural resources. Population was still by-and-large limited by food availability.

Then population growth went into overdrive. Especially in developing countries, mortality declined with the advent of modern hygiene, medicine, and agricultural fertilizers, but birth rates did not. By 1960, the world's population was 3 billion. Less than 15 years later, it had reached 4 billion. **Today, the world has a population of about 8 billion people.**

Human population growth has simply been mind-blowing. From 1900 to today, humanity has added over 6 billion people. The United Nations now expects population to level off at around 11 billion by the end of the century — still 40% more than we have today. (However, some other forecasts are as low as 10 billion.)

Unlike us, with our exponential growth, the planet does not grow. Human ingenuity has nevertheless made it possible to expand many resource

2. The Population Explosion

constraints. Yet we billions of people seem just about ready to push against other fixed planetary constraints sometime soon. As recently as 50 years ago, many scientists thought that humanity would soon run out of food or fossil fuels. This will almost surely not happen, although it is true that humans are ravaging our ecosystems and extinguishing species at a record pace (especially the oceans, which are a free-for-all, not owned or protected by anyone). Instead, it now looks as if we could run out of clean air, clean oceans, and healthy ecosystems first. Almost no one worried about these specific global environmental constraints just 50 years ago.

anecdote

> In a satirical commercial by The Onion, people can lower their carbon footprints by getting into the "Toyota Prius Solution," which then drives a stake through their hearts. The narrator exclaims "When you're dead, you can't pollute!"

Burgeoning population growth is also a principal cause of our climate problem. If our leaders had managed to curtail population growth in 1960 (at 3 billion), then stopping climate change would be much easier. If they could curtail it now, it would be as important a contribution to humanity's future as newer and cleaner technologies.

Indeed, it is not even the case that global emissions (especially in wealthier OECD countries) have grown a lot worse over the last half century *when measured on a per-capita basis*. (There are exceptions.) Instead, it is primarily the *global capita* that have increased — indeed, quadrupled. When calculating planet-wide energy consumption and emissions, we now have to multiply by 8 billion instead of 2 billion — and soon by 10-11 billion.

Note that we are not arguing that the planet cannot sustain 10 billion people. Indeed, most scientists believe that it can. But contemplate this: most population growth and increasingly negative environmental impacts occur in poor countries and not in rich industrialized countries. It is in Africa that humanity is experiencing the most suffering and not in Western Europe. This brings us to the next topic — regional variation.

3 Regional Population Variation

To understand global population growth, we need to look at where it has occurred and where it will continue. Figure 1.2 shows where humanity has settled. Contrary to casual Western impressions, most people now live on the Indian subcontinent, in South-East Asia, and in Sub-Saharan Africa.

Figure 1.2. World Population

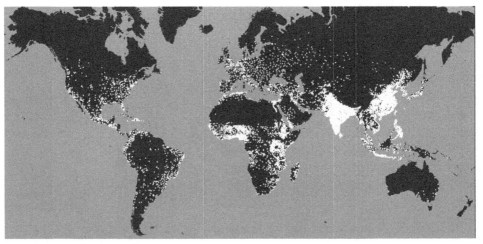

Source: World Population History, which also offers an even more interesting map animated over time. Each dot is 1 million people.

China, India, and Africa housed about 1.5 billion people each in 2022. Other Asian countries accounted for an additional 1.1 billion people.

The U.S. and Europe together accounted for only about 1 billion. Our Western perception of the world is too ethnocentric. We are less important than we like to think — both in our ability to cause problems and in our ability to fix them.

4 Rich and Poor: OECD and non-OECD

A useful categorization of countries is whether they are part of the *Organisation for Economic Co-operation and Development* (OECD) or not. The former are representative of the richest and economically most developed countries. This is not to say that there are rich countries that are not in the OECD (e.g., Quatar) and medium-income countries that are in the OECD (e.g., Mexico). However, the two categories are common enough to make them good standins for broader categories and their statistics are particularly easy to come by.

In 2022, according to the *U.S. Energy Information Administration* (EIA),

	Population (billion)	GDP (PPP) (US-$)	Energy (PWh)	Emissions (GtCO$_2$)
OECD	1.4	60	71	12
not OECD	6.5	75	116	24
% OECD	18%	44%	38%	33%

(We will cover the non-population aspects of this table in the next chapters.)

The important point here is that the OECD accounted for less than 20% of the population on the planet, but produced and consumed about 2–3 times its "fair" share. If the point of our book was to assign blame, it would be time to start finger-pointing. But it is not.

The point of our book is to consider viable solutions to reduce total global human greenhouse gas emission, as listed in the last column of the table. And even a casual glance at these numbers makes it clear that the OECD can no longer go it alone. OECD emissions are no longer the majority. In all respects, the OECD is already a minority player.

But it's actually worse than this: the non-OECD countries are still growing a lot faster in all columns compared to the OECD countries. Within one generation, the EIA estimates that the same table will look like this:

	Population (billion)	GDP (PPP) (US-$)	Energy (PWh)	Emissions (GtCO$_2$)
OECD	1.5	92	82	12
not OECD	8.2	191	177	31
% OECD	15%	33%	32%	28%

1. HUMANITY

Comparing the two tables, you can see that population and emission growth have essentially stopped in the OECD. (Population growth is being propped up only by immigration.) But population and emission growths have not stopped in non-OECD countries.

The numbers are stark. Even if the OECD unilaterally decided to extinguish itself, world emissions would still remain too high.

Figure 1.3. Population Growth

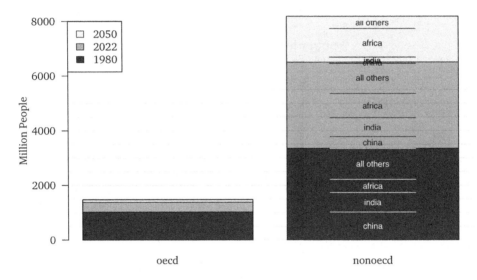

Source: *US Energy Information Association (EIA)*.

Where exactly will all the population growth occur? Figure 1.3 shows three snapshots of population.

In 1980, China and India were about one quarter each of the non-OECD population. The non-OECD population roughly doubled by 2022. China continued to increase its population but its share of population *growth* declined. India's population growth stayed the same at a very high level, and Africa's population growth increased.

Within one more generation, by 2050, the world will host another 1.8 billion of us. China's population will be shrinking. India's population will increase only modestly. Most population growth will come from Africa, a full

1.0 billion of the 1.8 billion (and even more in the generation beyond 2050). Africa's population growth is simply staggering. Nigeria alone had about 50 million people in 1965. It will have over 700 million people in 2050, almost as much as the USA and Western Europe combined! And about a third of the remaining 0.8 billion in population growth will occur in Pakistan, Indonesia, and Bangladesh.

In general, the poorer countries and regions are, the faster their populations grow.[2] It is here that the toughest environmental challenges will lie.

In fact, economists are often wondering whether accelerating the economic growth of the poorest countries would bring down their population growth rates quickly enough to reduce the environmental and other costs, relative to allowing them to remain poor for longer. (Unfortunately, regardless of what conclusion they might come up with, it is not easy to increase the wealth of poor nations. Despite decades of trying, the United Nation's foreign aid donation box has never managed to bring even one country out of poverty.)

5 Population Taboos

Population growth remains a vexing problem. Even if great steps were taken today, it would take many decades for it to stop. Population growth has self-momentum. In rapidly growing countries, the population is young. Their children today will often want to raise their own families within just another 20 years or so. This makes addressing the problem sooner rather than later even more urgent.

Yet even talking about how to slow population growth remains largely taboo. It raises difficult issues related to religion, imperialism, racism, ethnocentrism, culture, and parochialism. But discomfort does not make facts vanish: Humanity is still expanding and our planet is not. Earth remains in fixed supply. Africa and other poor regions are "overtaxed."

The most important shining beacon when it comes to curbing population growth have been improvements in womens' rights, especially greater access

---

[2]Ironically, richer countries will be struggling with the effects of shrinking population — and they generally do not like the partial solution of allowing more immigration from poorer countries.

to professional careers and birth control. (Other lights have been increasing education and secularization.) These factors have led to declining birth rates in many rich countries, which in turn have changed the character of families and societies.[3] Children have become fewer, and adults and seniors have been becoming the majority.

When countries undergo this change, it is called the "fertility transition." A glaring empirical fact is that — except for a few unusually oil-rich Arab states — only countries that have undergone the fertility transition enjoy high-quality lifestyles today, and vice-versa.

If we want civilization and our biosphere to remain sustainable, the world must get its population under control — the sooner the better. Countries that have not yet done so must undergo the fertility transition. To do so, they will have to become richer — and perhaps more "modern." But such economic development requires more resources and energy consumption per capita upfront, which in turn harms the planet sooner. It is a painful dilemma.

6 What Now?

In sum, for those environmentalists who long for a return to a time when humans lived in harmony with nature (those days *never* really existed) and consumed fewer resources and less energy (those days did exist), this time has passed. There is no going back.

Our civilization could simply not sustain eight billion people even at subsistence level without modern technology and energy — both agricultural and industrial. Short of unimaginably large human catastrophes, such as major nuclear wars or pandemics, our economies will have to find a way to support us all. We eight billion are already here. Another two billion will join us within one generation. And most of us live in poor non-OECD countries and deserve better.

Because we know of no ways to drastically reduce the human population, the best and most humane hope for our civilizations is now sensible and

---

[3] It is also why dire predictions of populations always expanding to the point of widespread poverty will likely not hold in the very long term. Population-growth induced poverty was imagined by the economist Thomas Robert Malthus and is the reason why economics is often called "The Dismal Science."

pragmatic environmentalism with painful compromises. That is what our book is about.

Before we move on from population to energy, you (our reader) should also ponder a novel political dimension here. The Internet now beams the West's rich lifestyles instantly to all parts of the globe. The global political situation is already a powder keg. It will only add more powder if poorer people do not see their standards of living increase — and *soon*. This holds even more strongly for those countries and people who will be suffering harmful consequences of a planet that will be warmed perhaps not primarily but surely overproportionately by richer countries' energy consumption. Many young poor individuals are likely to become radicalized.

A Terrible Tragedy: A mass suicide of extinction rebellion protesters after realizing they were all breathing out CO_2.

anecdote

> Some prominent environmentalists are on record having made some rather "interesting" population proposals. For example:
>
> - Ted Turner, billionaire, founder of CNN and major UN donor: "A total population of 250-300 million people, a 95% decline from present levels, would be ideal."
> - David Foreman, co-founder of Earth First!: "My three main goals would be to reduce human population to about 100 million worldwide, destroy the industrial infrastructure and see wilderness, with its full complement of species, returning throughout the world."
> - David Brower, a cofounder of the Sierra Club: "Childbearing should be a punishable crime against society, unless the parents hold a government license. All potential parents should be required to use contraceptive chemicals, the government issuing antidotes to citizens chosen for childbearing."

Yet, what can rich countries do? If poorer countries cannot reach the fertility transition (and they generally also resent interference by other countries), their standards of living cannot possibly increase quickly enough, even with all the development aid in the world. As already noted, the United Nation's donation box has also never helped countries escape poverty. And the last time that the Europeans intervened in Africa (the source of most future population growth), they killed millions. Belgium alone may have murdered 10 million Congolese. The locals will probably never again trust their good faith. Neither would we.

And who could blame poor individuals who want to join their wealthier neighbors? All they want is what we in richer countries assume to be our birth right.

7 We Are Most of The Problem

It is not an exaggeration to state that the number of humans on the planet is currently the most important underlying cause not just of global warming but of most environmental problems. These problems go way beyond the scope of our book. (And we also do not know how one could influence population growth.) Thus we limit the focus of our book "merely" to reductions in the emissions per capita for the sake of limiting climate change. It's more than big enough a problem!

A final word: it is all too easy for activists to point to, say, one million people for whom climate change will cause terrible suffering. Greenpeace has wonderful videos highlighting the fates that already are or will soon befall them — and Greenpeace is not lying, either. The fates of these people are not fair and you should not be callous about them. They don't deserve what they are getting.

But keep in mind that a million people are only about 0.1% of the human population. Almost 2,000 million people already live in abject poverty today, causing many to suffer and die from less spectacular causes. Economic growth is the only way out of this poverty. For the sake of humanity, even if the world could actively decide on a best policy (and we will argue that it really cannot), it should not base its policies only on the consequences for the 1 million and neglect the 2,000 million. Many things cannot be changed and for others, there are tough tradeoffs to be made. These are what our book is all about.

Further Readings

<u>Clarifications:</u>

Every chapter ends with further readings. Sources are listed in alphabetic order. When the title is not self-explanatory, we briefly try to explain the relevance. We do not necessarily share or endorse the views of further readings. It is sufficient if we find these readings interesting and relevant and not particularly badly biased, too polemic, or too political.

We will commonly refer to stories in certain news outlets in our book. <u>The Guardian</u> and the <u>Washington Post</u> are reliable center-left news outlets. The <u>Wall Street Journal</u>, <u>The Economist</u>, and <u>Bloomberg Businessweek</u> are reliable center-right news outlets. Our assessment of their centrism does not necessarily apply to their oped pages, which are run by completely different teams. Other outlets have little detectable bias in their climate coverage, such as <u>Ars Technica</u> (or its sister site <u>Clean Technica</u>) or the <u>MIT Technology Review</u>.

Books

- <u>Ehrlich, Paul</u>, 1968, <u>The Population Bomb</u>. Outdated assessment of <u>food shortage</u> fears. See also <u>Malthus</u>.
- Kaya, Yoichi; Yokoburi, Keiichi (1997). Environment, energy, and economy : strategies for sustainability. Tokyo [u.a.]: United Nations Univ. Press. ISBN 9280809113. This motivated the four factors described in the introduction.

Academic Articles

- <u>Bradshaw, Corey, J.A.</u>, et al., 2020, <u>Underestimating the Challenges of Avoiding a Ghastly Future</u>, Frontiers in Conservation Science. This paper estimates that in 1960 humans took about 0.75 of what the planet could replenish. By 2016, it was 1.7, mostly due to our ballooning population.
- <u>Marchetti, C.</u>, 1993, <u>$10^1 2$ — A Check on Earth Carrying Capacity for Man</u>, Global Bioethics, argued that the planet could conceivably sustain as many as 1 trillion people under an idealized reuse of resources — 100 times as many as there are today. This was to push back on the Club of Rome, which had argued that civilization would soon collapse under the burden of too many people. We would not take eitherthe Club of Rome's pessimistic or Marchetti's optimistic estimates too seriously.

Shorter Newspaper, Magazine Articles, and Clippings

- <u>Spinney, Laura</u>, 2021, <u>Are there too many people? All bets are off</u>, The Guardian.
- <u>Hodges, Glenn</u>, 2021, <u>Humans have 'stressed out' Earth far longer, and more dramatically, than realized</u>, National Geographic.

Websites

- `https://www.eia.gov/outlooks/ieo/` is the main source of our information.
- `https://ourworldindata.org/` curates important data used repeatedly throughout our book.
- `https://www.iucn.org/`, Union of Concerned Scientists.

Chapter 2

Energy

Our book is just about one consequence (climate change) of just one consequence (emissions) of (mostly) just one consequence (energy use) of humanity's huge population.

You read this right. Our book is "just" about byproducts. Human emissions and climate change are just byproducts. They are a sideshow of a sideshow. And if there were a lot fewer of us and/or each of us needed a lot less energy, civilization could tackle climate change much more easily. Indeed, until a few decades ago, it wasn't even fully appreciated that human emissions could cause meaningful climate change on a planetary scale in the first place.

What is so special about energy? It is that energy is a necessary input into almost all economic activity. It permeates every aspect of our lives. It is the life blood of modern economies. It heats and cools our homes, moves us and our goods within cities and across continents, powers our appliances and gadgets, and facilitates modern industry and agriculture. Cheap, reliable energy on demand is one reason why lives in rich countries have been transformed over the last two centuries. It should come as no surprise, then, that rich countries are prodigious users of energy (we will get to some numbers shortly) — and that poorer countries want in on the game.

With energy so central to our world, emissions, and global warming, it is important to explain more of its details. Thus we first have to take a detour and explain how to measure power and energy — especially but not only at large scales — before we return to a global economic analysis.

2. ENERGY

1 How To Measure Power and Energy

For starters, the description that we are "using" energy is misleading. One of the basic laws of physics is that energy is conserved and therefore cannot be used up. What we are actually using up is not energy, but higher-quality fuels.[1] You can think vaguely of widely dispersed lukewarm heat as being the lowest quality of energy and concentrated electricity as being the highest quality of energy. Our devices use high-quality forms of energy to perform productive work and in the process convert high-quality into low-quality energy.

An example can make this abstract idea more concrete. The chemical bonds in the molecules that comprise gasoline are a form of relatively high-quality energy. When an internal combustion engine burns gasoline, about 25 percent of the chemical energy is converted into useful kinetic energy (car movement). The remaining 75 percent is converted into useless heat and radiated into the environment. When the car is slowed or stopped by friction (for example, from applying the brakes), the kinetic energy is also converted into useless heat. Therefore, the net effect of driving a car is to convert all the energy in the chemical bonds of the gasoline into random atmospheric heat. No energy is lost, but unlike gasoline, the heat is no longer useful.

In this sense, our "energy needs" are not really about lacking energy. Instead, they are about finding high-quality energy that we can eventually convert into useful work, before it ends up as environmental heat. Although misleading, everyone just calls this "using energy," and we will thus do the same.

Measurement Challenges

Most vague statements about power and energy are platitudes. (Sometimes a little knowledge is a dangerous thing.) You (our reader) have to comprehend magnitudes if you want to understand climate change.

For example, everyone (including us) is excited about the progress in battery technology. But cursory knowledge cannot tell you whether batteries can or cannot plausibly satisfy global needs. (Spoiler: the answer is *not yet*,

---
[1]Physicists call high-quality fuels by the moniker of *low-entropy* fuels. Entropy is a fancy word for disorder and randomness.

1. How To Measure Power and Energy

but hopefully sooner rather than later.) You have to understand the energy storage problem not only in principle but also in scale. You need appropriate perspective.

Physicists themselves have tried hard *not* to make this easy. The famous physicist Richard Feynman once quipped that "if energy is one thing that is conserved, why do we need so many names for it"? Table 2.1 shows what Feynman was talking about: energy can be measured in terms of joules, calories, watt hours (Wh), British Thermal Units (BTUs), and (metric) tonnes of oil equivalents, to name just a few of the possible units.

Moreover, nobody normal can understand numbers that have a dozen zeros at the end. To make it "easier," the metric system uses standard abbreviations. 1 KWh ("Kilo") is 1,000 Wh; 1 MWh ("Mega", i.e., million) is 1,000 KWh; 1 GWh ("Giga", i.e., billion) is 1,000 MWh; 1 TWh ("Tera", i.e., trillion) is 1,000 GWh; and 1 PWh ("Peta", i.e., quadrillion) is 1,000 TWh.

Energy is power applied for a unit of time. A lightbulb has a certain power rating, and energy is running it for a certain time period. The Kilo-Watt-hour (KWh) is perhaps the most familiar unit of energy, because it is used for pricing electricity. However, watt-hours should not be thought of as merely an "electrical" measure. It makes perfect sense to speak of the number of kilowatt-hours of chemical energy in a gallon of gasoline. To make it easier, we will use only the standard "metric" measure. We will quote all energy in Watt-hours (Wh), appropriately modified by the metric zeros prefixes, like "Kilo" or "Mega."

Energy and power are also sometimes confused.[2] Power is the rate at which energy is being used. It is typically measured in Watts. A second prominent power measure is the horsepower (hp), which equals 746 watts. A horse can deliver much more than 1 hp — ironically, a mistake made by none other than James Watt in the 18th century. Most of the world is now abandoning horsepower in favor of the metric standard unit, the Watt. (Maybe Watt made the horsepower mistake intentionally to get his name onto the correct unit?!) As usual, the United States remains a laggard in adopting international measuring standards.

---

[2] It's also common to get the units mixed up. Even some scientific papers use phrases such as "a battery holds 100 KW of energy." The battery may be able to release power at a rate of 100 KW, but it does not hold 100 KW. Battery capacity must be measured as units of energy (as in KWh), not in units of power (as in KW).

2. ENERGY

Table 2.1. Energy and Power Conversion Factors

Power		Energy	
1 Joule Per Second	1 W	1 Tonne of Oil Equivalent	11,630 KWh
1 Horsepower	746 W	1 Barrel of Oil	1,700 KWh
		1 Therm	29.3 KWh
		1 Cubic feet of natural gas	1/3.6 KWh
		1 (Kilo) calories	1/860 KWh
		1 BTU	1/3,412 KWh
		1 Kilojoule	1/3,600 KWh
		1 Exajoule	278 TWh
		1 "Quad" (quadrillion BTUs)	293 TWh

Note: This is a reference table. It is not necessary to remember any details. For more units and more conversions, see www.convert-measurement-units.com. Our book primarily uses W and Wh as measures for power and energy, respectively.

In an electric car, the quoted power is the rate at which batteries can supply useful electricity. For instance, a 2020 Tesla Model 3 has a battery pack that holds 75 KWh of energy. Driving at a steady speed of 55 mph requires a power output of 15 KW. Therefore, a fully-charged battery can power the car for about 5 hours at 55 mph, giving it a range of about 275 miles.

The fact that energy conversions invariably involve losses (usually to heat) is important. You can think of electricity as the jack-of-all trades when it comes to energy. It can be transported instantly over wires and converted into other forms of energy with relatively high efficiency. (Admittedly, it is not cheap to store — *yet*!)

However, converting other types of energy into electricity often incurs severe conversion losses. For example, when natural gas is burnt to generate electricity, only about 30–50% of the energy in the molecular bonds of the gas is converted into electricity. The remaining 50–70% is lost to environmental heat. In this example, the natural gas bonds are called *primary energy*. Secondary energy is the useful electricity that is left after converting primary energy. Most electricity is secondary energy, having been derived from other forms first.

1. How To Measure Power and Energy

The difference matters. For example, running a 10 W light bulb for two hours per day for a whole year consumes about $10W \times 2h \times 365 \approx 7.3$ KWh of electricity. This 7.3 KWh is secondary energy. If the plant generating electricity is natural-gas based, then it requires about 15–25 KWh of primary chemical energy to generate this 7.3 KWh of electricity.

To avoid double counting, national and global energy usage is almost always measured in terms of primary energy. And because most energy in use today is from fossil fuels, and because their conversion into useful energy is mostly quite inefficient, primary energy use figures are markedly higher than the secondary energy that consumers actually end up using.

Typical Power Magnitudes

Tables 2.2 and 2.3 provide a more intuitive perspective on what power and energy scales mean. They go from the very small to the very large. They are not meant to be memorized but admired (or at least inspected).

Table 2.2 is all about power. It shows that human civilization can already generate and use primary power at a rate of about 18 TW. Future power plans must either provide the equivalent of 18 TW of primary fossil fuel energy, plus whatever is required for future growth (appropriately adjusted for conversion losses) — or induce civilization to get by with less power — <u>something that has never happened in the past.</u>

2. Energy

Table 2.2. Typical Power (Approximate Numbers)

Lifting 1kg at a rate of 1 meter per 10 seconds	1 W	
Light bulb, 800 lumen, LED	10 W	
Incandescent	60 W	
Human		
at rest (metabolism)	100 W	
Cycling (metabolism)	600 W	
Output at Pedals	150 W	
Automobile engine, 100hp, max power output	75,000 W	(75 KW)
cruising 65mph, typical power	15,000 W	(15 KW)
Direct Solar Power, Noon Clear Day, Earth Average, Per m^2 (about 10 sqft)		
Solar Cell Harvest, per m^2	150 W	
Ground, per m^2	1,000 W	
Above Atmosphere	1,360 W	
Typical Roof Solar Electricity peak output	6,000 W	
(20 panels each 300W, about $20k installed in 2021)		
Typical New Wind Turbine, 2021, 50m	2,500,000 W	(2.5 MW)
Typical Coal Plant	600,000,000 W	(600 MW)
Typical Nuclear Plant	1,500,000,000 W	(1.5 GW)
Worldwide Bitcoin Mining, early 2022	16,500,000,000 W	(16.5 GW)
USA, Consumption Rates		
Average US Electricity, circa 2021	500,000,000,000 W	(500 GW)
Peak US Electricity, circa 2021	850,000,000,000 W	(850 GW)
Installed US Electricity Power, circa 2021	1,200,000,000,000 W	(1.2 TW)
Average US Primary Power, circa 2021	3,000,000,000,000 W	(3 TW)
World, Average Consumption Rates		
Global Electricity, circa 2021	2,600,000,000,000 W	(2.6 TW)
Average US Primary Power, circa 2021	18,000,000,000,000 W	(18 TW)
Consumption rate for all life	130,000,000,000,000 W	(130 TW)
Sunlight Striking Earth	174,000,000,000,000 W	(174 TW)
World Population, 2023	8,000,000,000	(8 B)

Note: It is not important to remember these numbers, but it is important to look at them to understand the relative magnitudes involved. Power is always fluctuating, which strictly speaking means it requires modifiers on U.S. and World power use rates — average, peak (ever), or a hypothetical fully installed and running rate. We quote power for wind and solar generation at nameplate capacity, i.e., when operating at maximum. It is important to read Section 6 for appropriate qualifications.

1. How To Measure Power and Energy

Typical Energy Usage Magnitudes

Table 2.3. Typical One-Time or Non-Annual Energy Uses

Lift/drop 60 kg meters		1 Wh	
Light bulb, 800 lumen, LED	1 hour	10 Wh	
incandescent	1 hour	60 Wh	
Cyclist Pedal Output	1 hour	150 Wh	
Food Diet, 2000 (k)cal	1 day	2,300 Wh	
Automobile, 100hp/2, One Commute	1 hour	35,000 Wh	(35 KWh)
Tesla Battery for Model 3	Full	75,000 Wh	(75 KWh)
Primary Energy Use, Per Person, 2022			
Africa	1 Day	14,000 Wh	(14 KWh)
OECD Europe		109,000 Wh	(109 KWh)
USA		232,000 Wh	(232 KWh)
Electricity Component, USA		30,000 Wh	(30 KWh)
One Roundtrip Flight, LA to London	11 h×2	10,000,000 Wh	(10 MWh)

Table 2.3 is about energy. Remember that energy measures how long power is applied. For example, the average American adult typically consumes about 2.3 KWh in calories per day, about 30 KWh in electricity, and 230 KWh in total energy. A round trip flight to Europe or Asia consumes about 10,000 KWh. Although flying is among the most efficient forms of transportation per mile, flying quickly racks up a lot of miles!

Table 2.4 sums the numbers over a typical year. For example, the seventh line in Panel B calculates that if commuting uses about 35 KWh per day, then driving would consume about 365 × 35 KWh/day ≈ 13 MWh per year.

As with the power data, the energy numbers are really just for gawking at, more professionally called "perspective." They contain many interesting tidbits.

For example

- The round-trip airplane vacation to Europe or Asia (from Los Angeles) uses roughly as much energy (10 MWh) as a whole year's worth of either all of a household's electricity (10.6 MWh) or all of someone's car driving (13 MWh).

2. Energy

Table 2.4. Typical Annual Energy Use

	Use Pattern		
LED Lightbulb, 800 lumen	2h/day	7,500 Wh	(7.5 KWh)
	24h/day	90,000 Wh	(90 KWh)
Incandescent Lightbulb, 800 lumen	2h/day	44,000 Wh	(44 KWh)
(60 Various Device) "Wall-Warts"	24h/day	189,000 Wh	(189 KWh)
Air Conditioning	3h/day	3,500,000 Wh	(3.5 MWh)
All Household Electricity (US)	avg/day	10,600,000 Wh	(10.6 MWh)
Automobile, 100hp/2, Commute	1h/day	13,000,000 Wh	(13 MWh)
Roof Solar, 20 panels	5h/day	13,000,000 Wh	(13 MWh)
Typical Wind Turbine	6h/day,	3,285,000,000 Wh	(3.3 GWh)
Typical Coal Plant	20h/day	4,380,000,000,000 Wh	(4.4 TWh)
... typical utilization rate in 2019	12h/day	2,628,000,000,000 Wh	(2.6 TWh)
Average US Nuclear Plant	22h/day	10,000,000,000,000 Wh	(10.0 TWh)
Global Annualized Bitcoin Mining, early 2022	24/7	150,000,000,000,000 Wh	(150 TWh)
U.S. Electricity Consumption, circa 2021	Annual	4,500,000,000,000,000 Wh	(4.5 PWh)
World Electricity Consumption, circa 2021	Annual	27,000,000,000,000,000 Wh	(27 PWh)
U.S. Primary Energy Usage, circa 2021	Annual	26,000,000,000,000,000 Wh	(26 PWh)
World Primary Energy Usage, circa 2021	Annual	165,000,000,000,000,000 Wh	(165 PWh)
Approx World Population, 2023		8,000,000,000	(8 B)

Note: Sources are varied. It is not important to remember these numbers. It is important to read Section 6 for appropriate qualifications. The quoted numbers here are from different sources. The most prominent sources are the British Petroleum Annual review of World Energy and the U.S. Energy Information Administration. See the references for more explanations.

- A U.S. household uses about 32 MWh of electricity per year, so one typical wind turbine generating 3.3 GWh can supply the energy-needs equivalent of about 100 households, i.e., a small village. This ignores the discrepancy between when households need electricity and when wind turbines can supply it.

- Dividing 170 PWh by the population of 8 billion in 2022 tells us that the average human consumed about 21 MWh/year of primary energy. This is about 60 KWh/day — about the equivalent of 250 LED 10 W light bulbs continuously burning, or a 50–70 one-square-meter panels of roof solar working for about 5–6 hours per day.

Our nerdish side tempts us to look at these figures all day long, but we need to move on.

1. How To Measure Power and Energy

Our book is about the very largest scales of energy usage. The United States generates about 4–4.5 PWh (or trillion KWh) of electric energy from about 10 PWh of primary energy (mostly chemical bonds in fossil fuels) — the remaining 18 PWh of primary energy are used for transportation, heat, etc.

Although 4 PWh is a huge number (as the many trailing zeros make clear), electricity is only about 35–40% of the total primary energy consumption of the United States (28 PWh in 2022) and 10 PWh is only about 5% of the total primary energy consumption of the world. Think about that. Even if the United States managed to eliminate all fossil fuels from *all* of its electricity generation — which is an impossible feat for many decades — the world would have *barely* moved towards zero net emissions! The world's challenge is not to clean up the U.S. electricity grid consuming 10 PWh but to clean up the entire 187 PWh in 2022 (plus an additional ≈70–75 PWh that it will also consume by 2050).

I want to offset my carbon footprint, Jenkins...cancel your holiday flights

Our job is to convey to you, our reader, this quantitative information in an understandable fashion. You job is to grapple with comprehending these huge magnitudes if you want to understand climate change and be qualified to discuss possible solutions. Don't be easily swayed one way or another by half-truths.

2 Where Does All Our Energy Go?

How can humanity possibly use so much energy? Where does it all go?

Table 2.5 shows two aerial snapshots of what civilization is doing with all this energy. The two panels are somewhat disjointed, because they come from different sources, and it is unclear how they reconcile. Nevertheless, together, they convey a good overall impression.

Panel A shows that the largest use category is home and work, followed by transportation and industry. Within these broad categories, home heating and cooling (including refrigerators and water heaters), cars, and the production of cement, metals, and chemicals loom large. But even if we could eliminate those altogether, the remaining uses would not be trivial, either. For example, it would leave a lot of energy needed to fly around the world and to grow plants.

Panel B provides another view of energy usage. We use a lot of energy making stuff and plugging devices into electrical outlets. But agriculture and travel are large contributors, too, as is our need to heat or cool our buildings.

The panels make it clear that it is not enough to clean up just any one category (e.g., the electricity sector) and ignore the rest. There are too many big contributors to the problem. Humanity will need many solutions to many problems — many reductions by many different emitters.

2. Where Does All Our Energy Go?

Table 2.5. Estimating Primary Energy Use By Activity

Panel A: Classification 1.

	USA	World BP	World EIA
Home and Work		40%	30%	40%
Heating (Water, Air), A/C	50%			
Transportation		28%	20%	30%
Cars	60%			
Trucks	20%			
Aircraft	10%			
Boats or Buses, each	5%			
Industry		32%	50%	30%
Chemical	27%			
Petroleum Refining	22%			
Paper	17%			
Metals	17%			
Cement	4%			

Panel B: Classification 2 (Worldwide).

Making Things (cement, steel, plastic)	31%
Plugging In (electricity)	27%
Growing Things (agriculture)	19%
Getting Around (planes, cars, ships)	16%
Keeping Warm and Cool (heating, cooling)	7%

Source: For USA, NAS.edu and U.S. Energy Information Administration (EIA). For world, British Petroleum (BP) Statistical Review of World Energy [often named the Factbook] and Wikipedia World Energy and Supply (bottom right table), itself based on dated International Energy Agency data. Refining, metals, and paper are educated guesses. Component percentages are also stated in terms of total primary consumption (e.g., heating is 50% of home and work consumption). Of course, like most estimates in our book, these are just approximations. However, these particular estimates are also a little unusual in that they are a case in which some of our primary information sources disagree. Thus, we present two versions. Most of the time, though, the data sources align well. Panel B is from Gates (2021, p.55).

Table 2.6. Total and Per-Person Energy Consumption, 2022

	Population (million)	×	Primary Energy Consumption		
			Per Capita Day (KWh/P/Day)	=	Total Per Year (PWh/Year)
OECD	1,380	×	141	=	71
USA	335	×	232	=	28
Europe	593	×	109	=	24
Not OECD	6,502	×	49	=	116
China	1,449	×	90	=	48
India	1,408	×	23	=	12
Other Asia	1,177	×	32	=	14
Africa	1,367	×	14	=	7
Sub-Sahara	≈1,100	×	≈5	=	≈2
World	7,882	×	65	=	187

Note: The product omits the conversion factor from days to years. Sub-Sahara excludes South Africa and was inferred from equivalent shares in EIA data from 2019.

Source: US EIA International Energy Outlook, Oct 2021. The input series were Primary Energy and Population.

3 Rich and Poor Today

Obviously, air-conditioning and jet travel are not as common in Africa as they are in Florida. Is world energy consumption then primarily a luxury problem? Is it about too much air-conditioning and jet travel? Is it about the careless wasting of energy by richer people? Where could we cut its use? To answer this question, we need to look at energy consumption in different regions.

Table 2.6 shows the population numbers from the previous chapter on the left, their per-person per-day primary energy consumption in the middle, and the total annual primary energy consumption on the right. We will first discuss the per-person per-day numbers, measured in KWh, because they help us understand what luxury and what poverty consumption is.

3. Rich and Poor Today

The table shows that residents of rich OECD countries consumed almost three times as much energy per person (141 KWh per day) in 2022 as residents of poor non-OECD countries (49 KWh per day).

The average American was especially profligate, consuming about 230 KWh of primary energy per day (pPpD). For a meaningful economic perspective, if 230 KWh of primary energy were used only to generate about 100 KWh of electricity (and it is not) and if this electricity were sold at the typical electricity retail price of $0.15 per KWh, then the per-person energy bill would be about $15 of energy per day or $5,000 per year. In context, this would be the equivalent of (expensive) cappuccinos at Starbucks every morning for a family of four.

The average European consumed about half as much energy as the average American, at 110 KWh per person per day. This is commonly referred to as the *European standard*, but it also applies to other non-English speaking OECD countries like Japan and Korea.

China (with 90 KWh per person per day) has almost reached the European standard. However, because China burns more coal, its per-person greenhouse gas emissions have already exceeded those of the OECD. (We will describe emissions in the next chapter.)

Residents of most other poorer non-OECD countries were much more frugal. At 23 KWh per person per day, Indians consumed only about one quarter of what Chinese consumed. Hundreds of millions of Indians still don't have access to regular electricity. Many millions more suffer regular electricity outages even when they are connected to the grid. Residents of other Asian countries consumed about a third as much per-capita electricity as the Chinese, at 32 KWh per person per day.

Africans consumed only 14 KWh per person per day. The numbers are even lower in the sub-Saharan regions (South Africa excepted), with estimates of under 5 KWh per person per day. Such low energy consumption is a symptom of a subsistence economy with widespread extreme poverty. It is inconsistent with a healthy modern economic living standard. Many Africans still spend much of their day walking just to obtain and carry the necessities of life.

Are Americans and Europeans not only richer (in terms of income) but also more wasteful in their energy use? Do they use energy for careless luxuries, not paying attention to energy costs because they are so rich? Can we make progress by pushing them towards dealing with energy as efficiently and frugally as residents of poorer countries?

To assess this, we want to determine how efficient regions are in converting energy into economic output (which, for the most part, translates into income). The average American earned a gross income of about $61,000 per year in 2022. (Many households have of course multiple earners, and the average income even ascribes income to children. This means that the average gross income per household is considerably higher.)

Table 2.7. Per-Person Energy Use and Efficiency, 2022

	Per Capita Day (KWh/P/Day)	×	GDP/Person ($-1,000/Yr)	=	Efficiency (KWh/$)
OECD	141	×	43	=	1.19
USA	232	×	61	=	1.39
Europe	109	×	42	=	0.95
Not OECD	49	×	12	=	1.54
China	90	×	19	=	1.75
India	23	×	7	=	1.18
Other Asia	32	×	11	=	1.04
Africa	14	×	5	=	1.06
World	65	×	17	=	1.38

Source: US EIA International Energy Outlook, Oct 2021. The input series were Primary Energy, Population, and GDP in purchasing power parity.

Table 2.7 shows that OECD countries required about 1.19 KWh in energy to produce each dollar of income. Europeans earned less than Americans ($42,000/person), but they also worked about 20% less (mostly voluntarily!) and were more energy efficient per dollar earned. There was quite a bit of heterogeneity, though: Germans earned more and worked less than their American counterparts; Poles worked more and earned less.

3. Rich and Poor Today

Unfortunately, it is the poorer non-OECD countries that produced less with more energy. They required about 1.54 KWh of primary energy for each dollar of output. The Chinese were particularly inefficient in their energy use, using 1.75 KWh. (However, China's economy produced relatively more manufactured goods than, say, Europe. This may be the reason for China's lower energy efficiency.)

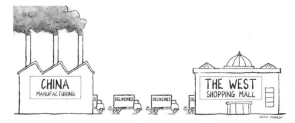

Why China's carbon footprint is so large...

The big picture does not seem to be that rich OECD countries care less about wasting energy. If anything, wealthier countries are producing more with less energy. They have spent more money on insulation, bought more energy-efficient machines, and reoriented themselves towards less energy-intensive activities. Rich countries have not been particularly wasteful. Their primary responsibility for the world's large energy consumption is that they produce more economic output, which has given them higher incomes that needed more requisite energy. Bringing their efficiencies into line with those in poorer countries would be counterproductive.

After you have gotten over the natural urge to point fingers — at Americans, Europeans, or Chinese for using too much energy per head, or at China for using energy too wastefully — you should reflect again that the global problem today is not about blame. It is not about *per-person* emissions (or energy consumption).

Instead, the global problem is about finding a way out of our collective malaise. It is about *total* emissions and energy consumption, i.e., per-capita resource use multiplied by population. As we already mentioned, despite its large per-capita consumption, the OECD contains only a small part of the planet's population. More than four out of five people in the world live in non-OECD countries. If we want to solve global problems, we need to contemplate primarily the world's total energy consumption and not just the energy consumption of the one in five.

With only 20% of the population living in the OECD, Table 2.7 shows again that OECD emissions are no longer as important as you may have thought. At

71 PWh per year, the OECD is already responsible for only about one-third of total world emissions. With its much higher population, China is responsible for about one-quarter of the world's energy use. China's energy consumption is already almost twice as large as that of the United States. In fact, China's consumption is already almost as large by itself as that of the U.S. and Europe *combined*. And China is not alone. If the OECD vanished tomorrow, primary energy consumption would still be a gigantic 116 PWh, growing fast.

4 The Future

Looking at longer-term trends can help us understand where we have come from and where we are going. We can still use both historical and forecast data from the U.S. Energy Information Administration. These are the best that we are aware of. (Pointing to them will also allow us to shift blame to them if — as will be inevitably the case — the forecasts will prove to be less than perfect.)

Per-Capita Consumption

Figure 2.8 shows generational trends in per-capita energy consumption.

OECD residents have been reducing their energy consumption since about the turn of the millennium. This was primarily due to the three English-speaking members of the OECD — the USA, Australia, and Canada — which have reduced their consumption more than other OECD countries. However, this was also easy for them, because they have been the most wasteful OECD countries for decades and continue to be so — and not by a small margin. (Australians consumed a little less, Canadians a little more than Americans.)

Looking forward to 2050, per-capita energy consumption in the OECD will likely remain stable. The predicted modest per-person increase to 2050 is well within the margin of prediction error.

In contrast to the stagnant OECD, non-OECD countries have been increasing their per-person energy use since the turn of the millennium. This was mostly due to China. The average Chinese consumed under 10 KWh per day in 1980 compared to about 90 KWh in 2022. Although Chinese energy consumption has been decelerating, it is still growing faster than those in the

Figure 2.8. Energy Use Per Person Per Day (pPpD)

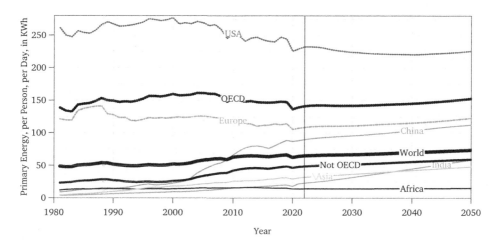

Note: The average American consumes about twice as much energy per capita as the average European and three times as much as the average Chinese. Per-capita consumption in Africa and India remains tiny.

Source: Pre-2020 figures are from the EIA. Post-2010 numbers are from the US EIA International Energy Outlook, Oct 2021, specifically the EIA World total primary energy consumption by region and the EIA World population by region. We aligned the data series to minimize discrepancies.

OECD countries. Thus, its *per-person* energy consumption will soon equal that of Europe's — and there are a lot more *persons* in China than in Europe. Fortunately, it does not seem as if China is aspiring to reach the U.S. energy standard anytime soon.

Of course, it would be hypocritical for the U.S. to complain to China about its high per-capita energy growth. China was only escaping abject poverty, and our own per-capita energy use remains twice as high. This is also why China's leaders usually bristle when U.S. leaders request that China reduce its emissions.[3]

Fortunately for Indians and unfortunately for global emissions, India's economy is already taking off as we are writing this book. Its energy use

---

[3] In the (now superseded) Kyoto protocol of 1997, China was even explicitly exempted from *any* obligation to curtail its emissions.

2. ENERGY

is predicted to reach half of the European standard by 2050 and more a generation later. Some population predictions further suggest that there could be more people on the Indian subcontinent by 2100 than in China, Europe, and the US combined. Multiply India's population count by European per-capita energy standards and you can see the problem.

High per-capita energy growth is also the case for other developing regions — except Africa. Africa is economically still in the poorhouse and its per-person energy consumption is not expected to increase much even by 2050. Simply put, the average African is poor and predicted to remain so. (And it's even worse in sub-Saharan Africa.) Yet the population of Africa is also expected to surpass India's, China's, and the OECD's *combined* by the end of the century. Pent-up demand for energy will explode if and when Africa develops.

Total Consumption

As we already stated, the world's energy problem is not about finger-pointing or per-capita consumption. It is about finding solutions to reduce total energy consumption and emissions.

Figure 2.9 catches the situation beginning in 1980. It plots OECD and non-OECD total energy consumption on the same scale.

In 1980, OECD emissions were still twice as large as non-OECD emissions. Europe and the US each easily exceeded China and India combined. After some modest growth, OECD total energy consumption stabilized around the turn of the millenium.

Yet just when OECD energy use plateaued, non-OECD energy use took off. Many countries were escaping poverty. China's transformation into a market economy under Deng Xiaoping had raised not just millions but hundreds of millions of Chinese out of poverty. (Moreover, China's population doubled from about 700 million people in 1965 to about 1.4 billion people today.)

Thus, today, the world's shares of energy use by OECD and non-OECD countries have reversed. The majority of the world's energy is now consumed by non-OECD countries. China now easily exceeds the energy consumption of the USA and Europe combined, and India will exceed either the USA or Europe by 2050, and both combined another generation later.

Fortunately, China's energy growth is now decelerating. Its population is no longer growing. The same cannot be said for other non-OECD countries.

4. The Future

Figure 2.9. Primary Energy Use

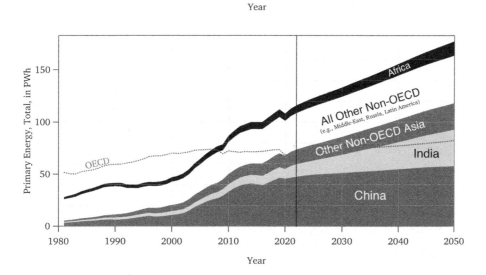

Note: The two graphs are drawn on the same scales. OECD energy consumption has been stable since the turn of the millennium. Non-OECD consumption has not been. China accounted for most of the dramatic growth in world energy use over the last two decades. Its consumption is now larger than that of North America and Europe *combined* but has also largely stabilized. The energy consumption of India remains small but is about to increase greatly.

Source: US EIA. Predictions are from the US EIA International Energy Outlook, Oct 2021. The input series was Primary Energy. We aligned the EIA data series to minimize discrepancies.

2. ENERGY

India's population is still growing, as are those of other countries. Sub-Saharan Africa is a rounding error as far as global energy consumption is concerned, but if and when it escapes poverty, its energy increase will be dramatic. This seems unlikely to occur within one generation, but it should not be discounted over longer timespans.

Summary Forecasts

As we explained in our preface, our main concern is about what we can do to move the needle in the next 10-30 years, not in the next 100-200 years. Table 2.10 puts together the most important numbers for the EIA's 2050 world energy consumption forecast.

Table 2.10. Total and Per-Person Energy Consumption, 2050e

	Population	×	Primary Energy Consumption		
			Per Capita Day	=	Total Per Year
	(million)		(KWh/P/Day)		(PWh/Year)
OECD	1,478	×	153	=	82
USA	386	×	226	=	32
Europe	613	×	123	=	28
Not OECD	8,177	×	59	=	177
China	1,402	×	113	=	58
India	1,640	×	59	=	35
Other Asia	1,432	×	48	=	25
Africa	2,413	×	15	=	13
World, 2050e	9,655	×	74	=	260
World, 2022	7,882	×	65	=	187

Note: This table is the equivalent of Table 2.6 but 30 years into the future.

Source: US EIA International Energy Outlook, Oct 2021. The input series were Primary Energy and Population.

4. The Future

Here is our summary:

- In the OECD, population numbers have largely stabilized.
  Beyond the OECD, population is still growing. In particular, Africa and Asia (but not China) will continue to grow — and from higher baselines than they had just 50 years ago. Their growth will continue trends that have existed at least since the 1980s.

- The world has been improving its energy efficiency, producing more output with less energy. Compared to 30 years ago, the world now produces its output with one-third less energy. Scientists are expecting this efficiency growth to remain as strong over the next 30 years.

- Over the last 30 years, per-capita energy consumption in the OECD has shrunk. This was due to large efficiency gains (more output given the same amount of energy). Beyond the OECD, per-capita energy consumption has increased greatly. This was due to much faster increases in the standard of living and not fully offset by (relatively smaller) efficiency gains. In particular, China, India, and the rest of Asia have been climbing out of poverty, with China having led the way. Sadly, most Africans will remain poor, consuming 15 KWh per person per day. In sub-Saharan Africa, it is even worse: 5 KWh per person per day.

- Over the next 30 years, efficiency gains are likely to continue, though they may or may not be enough to keep OECD per-capita energy consumption exactly constant over the next 30 years. A reasonable forecast is energy use growth of 15% over 30 years in the OECD.
  Beyond the OECD, total energy consumption will grow dramatically — at three times the OECD rate. In terms of living standards, the Chinese people are about halfway between Europeans and other non-OECD nations. Thus, they will likely still increase their per-capita energy consumption faster than Americans or Europeans, but no longer as fast as they have in the past. China will also improve its energy productivity, creating more GDP with less energy — indeed likely improving its efficiency more so than other regions of the globe.
  India and the rest of Asia are a little behind. They are thus expected to grow faster than China in per-capita consumption over the next 30 years. Indians are likely to double their energy consumption per person in one generation. Africa's consumption will grow by about 60% due to its population growth, but from so low a base that it barely matters.

- Thus, the OECD share of energy consumption will fall from about one-third of the world to about one-quarter.

5 Clean and Dirty Energy

Let's take stock. In the previous chapter, we explained why it is misleading to blame only the industrial revolution and capitalism for our increasing use of energy (and emissions). It was not just the industrial revolution, but also the second agricultural, hygiene, and medical revolutions that facilitated our population explosion.

- Without the population explosion, fossil fuel use — with all its nasty emissions (discussed in detail in the next chapter) — would not have endangered the climate (discussed in the chapter thereafter).
- Without fossil-fuel use, the population explosion would not have caused so many emissions and changed the climate.

The two needed each other. Population growth and fossil fuels have become the two horsemen of the climate-change apocalypse.

Realistically, our leaders will not be able to do much to curtail population growth. Nor do we see how they could do much to curtail the world's energy consumption.

Humanity's best hope, then, is to improve its efficiency — either the efficiency with which it uses energy to make its living or the efficiency with which it creates energy with fewer emissions. (Emissions and their effects are the subjects of our next chapters.)

Can we produce the same energy with fewer emissions? That is, can we switch from dirty fossil fuels to clean energy sources on a sufficient scale to meet global energy demand? And how quickly could we switch? This is the trillion dollar question.

Figure 2.11 is perhaps the most important illustration in our book. It stacks energy sources in order of dirtiness.

Biomass is the layer at the bottom, because it has the dubious distinction of being the dirtiest fuel in wide use. It consists primarily of the burning of wood plus agricultural waste (by farmers). Used since pre-biblical times, biomass still accounts for about 7% of humanity's energy use today. Healthwise, it's terrible. Its particle emissions create some of the worst health hazards among common pollutants studied by scientists. Its greenhouse gas emissions per useful energy beat even fossil fuels (in a bad way). Even in the United States, the EPA reports that about 10 million homes (30 million people) still use

5. Clean and Dirty Energy

Figure 2.11. Energy Breakdown, 1900 to 2020

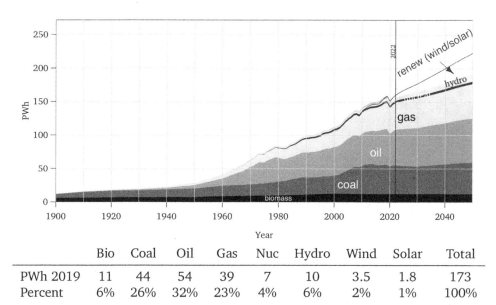

	Bio	Coal	Oil	Gas	Nuc	Hydro	Wind	Solar	Total
PWh 2019	11	44	54	39	7	10	3.5	1.8	173
Percent	6%	26%	32%	23%	4%	6%	2%	1%	100%

Note: This figure is based on the "fossil-fuel" equivalent way of counting non-fossil fuels as primary energy. (It grosses them up as if they had similar conversion losses.) About 80% of the world's energy is still supplied by the three fossil fuels. Wind and solar are rounding errors. The projection of energy use is from the EIA. Energy sources are stacked by order of dirtiness when burned (CO_2 emissions per Wh over 170 years). However, if Natural Gas production leaks are accounted for, natural gas would be as dirty as coal.

Source: Ritchie and Moser, Our World in Data, Energy Mix, originally based on Smil and British Petroleum. Forecasts are from the US EIA International Energy Outlook, Oct 2021. The input series were Forecasts by Fuel and Forecasts of Hydro. We aligned the data sources to minimize discrepancies.

wood as their primary or secondary heat source, causing an estimated 10,000 premature deaths per year. However, biomass is renewable, which is why it has (unfortunately) been exempted from many global emission treaties.

There are three fossil fuels that we are digging out of the ground: coal, oil, and natural gas. Like biomass, fossil fuels are primarily used in combustion processes. The figure shows that they are responsible for powering about 85 percent of the world today. Thus, they are also responsible for the majority of harmful emissions of all kinds. Reducing emissions will require tackling fossil fuel use. It will be a huge task.

2. ENERGY

Coal made a strong entry around 1850 and overtook biomass around 1900. We will cover coal in Chapter 12, but here is a basic description. Today, coal provides about four times as much energy as biomass. Unfortunately, coal is also only mildly less polluting than biomass.

You should have been here last week. *This* is clean coal.

If you look carefully at Figure 2.11, you will see a jump in coal consumption starting around 2009. This was due to the dramatic economic expansion in Asia, particularly China, where coal still accounts for 60% of electricity generation. Even today, despite claims that it wants to decarbonize by 2060 and despite ongoing *percentage* reduction in reliance on coal, China is building new coal replacement plants at a record pace. They will have an expected life span of about 30-50 years. From humanity's collective emissions perspective, this is a lost opportunity.

Fortunately, coal use has leveled off since around 2014 (though it has experienced a small and hopefully temporary renaissance in 2021). And even more fortunately, coal is not only universally despised for its more localized toxic non-CO_2 pollution (like smog and soot), but it is also expensive to mine and ship. Thus, coal has been becoming increasingly uncompetitive. No new coal plants have been built in the West for over a decade.

We are a little more optimistic than the EIA forecasts in Figure 2.11. We believe (or maybe just hope) that coal will decline under economic and political pressure from wealthier populations even in non-OECD regions where coal is still heavily used. Yet once a plant is built, it will be difficult not to use it. The decline of coal would be faster if politicians gave it a well-deserved push out the door with appropriate pollution taxation. What's delaying them? Remarkably, in many places, it is no longer the lack of cheaper and better economic alternative energy. Instead, it is the large employment in the coal sector. This is especially the case — where else? — in China and India.

Oil is a cleaner fossil fuel than coal. It is used mostly in transportation, secondarily for heating. Its use is still growing — despite all the Tesla cars in the West. Oil also fuels aircraft, ships, trucks, and so on. In the developing

5. Clean and Dirty Energy

world, oil is also sometimes used to power diesel-electric backup generators, because the electric grid is so unreliable.

Natural gas ("Natgas") is mostly methane. It is cleaner than coal and oil *when burnt*, but this is highly misleading. From a more complete supply-chain perspective, Natgas may well be as polluting as coal. The reason is that it leaks left and right. Bloomberg quotes estimates that about 2.3% of Methane escapes during its extraction and transportation. Methane is about 30–100 times more potent than CO_2 as a Greenhouse gas, so Natgas' true pollution may be *twice* of what it would be without leaks. At this point, given what scientists have learned in the last decade, we should no longer believe that Natgas is our clean "transition" fuel.

Despite all their drawbacks, both oil and gas are still projected to grow — at least for another generation, if not longer.

Other energy sources remain small — in relative terms, of course. In absolute terms, they are big. Nuclear and hydro-electric plants together can account for only 10% of the world's energy needs. Solar and wind power are about 3%, despite a full decade's worth of installations. They are becoming more important every year, though.

What is the outlook for the next generation? Unfortunately, not too different from the past. Fossil fuel use is still increasing, although fortunately now at a decelerating rate. This is mostly due to clean renewables coming online. But clean renewables are not coming on strongly enough even to arrest the growth of fossil fuels — much less to reduce them.

It is fairly straightforward to make good predictions for the world's energy use. Even with the greatest of effort, fossil fuels — the main source of our emissions — will play a key role for at least a few more decades, whether we like it or not. And fossil fuels will *not* play a key role in one to two centuries or so, simply because we will likely have exhausted most of the cheap-to-mine higher-quality fossil fuels. The age of fossil fuels will come to an end. The only question now is how quickly.

Why can't we phase out fossil fuels within, say, one generation? Probably because there are physical limits to the speed of change. (We describe the various challenges clean tech faces in upcoming chapters.) As in all things economic, there are costs and benefits. On the one hand, we know that we cannot eliminate emissions too quickly. Rapid reduction could impoverish the lives of billions of others now and maybe even kill millions. On the other hand, we also know that it is also not a good solution to push emissions down too slowly. The environmental impact — and not only from global warming but also local particle emissions — could again impoverish the lives of billions and kill millions of people. Civilization should try to find a good middle path between the two extremes. We should move the needle toward clean solutions as soon as doing so becomes reasonable.

anecdote

> Some prominent thought leaders are on record having made some "interesting" energy proposals. For example:
> - Paul R. Ehrlich, Professor, Stanford: "Giving society cheap, abundant energy would be the equivalent of giving an idiot child a machine gun."
> - Jeremy Rifkin, Greenhouse Crisis Foundation: "The prospect of cheap fusion energy is the worst thing that could happen to the planet."
> - Barack Obama, Presidential Candidate 2008: "Under my plan of a cap-and-trade system, electricity rates would necessarily skyrocket. Coal-powered plants, you know, natural gas, you name it, whatever the plants were, whatever the industry was, they would have to retrofit their operations. That will cost money. They will pass that money on to consumers."

6 Important Details and Clarifications

Before we leave energy, we want to quickly cover a few more details.

Clean and Renewable Energy

First, some clarifications. "Renewable energy" is not the same as "clean energy" and vice-versa. Renewable energy includes solar power (mostly solar photovoltaic cells), wind power (mostly giant turbines standing around the landscape), geothermal power (think of giant pits that tap heat from deep underground), hydro-power (think dams that refill from precipitation when emptied), but also biomass (the aforementioned burning of wood and nastily dirty affair), but not nuclear power. In contrast, clean energy includes solar, wind, geothermal, hydro-power *and* nuclear power (but not biomass). Although "clean" would have been the shorter word when discussing wind and solar, the adjective more commonly used is "renewable." (Maybe it sounds more sophisticated?) We can only hope that speakers don't mean biomass.

The two brightest beacons on the horizon to replace fossil fuels are indisputably solar and wind power. They are both clean and renewable. We are not yet sure about other renewable and clean energy sources — it will depend on their technological progress relative to that of wind, solar, and batteries. (Batteries required but not included.) Civilization could certainly use other energy sources, too, and more horses to bet on. Technology will be the subject of Part III of our book. Here is a very brief preview.

The problem with wind and solar is that their contribution to today's power generation is so small that it is difficult even to see their slivers in Figure 2.11. Thus, you need to keep perspective. Although they have indeed been growing more rapidly in percentage terms than any other sources of energy, they still account for less than 5% of primary energy as of 2020. This is also why they are not growing more rapidly in absolute terms than fossil-fuel plants. In 2020, the world was still installing about 2 PWh of fossil-fuel energy compared to only about 1 PWh of wind and solar energy. In a few years, wind and solar will overtake fossil fuels — but we are not there yet.

Furthermore, solar and wind have so far made their appearance overwhelmingly only in the electricity sector. As we noted in Table 2.3, electricity itself accounts for only about one-third of the world's energy consumption today, although it will account for more in the future. Thus, despite a lot of (warranted) hoopla, wind and solar are nowhere near where they will have to be in order to significantly reduce fossil fuels and associated emissions. We have little doubt that they will get there — the only question is how long it will take.

Nameplate Power, Conversion Losses, and Intermittency

It is difficult to compare energy sources that are so different and that have to be converted to end uses with varying efficiency losses.

The EIA *energy figures* already include adjustments — in particular, they try to put renewables and nuclear energy on an even footing with fossil fuels by grossing up delivery as if its conversion to electricity was instead delivered by fossil fuels with standard inefficiency.

The problem the adjustment fixes is that primary energy is not secondary energy. When we generate electricity from fossil fuels, we lose more than half in the conversion. (Electricity will be the subject of Chapter 12.) If all fuels were used for electricity generation (and they are not!), about 100 PWh in secondary energy (instead of 260 PWh of primary energy) would be sufficient. Solar and wind power, the two most prominent candidates for a clean and renewable energy future, do not have the same large conversion heat waste losses as fossil fuels. Solar photovoltaics in particular generate electricity (nearly) directly.

Unfortunately, solar and wind have a different problem — a problem especially for *power figures*. They work only intermittently. Their so-called capacity factor is low. They cannot operate 24/7.

	Nuclear	Fossil Fuel	Wind	Hydroelectric	Solar PV
Capacity Factor	90%	50–60%	35%	25–50%	20%

Much worse, it is not at the operator's discretion as to when they work. They are at the mercy of the local weather. Consequently, on average, wind farms typically generate about one-third of their so-called nameplate capacity (i.e., their maximum output), more in some places, less in others. Solar farms produce even less. Both are even often turned off when there is too much electricity on the grid already. Nevertheless, wind or solar is usually available in most places on earth in abundance, even near population centers.

To satisfy electricity demand when the sun does not shine and the wind does not blow, wind and solar have to generate even more power to charge energy storage devices while they are operating. This also wastes some energy and costs *a lot* more money. Energy storage is the one critical clean-tech aspect that has not yet been solved. Once storage costs drop far enough in price, the fossil-fuel age will quickly come to an end.

For now, roughly speaking, if energy comes from wind and solar plants and the end product is electricity, civilization will need more nameplate power in wind and solar plants than it needs in primary input power from fossil fuel plants. Intermittent generation may lose more power relative to nameplate power than fossil fuels lose in power in their efficiency conversion, but the two are not too far off in terms of order of magnitude.

It is not a bad approximation to think that we will need about 300 PWh of wind and solar nameplate capacity instead of 220 PWh of primary fossil-fuel energy.

7 The Situation Today

So where is the world in 2022?

We have said this a few times already: Civilization should try to move away from fossil fuels as fast as is reasonable, but no faster. This may sound like a vague statement — and it is — but it is also true.

The world seems to be at the start of the fastest energy transition in its history. We venture to guess that this is not because of increasing environmental conscience but because of declining clean energy costs. Many fossil fuel plants have been shut down by their operators even before they have reached the ends of their lifespans, because they could no longer compete economically against wind and solar farms. In many OECD countries, it's been over a decade since a new coal plant has been built. Even natural gas deployment is no longer growing greatly, despite their remaining economic advantages. Clean wind and solar power are growing at an accelerating record pace.

But not all news is good from an emissions perspective. Clean nuclear power is declining and is often replaced by natural gas. (In Europe 2022, nuclear power is not replaced by Russian Natgas but by old coal plants.) Worldwide, wind and solar are growing faster than fossil fuels only *in relative terms* (about 50% per year), but this is from a very low base, where relative increases are both small and easy. Wind and solar are not yet even growing faster than fossil fuel *in absolute terms*. It will take decades just to arrest the installation of new fossil fuel plants, and then some more decades to retire all existing ones. This is why we are so confidently predicting that fossil fuels will play an important role for decades to come — whether we like it or not. Please don't shoot the messenger.

What about all the wonderful news in the press about how the United States is making progress in cleaning up its electricity grid? Yes, it is wonderful, but the fact is that the entire U.S. electricity grid today transmits "only" 4 PWh/year. Cleaning up American electricity *quickly* will be difficult, but it is *relatively* easy compared to cleaning up other sources of emissions, future demand, and many more other countries, too.

We would love clean energy to limit fossil fuel use to today's consumption of ≈150 PWh/year. But it likely won't happen. The ongoing growth in fossil fuel usage will lead to growing emissions. By 2050, clean energy will likely cover 30–50 PWh that would otherwise have been filled by fossil fuels — about a third of today's energy consumption and about half of the growth of energy from today to 2050. This will be a great success — but it will also be a failure in not going far enough.

We began our chapter with the observation that energy demand and supply is not just a technological problem (Part III of our book) but also a social and economic problem (Part II). And it is not the case, as some allege, that the main reason for global emissions is that evil, unscrupulous big-oil capitalists are willing and eager to destroy the environment in order to satisfy their own greed. If that were the case, it would be easy to fix the problem the Shakespearean way — "first kill all the lawyers." But evil capitalists are not the main problem.

Instead, we have to repeat (again and again) that the global emissions problem presents a harsh dilemma. Turning off the fossil-fuel spigot too quickly would result in economic chaos and condemn billions of people to continued long-term poverty. The 1–2 billion people in rich countries and regions might be upset, but they would come out okay. Not all the remaining 6 billion people in poorer countries would. The world shouldn't cut off fossil fuels too quickly. More importantly, even if the world should have done so (say, if the harm was far worse than it is projected to be), realistically, it also wouldn't do so. Those 6 billion people would not tolerate it.

Further Readings

BOOKS

- Gates, Bill, 2021, How to Avoid a Climate Disaster: The Solutions We Have and the Breakthroughs We Need, Penguin Books, New York.
- MacKay, David J.C., 2009, Sustainable Energy — Without the hot air, UIT Publisher, Cambridge, England. Free online. This book is the classic explanation of the energy dilemma, but it is about the United Kingdom around the turn of the millennium. Many good subsequent books (including our own) have borrowed heavily from MacKay's much more original ideas.
- Smil, Vaclav, 2017, Energy and Civilization — A History, MIT Press. A comprehensive history of energy usage.

REPORTS

- The British Petroleum (BP) Statistical Review of World Energy is a remarkably accurate and reliable source of information — despite the fact that the data is collected by a fossil-fuel company. Kudos where it is deserved.
- US EIA International Energy Outlook, Oct 2021.

WEBSITES

- https://www.eia.gov/ (Energy Information Administration), our primary source.
- http://www.bp.com/statisticalreview is another authoritative (and unbiased) British Petroleum (BP) source.

The BP numbers can differ by about 10% from the EIA numbers, making it difficult to cross-compare numbers. However, the series are internally consistent and they typically move closely together.

We prefer the EIA data primarily because it provides projections up to 2050. However, even within the EIA data, it is possible to find different numbers for the same series. For example, in February 2022, two series were quoting primary energy consumption for the world in 2019 as either 632.9 quads or 601.0 (due to different treatments of biomass).

- https://www.eia.gov/outlooks/ieo/ is the main source of our information.
- https://ourworldindata.org/ curates important data we use repeatedly.
- https://www.iucn.org/, International Union of Concerned Scientists.
- https://www.rff.org/geo/, Resources for the Future, offers a meta outlook based on different projections from different sources.

Chapter 3

Greenhouse Gases

The previous chapter showed that the combination of population growth and economic development has translated almost one-to-one into increased primary energy consumption. In this chapter, we show that this increase in primary energy use has in turn translated almost one-to-one into increased emissions of greenhouse gases. This is because civilization has relied so strongly on fossil fuels.

As in the previous chapter, we begin with an explanation of the underlying science —- here, the science of greenhouse gases and the planet's carbon cycle. We then describe which regions have been and will continue to be most responsible for their emissions.

In the next chapter, we will describe the effect of greenhouse gas emissions and concentration in the atmosphere on the planet's climate.

The next decade of indecision could be decisive.

1 Measuring Human Emissions

There are four important long-lived greenhouse gases that scientists have identified as the most troublesome. The most important one by far is carbon-dioxide (CO_2). Methane (CH_4) is a distant second. The other two, nitrous oxide (NO_x) and "F-gases" (containing Fluor), are even less important. We will discuss each gas in turn, but a quick overview makes the abstract more real. Figure 3.1 shows the main emitters of the different greenhouse gases.

Figure 3.1. Global Annual Greenhouse Gas (GHG) Emissions

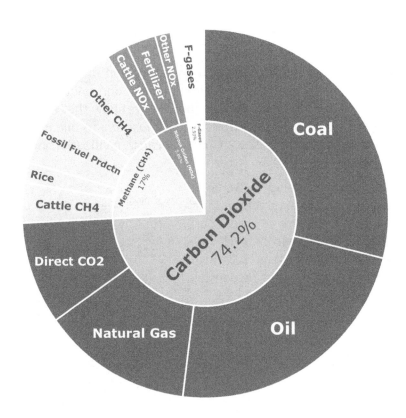

Note: The figure is for emissions around 2020. The detailed explanation is in Table 3.2.

1. Measuring Human Emissions

Carbon Dioxide

Carbon-dioxide is a colorless, odorless, non-toxic gas. It is the most natural of all emissions. All animals create it when they breathe. And all plants need it to photosynthesize.

The problem is not that CO_2 is intrinsically bad, but that too much CO_2 is bad. From our perspective, just as it was for energy, it's not enough to understand that humanity has been emitting more CO_2. Instead, it's a numbers game. It's important to know *how much more* we have been emitting and *how much more* has accumulated in the atmosphere. Therefore, we first need to explain how to measure it.

One cubic yard of anthracite coal weighs about 1,540 pounds or 0.70 metric tons (tonne). When burned, the added oxygen transforms it into 2.57 metric tonnes of CO_2. 1 tonne of carbon is therefore equivalent to 3.67 tonnes of CO_2. The "metric tonne of CO_2 (tCO_2)" is the principal unit of CO_2 emissions. When we want to put the scale of civilization's emissions into perspective — with billions of humans (consuming trillions of KWh of energy) — we have to measure global emissions in billions, too — specifically, in Giga-tonnes of CO_2 (**GtCO_2**).

As was the case with energy, you may be dismayed to learn that the scientists also love confusing their audiences about emissions. (It's almost as if they are having some devious fun at our expense.) The most important and painful ambiguity is their common, casual equating of carbon and carbon-dioxide: Many experts talk about emissions in terms of "tonnes of carbon," when they really mean "tonnes of carbon-dioxide."[1] We will try to stick to tonnes of CO_2 in our book *and* spell it out, but be aware of this common ambiguity when you read other books or articles.

Remarkably, our human population has grown so large that even our breathing now matters to the planet! The average person exhales about 1 kg (2.3 pounds) of CO_2 per day. Multiplying that amount by 8 billion people and by 365 days in a year implies that human metabolisms emit about 3 GtCO_2 per year. About 8% of humanity's emissions (of 38 GtCO_2) is just breathing.[2]

---

[1] This is especially problematic when they discuss "carbon taxes" (which we will cover in Chapter 5). It makes a big difference whether they mean a tax of $50/tonne of carbon or $50/tonne of carbon-dioxide — a dollar difference of 3.67 times!

[2] However, most of our breathing is 'just' re-emitting carbon that already was in our food. In some sense, humans were also responsible for removing this CO_2 from the atmosphere by

Table 3.2. Global Annual Greenhouse Gas (GHG) Emissions (circa 2020)

Carbon Dioxide (CO_2), 74.5%	38.0 $GtCO_2$
Coal, 39%	14.8 $GtCO_2$
Oil, 31%	11.8 $GtCO_2$
Gas, 18%	6.8 $GtCO_2$
Not Fossil Fuel Combustion, 12%	4.6 $GtCO_2$
Methane (CH4), 17.0%	8.7 $GtCO_2e$
Cattle, 21%	1.8 $GtCO_2e$
Rice, 10%	0.9 $GtCO_2e$
Fossil Fuel Production, 33%	2.9 $GtCO_2e$
Nitrous Oxides (NOx), 5.9%	3.0 $GtCO_2e$
Cattle, 23%	1.0 $GtCO_2e$
Fertilizer, 42%	1.3 $GtCO_2e$
Other (fluorinated) Greenhouse Gases	1.5 $GtCO_2e$
All GHG Emissions, Gates (2021)	51 $GtCO_2e$
Plus Land Charge (GCP via NL), 3.8 $GtCO_2e$	55 $GtCO_2e$
(reduction of green land caused by humans, ≈ 4 $GtCO_2e$)	

Note: These numbers were patched together from multiple sources and years and extrapolated to 2018–2022. The primary source was Olivier and Peters (2020), Netherlands EAA 2019 Report. We adopted Gates' (2021) overall GHG estimate of 51 $GtCO_2e$, and used CAIT/PIK/Olivier-Peters percentage estimates to extrapolate gas ingredients. The detailed subcategory estimates are also scaled from the EAA report and do not add to 100% for each category, because they omit some components.

Not shown, the same sources state that GHG emissions were about 34 $GtCO_2e$ in 1990, compared to 51 $GtCO_2e$ today, a growth rate of about 1.4% per annum.

There are also other greenhouse vapors that are not listed in this table. The most important GHG is water vapor (think humidity), responsible for about 70–90% of the greenhouse effect of our atmosphere. There are also soot and other less common substances. Global warming will be the subject of the next chapter.

1. Measuring Human Emissions

Table 3.2 (and Figure 3.1) show the main sources of our CO_2 emissions. (Un-)naturally, our global industrial activities and the burning of fossil fuels (circa 2021) emit a lot more CO_2 than just breathing — about 35 $GtCO_2$ in total.

As with the energy data in the previous chapter, it is common to see different emission estimates quoted. This means estimates are usually perfectly consistent only if you stay within the same data source.[3] For CO_2, there are also more reasons. First, reasonable measurement estimate variations are about ±5%. Second, some sources quote only CO_2 from fossil fuels (such as the Global Carbon Atlas with 34 $GtCO_2$ for 2020), others quote only CO_2 from combustion or emitting agriculture, etc. Thus, it is not uncommon to see CO_2 emission estimates anywhere from 34 $GtCO_2$ to 40 $GtCO_2$. Third, emissions have been increasing. Quoting 2018 gives a lower number than quoting 2021 — and then the Covid year of 2020 has caused all sorts of strange blips, leading some to prefer the earlier 2019 number. (We try to skip straight to 2022 estimates when we can, even though the year is not yet complete.) Usually, the data are not different enough to change the basic points we make.

Furthermore, civilization emits CO_2 not only by burning fossil fuels but also through some agricultural and chemical processes (principally CO_2 outgases in cement production) — roughly accounting for another 5 $GtCO_2$.

If we want to hold humanity responsible for increased CO_2 in the atmosphere, we also have to take into account that humans have reduced green land. This depletion mainly has to do with forests, which previously removed and sequestered CO_2 from the atmosphere and stored it, primarily in the form of wood. The current state of planetary deforestation is accounting for a loss of "CO_2 scrubbing" equivalent to about 10–15% of all human CO_2 emissions — somewhere between 3–5 $GtCO_2$ per year.

---

growing plants (possibly feeding them to food animals) in the first place. Thus, our breathing "cost" can be viewed as being attributable to our agricultural land-charge.

[3]It can become an issue when one attempts to patch together figures from different sources. For any given table, we try to use the authors' sets rather than patch in figures from other sets in order to continue comparing apples to apples.

3. Greenhouse Gases

The Other Long-Lived Greenhouse Gases

Carbon-dioxide is not the only important greenhouse gas. Table 3.2 also describes the effects of three other greenhouse gases. The two more important ones are Methane (CH4) and Nitrous Oxides (NOx). Methane is essentially the odorless natural gas that burns easily and runs most stoves and heating systems in the United States today. Nitrous oxide is also called laughing gas and was quite popular with dentists before there were good local anesthetics.

Methane and nitrous oxides are at least a thousand times rarer than CO_2 in the atmosphere, but they also have more potent warming effects. (Therefore, curbing methane — even simply as flaring it off — tends to be more cost-effective than reducing carbon-dioxide.) To make it easier to measure the entire sum of human greenhouse gases in terms of warming contribution, their emissions are often quoted in terms of CO_2 equivalents (**CO_2e**). There is some subjectivity regarding how CH4 and NOx should be counted with respect to their lifetime planetary warming contributions, but the standard approximations are good enough for our needs. Based on these standard equivalents, the most common estimates are that anthropogenic CO_2 is responsible for about 75% of global warming, Methane for about 15%, and Nitrous oxides (NOx) for about 6%.

In sum, direct human CO_2 emissions now run to about 40 GtCO_2, and greenhouse gas emissions now run at about **50 GtCO_2e per year**. They increase to about **55 GtCO_2e/year** when we add the land charge — the reduction of green land. Reasonable estimates can be 5% higher or lower.

Sources of Greenhouse Gases

Figure 3.3 provides a breakdown of energy use that is perhaps too detailed but again interesting to gawk at. The two overwhelming emitters are **energy** — the subject of our previous chapter — and **agriculture**.

Different activities produce different GHG pollution mixes. Burning coal produces relatively more nitrous oxides than burning natural gas. Agriculture produces relatively more methane than carbon-dioxide (mostly from cow and rice farming). Nevertheless, it is generally the case that where there is more CO_2, there are also more other GHGs.

More systematically, where do all these greenhouse gases come from? Table 3.4 breaks the sources into broad categories. Unsurprisingly, fossil-fuel

Figure 3.3. Source of Greenhouse Gases, 2016

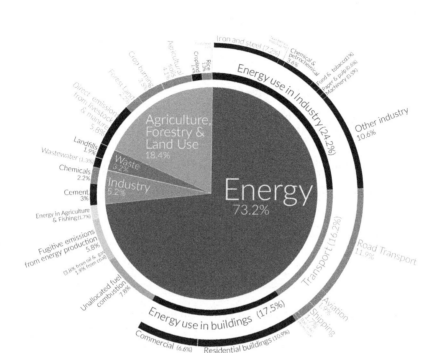

Source: Hannah Ritchie, 2021, Our World In Data.

combustion looms large. However, agriculture and land use are important non-combustion sources of greenhouse gases as well.

When reading about climate change, we are often struck by how easy it is to misunderstand authors. For example, many articles discuss CO_2 emissions, but that misses one-quarter of all effective GHG emissions (and it is rarely clear if the authors' figures include the land use charge). Fortunately, because CO_2 emissions are generally reasonably in line with GHG emissions (except for agriculture), and authors quote percentages, one can often mentally scale up the CO_2 picture. Countries and activities that emit more CO_2 typically emit more GHGs as well.

3. GREENHOUSE GASES

Table 3.4. Annual Emissions, circa 2018–2022

Power, Heat, Agriculture, All GHGs		51 GtCO$_2$e
Agriculture, 19% (of 51 GtCO$_2$e)	9.7 GtCO$_2$e	
Non-Ag Emissions, All GHGs		41 GtCO$_2$e
Combustion, CO$_2$ Only	33.4 GtCO$_2$	
Combustion, NOx Only	0.5 GtCO$_2$e	
Transport, 16% (of 51 GtCO$_2$e)	8.2 GtCO$_2$e	
Electricity, 27%	13.8 GtCO$_2$e	
Heating, 7%	3.6 GtCO$_2$e	
Industrial, 31%	15.8 GtCO$_2$e	
Reasonably Electrifiable Emissions		25–35 GtCO$_2$e
(Heating=3.6, less fossil-fuel extraction=5, some industrial=5, cars/trucks=5.5.)		
Difficult/Costlier To Electrify		15–25 GtCO$_2$e
(Some high industrial heat and cement=10, agriculture=10, ships/airplanes=1.6.)		

Note: The primary source was Olivier and Peters (2020) and Gates (2021). Gates (2021) estimates are reasonably close. Both are based on similar sources. Our own estimates of potentially electrifiable emissions were a little less optimistic than Gates' but generally similar. (WRI reports that agricultural emissions are about 40% livestock, 25% fertilizer, 15% burning, and 10% rice cultivation.)

Table 3.4 also hints at a more serious confusion. It arises when articles discuss "total" decarbonization but refer only to electricity. As we explained in Chapter 2, electric power generation accounts for less than one-third of human primary energy use today. Even zero carbon emissions in electricity generation would not mean zero total emissions — far from it. In a few decades, electrification of ground transportation and heating could realistically increase the share of electric power to two-thirds. Unfortunately, the final third will be much more difficult to decarbonize — agriculture, airplanes, ships, industrial heat, etc. — and perhaps will never rely exclusively on electricity.

And, of course, humanity is still on the move. With continued population growth and economic development, the demand for energy is continuing to increase. As we explained in the previous chapter, humanity will use a lot

Other Fossil-Fuel Pollution

Fossil fuels emit not only greenhouse gases but also other more local pollution. Most importantly, coal and oil emit tiny aerosol particles (such as smog and soot). The negative health effects of these local emissions are enough to more than justify drastically curbing fossil-fuel use — even ignoring their global warming consequences.

Scientists have estimated that without local fossil-fuel pollution, the average life expectancy of the world's population could perhaps increase by 3 to 5 years, and global economic and health costs could fall by more than $3 trillion (out of a world GDP of about $90 trillion). On the high end of death estimates, as many as 5 million to 9 million people may die prematurely every year due to direct pollution caused by fossil-fuel combustion.[4] Between 1-in-5 and 1-in-10 deaths may be hastened by the same fossil-fuel processes that generate our energy and emit our greenhouse gases. Fossil fuels are murderous.

To be fair, these mortality figures are rough estimates and other reasonable scientists might halve them. But they are not outlandish and there is no doubt: The local adverse health effects and health care costs of fossil fuels are severe.

Ironically, not all fossil-fuel pollutants are bad from a climate-change perspective. The burning of sulphur-laden coal produces tiny sulfur dioxide (SO_2) aerosol particles — famous for causing acid rain. However, these SO_2 particles are also reflective and thereby enhance the planet's albedo.[5] The burning of dirty coal has therefore probably held down global temperature by about 0.6°C (out of a total of 1.5°C). An ongoing shift towards cleaner coal is about to reduce this cooling effect.

---

[4]It is not clear whether climate-change itself is already increasing deaths (e.g. through droughts and heat waves). Colder temperatures kill more people than warmer temperatures. For example, the UK Office for National Statistics (ONS) found that more climate change has saved more than 25,000 lives per year between 2001 and 2020. This is not generalizable to the rest of the world.

[5]Albedo means literally "whiteness" in Latin. Snow and clouds are the most important sources of planetary albedo.

3. Greenhouse Gases

2 Earth's Natural Carbon Cycle

Fortunately, not all human emissions accumulate in the atmosphere. Thus we now take a brief detour into the earth sciences to explain where they ultimately end up.[6]

Land, Sea, and Air

Carbon in its various forms, including carbon dioxide (CO_2) and methane (CH4), can be found on land, in the sea, or in the air. In the ocean, dissolved CO_2 acidifies the water. There is about 50 times more CO_2 (140,000 $GtCO_2$) in the oceans than there is in the atmosphere (3,200 $GtCO_2$). In addition, the oceans also store large amounts of frozen methane at their deepest bottom.[7]

In the ground, carbon is typically not a problem, because it is generally bound in stable solid or liquid forms. This carbon and its compounds are stored in biological matter (including not only in trees but also in us), in coal and oil, in weathered rocks, or in the deep underground (where both CO_2 and CH_4 become pressurized liquids). In total, the soil holds about 2,500 gigatonnes of carbon, equivalent to about 9,000 $GtCO_2$.

There is one big and one small exception to the general rule that carbon in the ground is no problem. The Arctic permafrost is comprised of the regions of northern Canada and Russia where the ground has not melted even in summer for millennia. It now contains a lot of undecomposed organic matter. If (or better when) the temperature in the high north increases to the point where the permafrost melts, microorganisms will turn this matter into atmospheric carbon-dioxide — or, worse yet, methane. Remarkably, there is more carbon buried in the permafrost (about 3,700 $GtCO_2$) than there is in total in the atmosphere today (about 3,200 $GtCO_2$). From a greenhouse perspective, the Permafrost is a live (though probably not a quick-release) time bomb. The smaller exceptions are other non-Permafrost peat lands, which emit CO_2 when they are drying out or being dried out by farmers — still to the tune of about 2 $GtCO_2$ per year.

---

[6] For a more detailed and yet readable discussion, we recommend David Archer's The Long Thaw.

[7] Scientists do not know whether it is 1,000 $GtCO_2e$ or 30,000 $GtCO_2e$. Fortunately, it seems highly unlikely that the planet will warm enough to release this Methane during the next few thousand years.

2. Earth's Natural Carbon Cycle

Oceans	Atmosphere	Permafrost	Other Terrestrial
140,000 GtCO$_2$	3,200 GtCO$_2$	3,700 GtCO$_2$	5,300 GtCO$_2$

Carbon Cycle Equilibrium

It is the CO$_2$ and other long-lived GHGs in the air that are the sources of humanity's climate-change problem. Their balance in the atmosphere is the main issue of this chapter. (We delay the discussion of *how* the atmosphere raises the planet's temperature through the greenhouse effect to the next chapter.)

Each year, about 1,000 GtCO$_2$ moves naturally into the atmosphere. Common sources are warm ocean surfaces (essentially bubbling out of dissolved CO$_2$, carbonic acid), fires, and volcanoes. Each year, an almost equal amount of 1,000 GtCO$_2$ moves naturally back out of the atmosphere. That is, carbon-dioxide flows out of the atmosphere into what are called "carbon sinks." This circulation is called the "carbon cycle."

The most important carbon sink is the ocean. Rain water captures and dissolves CO$_2$ and eventually flows into the ocean. This CO$_2$ is then integrated into plankton (which itself contains large amounts of calcium, Ca). It then turns into limestone (CaCO3) on the ocean floor and is finally subducted by tectonic forces beneath the ocean into the earth's interior.

Fortunately, there is more than enough calcium in the oceans to absorb all the CO$_2$ that humanity could ever dump into the environment many times over. Unfortunately, the speed with which the ocean can bring this new calcium online (and thus shuttle more CO$_2$ from the air to the ocean bottom) is (too) slow. Thus, when CO$_2$ accumulates in the atmosphere and presses into the ocean, there is not enough calcium to immediately react with the CO$_2$. The time lag reduces the ocean's ability to absorb and store CO$_2$. It is only in the very long-term that the oceans can bring enough calcium back online and expose enough cold ocean surface to the atmosphere that they can scrub out all excess atmospheric CO$_2$

When the existing calcium buffers become temporarily exhausted, excess CO$_2$ turns into increased carbonic acid, and the ocean's native alkalinity decreases.[8] Over the last 30 years, anthropogenic CO$_2$ has increased the

---

[8]The level measures alkalinity, of which acidity is the opposite. An acidity level of 1 is battery acid, of 6 is milk, an acidity level of 13 is bleach, an acidity level of 11 is Ammonia. Pure water is a neutral 7. Thus, the oceans are alkaline, but are becoming less so now.

3. Greenhouse Gases

ocean acidity from <u>a ph level of about 8.11 to about 8.08</u>. Given the giant size of the oceans, this is an impressive change.

Why does ocean acidity matter? Human CO_2 emissions will not make the ocean so much less alkaline (relatively more acidic) that it would poison sea creatures. The effect on marine life will work through a different channel. The same calcium that is now pulled out of the solution into sequestering more CO_2 was previously used by marine life (especially plankton) to build their shells. With less available calcium, many species will no longer be able to build effective shells and will go extinct. In turn, this could percolate up the food chain. The effects could be <u>deadly serious</u> far beyond the smallest ocean creatures.[9] Therefore some researchers are now investigating whether <u>lime</u> ($Ca(OH)_2$) could be added to the ocean in order to help speed up the slow natural calcium cycle *at an affordable cost*. (Yes, environmentalism is all about economics, too.)

The next two important sinks are terrestrial. The first are minerals that <u>weather</u>, i.e., change from one type of rock into another by absorbing CO_2. The most important such mineral is <u>Olivine</u>. It constitutes about half of Earth's crust. Fortunately, there is enough Olivine around to absorb human emissions a hundred times over. Unfortunately, like the ocean calcium process, the natural weathering process is also very slow, taking many centuries. Therefore other researchers are now investigating whether we can actively coax Olivine to absorb CO_2 faster *at an affordable cost*. (Once again, it is all about economics.)

The second terrestrial sink is life itself. Living organisms are estimated to contain about 550 Gt of carbon, equivalent to about 2,000 $GtCO_2e$. Wood is a particularly effective and valuable carbon sink, because it is long-lived and decays slowly after death; and young, growing trees are particularly efficient in pulling out CO_2, because they are growing trunks more aggressively. Some researchers are now investigating whether planting more trees can sequester CO_2 more quickly *at an affordable cost*. (Economics yet again. Are you detecting a pattern?)

---

[9]Humans will not realize the extent of this problem for a long time. Ironically, it will be difficult to ascertain the mechanism of our destructive influence, because humanity is doing so much harm on so many fronts at the same time. Humans are simultaneously wiping out fish at an unprecedented rate, changing the ocean currents through global warming, and acidifying the oceans — a veritable trifecta.

However, such schemes will work well only if the wood is buried under soil (to become fossil fuel in a few million years) or is harvested and used for lumber. If wood is allowed to burn or die-and-decay, the CO_2 is released back into the atmosphere. (Of course, environmentalists love to sue whenever timber companies are logging forests. They would probably love to sue when fires destroy trees, too, but fire is difficult to drag into court.)

(We will return to research underway to capture CO_2 via enhanced ocean absorption, weathering, or tree planting — called sequestration — in Chapter 14.)

3 Accumulating Human Emissions

The carbon cycle is often compared to a giant barrel, with a roughly equal inflow and outflow of water. The flows into and out of the barrel are never perfectly balanced, but small fluctuations do not matter much. It is a big barrel, and it takes large one-sided inflows or outflows to raise or lower the level. However, even modest unbalanced excess inflows or outflows can and do accumulate if they occur consistently over long enough time spans.

For many millennia, the natural atmospheric inflows and outflows were reasonably well-balanced. Popular belief to the contrary, even large volcanic eruptions have had only temporary and small influences over the course of the last few millions of years. The giant eruption of Mount Pinatubo in 1991 emitted about 0.05 $GtCO_2$. All global volcanic activity combined emits about 0.1 to 0.5 $GtCO_2$ per year. Much bigger supervolcano eruptions could emit 50–100 $GtCO_2$ or more — but the last modest supervolcano eruption (Lake Taupo) occurred in New Zealand about 25,000 years ago. It also emitted only about 2–3 times as much as humanity emits each year. Yellowstone is an even larger supervolcano, but it erupted most recently about 500,000 years ago.[10]

By comparison, humans keep pushing an extra 40 $GtCO_2$ per year *every year* into the atmosphere. This is roughly 50–100 times more than what volcanic activity or fires emit in a typical year. Of course, 40 $GtCO_2$ is also much less than the 1,000 $GtCO_2$ that move in and out of the carbon cycle every year

---

[10] The Siberian Traps did emit vastly larger amounts of CO_2 and other gases about 500 million years ago. This probably caused the *Great Dying* in which 97% of all species vanished.

3. Greenhouse Gases

or the 140,000 $GtCO_2$ that are already present in the ocean. And if humanity emitted 40 $GtCO_2$ for a year or two, it would not make much difference — the atmosphere is a very big barrel. The problem is that civilization has been emitting 40 $GtCO_2$ every year for many years now *and* it will emit a lot more soon *and* it will do so for many more decades — and this does make a big difference.

Atmospheric Carbon-Dioxide Readjustment Processes

David Archer, who researches the complex long-run and earth-state-specific changes of our atmosphere, characterizes the scrubbing process as it pertains to our human excess CO_2 emissions as follows: about half of our emitted CO_2 is scrubbed immediately (and of this half, equal parts disappear into the ocean and into the soil); another half of the remaining half will disappear within about 30 years; and the remainder will lurk in the atmosphere for thousands of years or more. What human civilization does in the 21st century will have long-lasting effects.

A longer description is that the annual absorption of greenhouse gases from the atmosphere into sinks is not directly linked to contemporaneous annual human emissions. Instead, it is determined by the momentary relative balance of CO_2 in its three reservoirs (air, ocean, and land) and influenced by many other aspects relating to the state of the planet — such as the planetary temperature, the current calcite level in the ocean, the availability of olivine on land (plus the rain necessary to allow olivine to weather), and so on.

Figure 3.5 sketches how inflows and outflows were linked (circa 2020). A little more than 10 Gt of carbon from human activity ultimately combined with oxygen to become about 38 $GtCO_2$ of human emissions. We should add a land charge (reduced CO_2 absorption) of about 4 $GtCO_2$, because humans were responsible for tree reductions, too. Call it about 40 $GtCO_2$ over one year. Simultaneously, over the same year, above and beyond the "base sink rate" of about 1,000 $GtCO_2$/year, the planet weathered about an extra 10 $GtCO_2$ into rocks and dissolved about an extra 10 $GtCO_2$ into the ocean due to the differences in the relative CO_2 pressure among the three reservoirs. Even if humanity went cold-turkey and stopped emitting CO_2 altogether, the land and ocean sinks would (likely) still continue to each scrub about 10 $GtCO_2$ per year from the atmosphere for many years. Eventually, these scrubbing processes would then slow down as the CO_2 pressure from the atmosphere into the water would drop.

3. Accumulating Human Emissions

Figure 3.5. Annual Human-Related CO_2 Flows, ca 2020

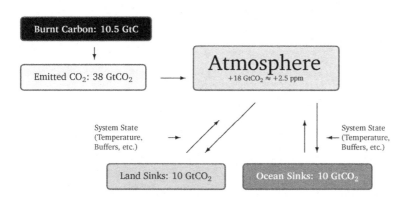

Note: The link between inflows and outflows is weak over human lifespans. Outflows are determined by the system state, not by that year's emissions. If human CO_2 emissions suddenly stopped, it would not instantly reduce the CO_2 outflow rate from the atmosphere into land and ocean sinks. Instead, the outflow rate would slowly start declining, e.g., based on the (relative) CO_2 in the atmosphere, the planetary temperature, the availability of rocks that can weather, the CO_2 concentration in the ocean, and so on.

Source: David Archer, The Long Thaw.

Figure 3.6 shows estimates of how the emissions and removal processes have worked year by year over the last 50 years. About 90% of our CO_2 charge were emissions from fossil fuels; the rest was from land use. The oceans have been taking up CO_2 very steadily, while land sinks and the atmosphere have been absorbing CO_2 with much year-to-year variation.

The Half-Life of Human Excess CO_2

The dependence of the CO_2 processes on many other state variables explains why there is no straightforward half-life of CO_2 in the atmosphere. Nevertheless, the concept of a half-life — how long it takes to remove half of any given emission of CO_2 — at least at the moment can still be a useful conceptual guide. Figure 3.7 shows a current educated guess about the speed of the removal processes of extra CO_2 in the atmosphere.

In sum, humans can be held responsible for adding about 20 GtCO_2 *net* to our atmosphere every year (ca 2020), i.e., CO_2 that the planet does not scrub

3. Greenhouse Gases

Figure 3.6. Sources and Sinks, 1959–2019

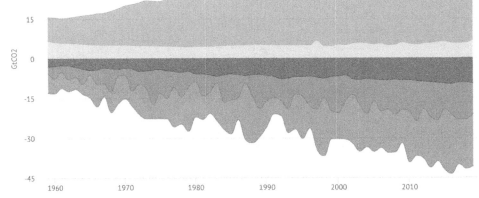

Source: The Global Carbon Project and CO2.earth.

Figure 3.7. CO_2 Time to Equilibrium

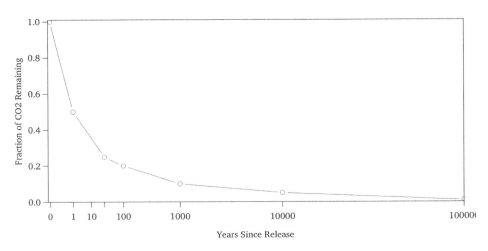

Source: Our interpretation of David Archer's, Long Thaw.

away in the same year. About 15 GtCO$_2$ will slowly disappear in a matter of decades or centuries. The final 5 GtCO$_2$ will remain in the atmosphere for a millennium or longer.

Don't worry. The planet will adjust. In the very long term — over a few hundred thousand years — natural earth processes will eventually scrub all the human-emitted CO$_2$ into sinks, where this CO$_2$ will no longer have much impact on the climate. You need to worry "only" if you are more interested in the next few thousand years than in the next few hundred-thousand years!

4 The Balance Sheet

You are now armed with the knowledge to understand the bigger picture.

The Historical Accumulation

Figure 3.8. Atmospheric CO$_2$ Concentration

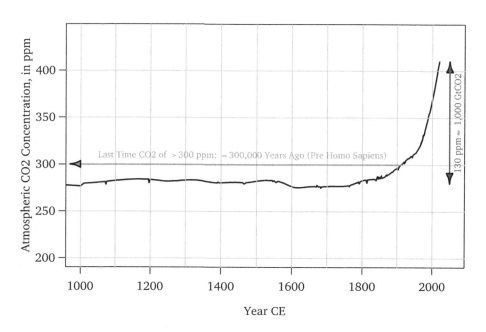

Source: Pre-1955 values based on smoothed Vostok ice core samples (ClimateData.Info). Post-1955 values based on direct NOAA CO$_2$ measurement on Mauna Loa.

3. GREENHOUSE GASES

Scientists have measurements of atmospheric CO_2 concentration going back a long time. These measurements are accurate enough to learn how the concentration of CO_2 in the atmosphere has changed. Figure 3.8 plots them over the past 1,000 years.[11] The planetary CO_2 concentration was stable between about 270 and 280 parts-per-million (ppm) until the 19th century. Until the 19th century, atmospheric CO_2 concentration changes due to human emissions were so small, going up or down year by year, that scientists cannot determine reliably whether they came from anthropogenic or natural processes. (And they were mostly scrubbed away by barrel Earth within a few years, anyway, although a little residual may have accumulated very slowly since about 1800.)

Beginning around 1800 but certainly after 1900, scientists observed a steady rise in atmospheric CO_2. They also observed corroborating chemical evidence that most of this CO_2 increase came from burning ancient fossil-fuel-based carbon and not from recent organic carbon or volcanoes. Thus, scientists know that humans are responsible for most or all of the increase in atmospheric CO_2 concentration since 1900.

Moreover, we have confirmation. Simple chemistry and math implies that 1 ppm of CO_2 over the entire planetary atmosphere is the equivalent of about 7.8 $GtCO_2$. Thus, the increase from the pre-industrial 280 ppm to the 410 ppm in 2020, i.e., 130 ppm, is equivalent to about 1,000 $GtCO_2$ in added CO_2 stored in the atmosphere. Scientists can compare this to human activities directly. National accounting estimates suggest that humanity ramped up its fossil-fuel based activities beginning with the Industrial Revolution. Adding up national emissions, humans have pushed out a total of about 1,700 $GtCO_2$. Coal was responsible for about 800 $GtCO_2$ (47%), oil for 600 $GtCO_2$ (35%), gas for 250 $GtCO_2$ (15%), and cement for 50 $GtCO_2$ (3%). This sums to about 1,700 $GtCO_2$ of human CO_2 emissions. Another 300 $GtCO_2$ are the land charge. The planet scrubbed a net of about 700 $GtCO_2$ of this; the remaining 1,000 $GtCO_2$ are in the atmosphere. Table 3.9 puts observed CO_2 concentrations and human emissions together in order to summarize how the world got to where it is today. (The numbers are continuing to change quickly, though. As of 2022, we are about to reach 420 ppm and accelerating.))

---

[11]The CO_2 estimates were smoothed to reduce measurement noise. Not shown, the CO_2 concentration over the last 300,000 years was stable. It looks just like the first part of the figure.

4. The Balance Sheet

Table 3.9. Human CO_2 Emissions and Atmospheric CO_2

Year →:	1870	1970	2000	2020
1. Annual Emissions, $GtCO_2$	0.5/y	14.8/y	25.1/y	36.5/y
2. **(Cumulative) Emitted, $GtCO_2$**	11.5	423	1,040	1,690
3. Change in Atmospheric CO_2 since 1770, $GtCO_2$	50	350	850	1,050
4. Atmospheric CO_2 (Total), $GtCO_2$	2,200	2,500	3,000	3,200
5. **Atmospheric CO_2 ppm**	280	320	380	410
6. Rate of Change, CO_2 ppm	+0.14/y	+0.9/y	+2.0/y	+2.2/y

Note: The primary point of this table is to show that planetary changes in CO_2 concentration were determined primarily by non-human sources before 1950 and increasingly by human sources thereafter.

Source: Cumulative and annual human emissions are from Our World in Data and NASA. The CO_2 concentrations can be found, e.g., at the EPA or Ahn et al (2012).

[1,2] Human cumulative emitted CO_2 are summed beginning in 1770. The retained change in atmospheric $GtCO_2$ levels since 1770 [3] are net of baseline [4] and obtained via simple translations of atmospheric CO_2 ppm estimates [5]. The rate of change [6] is estimated from single-year changes. This table excludes the land charge, which would add another 600 $GtCO_2e$ for which humanity is responsible. The starting year 1770 for the accumulation was chosen because the second agricultural revolution (and with it the industrial revolution and high population growth) began around 1800.

The Future

What will happen next? Scientists know how the ocean and land sinking processes have functioned in the past. Fortunately, they have not yet detected any visible deterioration in their absorbing capabilities. (They are very big sinks indeed!) Scientists also *believe* they will continue to function in the future. But they are not certain. Earth is a complex system and not fully understood. The scrubbing processes could hit snags.

A disruption in the carbon sinking processes is not entirely implausible, because these processes depend on other Earth state aspects — such as the planetary temperature which has not yet increased even half as much as scientists predict it will. Scientists have never observed the state configuration which Earth will soon be experiencing.[12] If the planetary temperature were

---

[12] Scientific Clarification: By state configuration, we do not only mean levels in various inputs but also their direction and rate of change. We are less concerned with 1,000,000 year

to rise in the future, it could alter or even reverse both the ocean and the soil carbon-dioxide sink rates. The oceans could start bubbling out relatively more CO_2 and absorb relatively less than they do in the cooler waters of today. Similarly, melting permafrost could start releasing more greenhouse gases. In addition, a less reflective ice layer could further heat the planet. But other processes could counterbalance such scary feedback loops, including <u>increased plant growth due to higher CO_2 levels</u> and increased rainfall.

Frankly, scientists cannot know for sure what will happen. They are making "well-educated guesses" — better than those made by pundits and interest groups. But consider this: many scientists are loathe to make "worst-case" predictions. The point of worst-case predictions is that they are not supposed to be likely to come true. Given the politicized "climate around climate change," making starker and more extreme predictions could easily lead to accusations of misrepresentation or even unscientific political bias. What serious scientist wants to risk this?

How many unknown unknowns could be out there? What is the probability that carbon sinks will become exhausted, and how dangerous would this be? Given the path that the planet is on, it looks as if we will find out all too soon.

5 Growth in Human CO_2 Emissions

If we want to tackle greenhouse gas emissions, we have to determine where they are originating. We already explained above that 73% of emissions are from energy provision, which are themselves 85% fossil fuels; and that the remaining 27% tend to be highly correlated with energy use, too.

In the previous chapter, we described how different countries and regions consume energy. In the remainder of this chapter, we explain how they have been and will be responsible for emissions.

Figure 3.10 plots a broad measure of global CO_2 emissions by country/region over the decades. You can see how emissions have grown alarmingly quickly and are still accelerating today. Before 1900, emissions were negligible — just like energy use. In 1900, Europe was still the world leader in emissions.

---

long-run equilibrium outcome than with short-term 100-year spikes that could "temporarily" devastate the biosphere.

5. Growth in Human CO_2 Emissions

Figure 3.10. Annual CO_2 Emissions By Area/Country

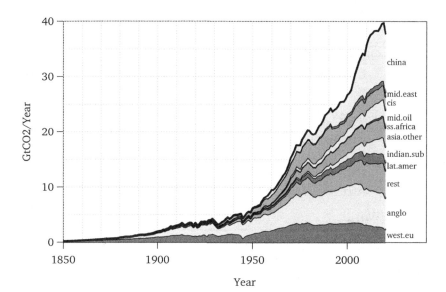

Note: Countries were grouped into regions and ordered by emissions in 1850. See appendix for precise classification. In brief, "anglo" are the US, Canada, and Australia. The "rest" of the world contains many countries that are not easy to classify — such as South Africa or Israel. The Middle East was split into oil-rich countries (such as Saudi Arabia) and others (such as Egypt or Jordan).

Source: Our World in Data. The figure does not include greenhouse gases other than CO_2 and the land charge, but it can be mentally scaled up proportionally.

By World War I, the dubious-distinction baton had passed to North America, primarily the United States. By the turn of the millennium, it had passed again, this time to China and Asia.

Even by 1950, total emissions were still running at the low rate of 6 $GtCO_2$ per year — only about twice what human respiration alone produces today. By 1988, our emissions had more than tripled to 22 $GtCO_2$ per year. By now, civilization emits about 40 $GtCO_2$ per year, 50 $GtCO_2$e including other greenhouse gases, and 55 $GtCO_2$e if we add the land charge.

Our by-now familiar region classification in Table 3.11 shows that emissions have stabilized in OECD countries (about 12 $GtCO_2$ per year) but continue to grow in non-OECD countries (about 24 $GtCO_2$ per year currently;

3. Greenhouse Gases

Table 3.11. CO_2 Energy-Based Emissions, in $GtCO_2$

	1981	2000	2020	2050e
OECD	11.3	13.6	12.1	12.1
USA	4.6	5.9	4.8	4.8
Europe	4.3	4.4	3.8	3.7
Not OECD	6.8	10.3	24.1	30.8
China	1.2	3.2	11.0	10.5
India	0.2	0.8	2.7	5.8
Other Asia	0.5	1.3	2.8	4.9
Africa	0.5	0.8	1.3	2.0
World	18.1	23.6	34.3	42.8

Source: US EIA International Energy Outlook, Oct 2021. Carbon Dioxide Emissions by Region. Other Asia includes, e.g., Indonesia, Pakistan, and Bangladesh.

about 31 $GtCO_2$ within one generation). Without a fundamental change, within our lifetimes, non-OECD countries will emit by themselves what the world emits *in total* today.

The majority of the global growth in emissions over the last 20 years (+10.4 $GtCO_2$) has come from China (+7.8 $GtCO_2$). China alone already emits 11.0 $GtCO_2$ — more than the United States (4.8 $GtCO_2$) and OECD Europe (3.8 $GtCO_2$) combined. It is mostly due to its larger population and its extensive use of coal that it already emits more CO_2 than all rich countries combined. It also emits more than the next four biggest country emitters combined.

With a per-capita income of about $20,000 per year, China is also still only about halfway between rich and poor countries. Fortunately, China's emissions are now stable. Unfortunately, they are stable at a very high level. (This is just like the OECD. Remember, these facts are not about finger-pointing or assigning blame.)

The EIA projects that emissions in the next 30 years will grow most in India and other Asian countries. India alone will soon emit more than either the USA or Europe. The same will apply to the rest of Asia (countries other

5. Growth in Human CO_2 Emissions

than China and India). That is, each of the three regions — China, India, Other Asia — will soon emit more than either the USA or Europe.

Africa, with its fast-growing population, remains far behind. It emits so little because it uses so little energy because it is so poor. Of course, Africans deserve no less than the humans in the rest of the planet. Contemplate this — how do you imagine will Africans attain reasonable standards of living?

sidenote

> A 2021 report by Carbonbrief comes to a surprising conclusion: CO_2 emissions may have already been flat since 2012, because increasing fossil fuel emissions were balanced by a declining land-use charge. Though good news, there is large uncertainty surrounding this estimate and the inference should be confirmed by other scientists.

3. GREENHOUSE GASES

6 Energy and Emissions

So far, across countries, regions and time, wherever and whenever economic growth has increased, so has primary energy use, so has fossil fuel consumption, and so have greenhouse gas emissions. (Efficiency improvements have mitigated this but not stopped it.)

6. Energy and Emissions

Global Trends

Is it possible to disconnect the world's emission growth from its energy growth (itself caused by population and income growth, especially in non-OECD countries)? Is climate activism an important part of the answer?

We can test this informally. The scientific and popular concern about the cumulative effect of greenhouse gases began only fairly recently. In 1988, global emissions and climate change entered the popular conscience through (bipartisan) landmark testimony by NASA scientist James Hansen to Congress.

Figure 3.12. Primary Energy Use and CO_2 Emissions

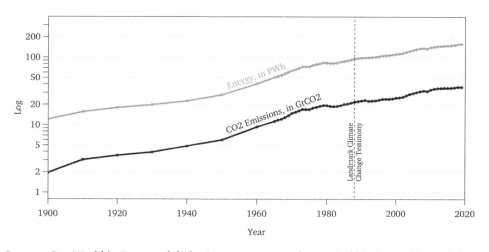

Source: Our World in Energy, global primary energy; and Our World in Data, CO_2 emissions. The numbers are annual measures. The scale is in log, which is visually less alarming. Don't look at the trends. Many variables have upward time trends. Instead look at how deviations from the trend have occured together.

Has the subsequent increased awareness of global warming made much of a difference in reduced economic activity or decoupling CO_2 emissions from economic development? Apparently not yet.

Figure 3.12 shows how world CO_2 emissions continued to grow in lockstep with energy just as (un-)healthily after 1988 as they did before. The reason is simple: most human emissions of CO_2 were still occurring in the energy provision sector, which was still mostly met by fossil fuels emitting greenhouse gases; and most population and consumption growth occurred in China.

3. Greenhouse Gases

Although the figure is not proof, it does suggest that climate activism has not made a large dent. If you place your faith in more activism, ask yourself this: who needs to become more aware of or concerned about climate change than they are already today? If it has not worked in the past, how exactly will it work in the future? We are not against climate activism. We are just skeptical about whether activism will be more effective in the future.

So what have climate activists *really* accomplished? Pundits have been lamenting year after year that no meaningful progress has been made — today's *Associated Press* headline reads "UN climate report: 'Atlas of human suffering' worse, bigger"). And then the pundits have repeated the same script the following year.

You may or may not agree with Greta Thunberg, but she does have one incontrovertible point: collectively, humanity has not done much so far to curb its emissions. Every year, the world is falling further and further behind on climate activists' aspiration.

Greta is right on her observation, but she is wrong about what the relevant problem is. If she is hoping for collective world action, we are predicting that it will not happen. (We will explain our reasoning in part II of our book.) The world has no central collective decision-maker. Instead, countries and people can only make decisions for themselves. It is not the collective decision problem that matters but the many individual decision problems that do.

Emissions and Emissions Efficiency

The previous chapter explained that primary energy consumption is determined primarily by (a) population, (b) economic activity (income) per capita, and (c) efficiency. It also explained how difficult it would be to tackle population and per-capita income, especially in non-OECD countries which want to escape widespread poverty. Finally, it described how humanity is making good but insufficient progress in energy efficiency all over the globe.

This means that in order to break the link between energy use and emissions, the only remaining lever is working towards higher emission efficiency — less CO_2 emitted for each unit of energy. This means using cleaner energy sources.

6. Energy and Emissions

Table 3.13. CO2 Per Unit of Primary Energy, 2022

	CO_2 (GtCO$_2$)	=	Energy (PWh)	×	Energy Efficiency (g/KWh)
OECD	12.1	=	71	×	170
USA	4.8	=	28	×	170
EUR	3.8	=	24	×	160
Not OECD	24.2	=	116	×	209
China	11.0	=	48	×	232
India	2.7	=	12	×	226
Other Asia	2.8	=	14	×	203
Africa	1.3	=	7	×	184
World	36.3	=	187	×	194

Note: Non-OECD countries generally use dirtier energy sources.

Source: US EIA International Energy Outlook, Oct 2021. CO_2 and Primary Energy.

▶ The Current State

Table 3.13 shows how energy and emissions are related in different regions. OECD countries emit less CO_2 for each unit of energy consumed than non-OECD countries. They rely relatively less on coal and relatively more on natural gas, hydro, nuclear, wind, and solar power. Europe is the most emission-efficient region; China is the least emission-efficient.

Although there are clear differences in the types of energy used and thus their emissions, these differences are not orders of magnitudes. If the goal is to reduce the emissions of CO_2, the efficiency gains need to improve by an order of magnitude. This can only be accomplished by using more clean energy and less fossil fuel.

▶ The Growth

Do trends suggest that the world can reach much greater emissions efficiency anytime soon? Table 3.14 shows what the EIA reports for historical growth and expects for future growth.

3. Greenhouse Gases

OECD countries have been and are likely to continue covering all their growth in (GDP and) energy use with cleaner energy — although they will remain at their high levels of per-capita emissions. Europe's per-capita emissions are already close to those in non-OECD countries. This is not the case for the United States, whose per-capita emissions run at twice the rate of Europe's. The United States could reasonably and plausibly go quite a bit lower in per-capita emissions with some effort.

Beyond the OECD, the EIA projects that clean energy is not likely to arrest emissions growth, either this coming generation (2050e) or the generation thereafter (2080e). Yes, non-OECD emissions will grow relatively slower in percentage terms than they have in the past, but there is little cause to celebrate. Their base level of emissions is now much higher than it was in 1994, so their growth in emissions is not even slowing down in absolute terms. And unfortunately Earth has not been growing in its ability to absorb more emissions.

China and India had the highest growth in emissions over the last 30 years, relying disproportionately on coal power. India is the only region which did not manage to improve emission efficiency, but its share of global emissions was small in the past.

Looking forward, China will continue to grow its per-capita income growth and energy consumption, but efficiency improvements will be able to cover it. Its population growth has slowed, its emissions efficiency is improving faster than its GDP growth, and it is installing clean energy sources faster than any other country in the world. But not all is well. China's more ambitious public climate-change promises are only set to kick in after 2050 — long after any contemporary promises will still be remembered. Its current policy is still to plan and build new coal plants at a record pace. Like the United States, China could do better on its emissions efficiency — not enough to stop the growth in world-wide emissions but important nevertheless. Currently, it's not happening.

Table 3.14. CO_2 Growth Per Unit of (Primary) Energy

	CO_2		Energy (E)		Efficiency (CO2/E)	
	94–22	22–50e	94–22	22–50e	94–22	22–50e
OECD	−0.08	−0.01	0.08	0.11	−0.16	−0.12
USA	−0.13	−0.03	0.05	0.16	−0.18	−0.18
EUR	−0.04	0.00	0.13	0.15	−0.16	−0.15
Not OECD	0.96	0.24	1.06	0.43	−0.10	−0.18
China	1.44	−0.05	1.74	0.19	−0.29	−0.24
India	1.51	0.77	1.51	1.08	0.00	−0.31
Other Asia	1.10	0.55	1.28	0.60	−0.18	−0.05
Africa	0.62	0.44	0.76	0.63	−0.14	−0.19
World	0.54	0.17	0.61	0.33	−0.08	−0.16

Note: These should be interpreted as (fractional) changes. However, because they are logged, the components on the right add up to the quantity on the left. The 2050e numbers are quoted in $GtCO_2$ in Appendix Chapter E.

Source: US EIA International Energy Outlook, Oct 2021. CO_2 and Primary Energy.

7 The Kaya Components

Remember that we started the book with the statement that there are only four ways to reduce emissions? The economist Yoichi Kaya first explained why. We can view human emissions through the lens of four components to summarize where we are:

1. We can reduce the number of people on the planet.

   Unfortunately, this is not happening. Human population is not decreasing but increasing. This is especially the case in poorer non-OECD countries — India and Africa, in particular.

2. We can reduce how much energy each of us consumes by reducing our economic activities — for example, by working, producing, and consuming less.

   This is also not happening. People in non-OECD countries want their economies to grow in order to escape poverty. Asian countries are making good progress.

3. Greenhouse Gases

3. We can improve the energy efficiency of our economic activities, reducing how much energy each of us consumes — for example, by insulating our buildings better, or producing and consuming less energy-intensive products.

   This is happening but not fast enough. China and the (much smaller) USA might be able to accelerate their efficiency gains in order to get closer to European standards, but it still won't be fast enough.

4. We can improve the emissions efficiency of our energy use — by switching from fossil fuels to clean energy.

   This is also happening but also not fast enough. Non-OECD countries still use dirtier energy sources than OECD countries. The EIA does not predict that they will be able to leap-frog wholesale over fossil fuels into clean energy.

These facts are collected and put into numbers in Table 3.15. Over the next generation, expect world emissions to grow as population grows and poor countries escape poverty — despite (insufficient) improvements in energy efficiency and emission efficiency.

Emissions By Population

Our discussion of emission associations is almost done. The last two questions that we want to address quickly are the following: How should we expect per-capita emissions to change with (1) population growth and (2) economic development.

Figure 3.16 stacks the CO_2 emissions by area. Admittedly, this figure is more tempting for finger-pointing than it is for finding a solution. If fingers are to be pointed, it should be at the countries with the highest bars. Anglo-Americans (USA, Canada, Australia) remain most profligate, followed closely by oil-rich middle-eastern countries and the former Soviet Union.

However, finger-pointing is unproductive. For example, when Wyoming residents emit 100 tCO_2/capita per year, the bar may be high, but it doesn't affect the planet much. It would only matter if there were hundreds of millions of Wyomans. Fortunately for the planet, there are only a few hundred thousand of Wyomans — negligibly few from a planetary perspective.

7. The Kaya Components

Table 3.15. Kaya Decomposition of Emissions, 2022

	emissions	=	population	×	income per person	×	energy inefficiency	×	emission inefficiency
	CO_2 (GtCO_2)	=	N (million)	×	GDP/N (1,000-$)	×	PE/GDP (KWh/$)	×	CO_2/PE (g/KWh)
OECD	12.1	=	1,380	×	43	×	1.2	×	170
USA	4.8	=	335	×	**61**	×	1.4	×	170
EUR	3.8	=	593	×	42	×	**0.9**	×	**160**
Not OECD	24.2	=	6,502	×	12	×	1.5	×	209
China	11.0	=	1,449	×	19	×	***1.8***	×	***232***
India	2.7	=	1,408	×	7	×	1.2	×	226
Other Asia	2.8	=	1,177	×	11	×	1.0	×	203
Africa	1.3	=	1,367	×	**5**	×	1.1	×	184
World	36.3	=	7,882	×	17	×	1.4	×	194
Forward-Looking Expected Growth Per Year									
2020-2050	+0.7%	=	+0.7%	×	+2.1%	×	−1.5%	×	−0.6%

Source: US EIA International Energy Outlook, Oct 2021: Primary Energy, Population, GDP in purchasing power parity, and CO_2.

If world emissions are to be reduced, it must be via reductions of the large areas. Reductions in thin slivers with higher bars just do not move the needle much. China's area in Figure 3.16 is already the biggest area. Even its per-person emissions — the height in the figure — are already larger than those in Western Europe. The blocks on the far right — the Indian subcontinent and sub-Saharan Africa — are likely to grow in both directions — in width and height.

Note also that the poorer non-OECD countries will not consent to global limits or shaming. They will always consider themselves not having emitted their "fair shares." The richer countries cannot help this either. Even if the richer countries on the left were to radically reduce their emissions in the future, there would still be their historical emissions to point to. Appealing to the global responsibility of poorer countries seems futile. Instead, for poorer countries not to emit more, they will need to come themselves to the conclusion that it is not in their own interests to emit their fair share.

3. GREENHOUSE GASES

Figure 3.16. Per-Capita and Total Emissions, 2019

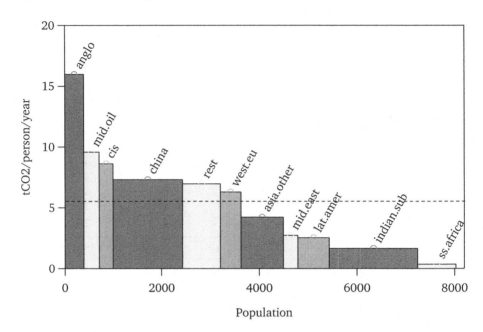

Note: The population size (in millions) is on the x axis. The CO_2 emission per person per year is on the y axis. The size of the rectangle is total emissions. The dashed line is the world average. Classification of countries is listed in the appendix to this chapter.

Source: Our World in Data. The idea was borrowed from MacKay, Chapter 1 (p12f).

No one will take me seriously in these shoes

Although the height of the bar in the graph is not the central concern when it comes to climate change, it does matter for global leadership. Realistically, with this high a bar, the United States will not be able to take on a leadership role. The rest of the world sees U.S. efforts largely as hypocritical — and subject to a "four-year fickle cycle."

The only credible region to lead the world could be the Europeans,

7. The Kaya Components

having more aggressively curtailed their own emissions than any other region of the world.

However, so far, the European efforts have been largely misguided. They have been willing to offer many declarations (mostly about their own CO_2 goals) and too little technology to help poorer nations. That is, to the extent that the Europeans have put skin into the game, it has been the wrong skin. For example, when Germany declares war on global warming by reaching for zero emissions by 2035, it's laughable — not for its aspiration, but for its misunderstanding. Germany emits under 1 $GtCO_2$ per year. Even if Germany eliminates all its emissions, it will still have a negligible effect on world emissions. Germany is simply too small to matter much in itself. Germany could however help far more effectively and cheaply in a different way. It could deploy its advanced science, technology, and industrial base to help figure out how it can make it in the interest of all nations (and especially poorer ones) to reduce their emissions. For instance, it could work on better energy storage solutions. (We will return to this theme in later parts of our book.) Germany's current efforts seem not only wildly expensive, but also rather misguided as far as global warming is concerned. The goal should not be to reduce guilt feelings, but to reduce CO_2 in the atmosphere.

Table 3.17 summarizes the emissions both in total and per-person for our familiar regional categories. Per capita, Americans emit an embarrassing amount. Europeans emit remarkably little — they are already near the non-OECD average. Ironically, the Chinese are so bad they are now ven "beating" the Europeans as far as emissions per person is concerned — not an accomplishment to be proud of. And there are hundreds of million more Chinese than Europeans!

3. Greenhouse Gases

Table 3.17. CO2 Per Person, 2022

	CO_2 (GtCO_2)	Population (million)	Per-Capita (tCO_2/year)
OECD	12.1	1,380	8.8
USA	4.8	335	14.4
EUR	3.8	593	6.4
Not OECD	24.2	6,502	3.7
China	11.0	1,449	7.6
India	2.7	1,408	1.9
Other Asia	2.8	1,177	2.4
Africa	1.3	1,367	1.0
World	36.3	7,882	4.6

add % change for world expected to 2050e?

Source: US EIA International Energy Outlook, Oct 2021. CO_2 and Population.

Emissions By Economic Development

How do emissions change with economic development? Figure 3.18 suggests (but does not prove) an interesting association. As countries climb the economic development ladder, per-capita emissions first climb steeply for the very poorest regions. Once countries reach middle-income levels, their per-capita emissions still climb but less steeply. They can afford to become more energy-efficient. Furthermore, not shown in the figure, as countries climb out of the low-income into the middle-income group, they also tend to have fewer children, so their population growth slows. The unabated population growth pattern suggests that allowing subsistence poverty to persist in the lowest-income countries may not just be ethically wrong, but it may ultimately harm the planet's climate as well.

7. The Kaya Components

Figure 3.18. CO_2 Per-Capita vs GDP, By Country in 2018

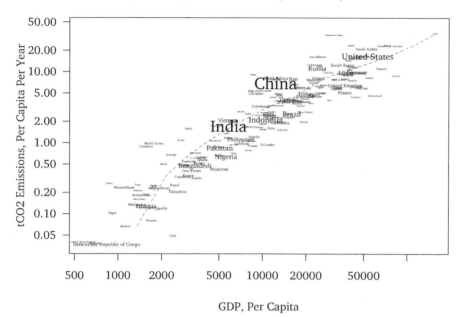

Note: The population size is indicated by the size of the text. The relation between GDP and CO_2 is roughly linear (both X and Y are in logs), although it is steeper for the poorest countries. Richer countries use more fossil fuels and thus emit more CO_2.

Source: Our World in Data.

Emissions Efficiency by Country

Our last subject is about emissions efficiency. It is the (inverse of the) product of Kaya's two efficiency components: (1) energy use per unit of GDP (energy efficiency) and (2) emissions per unit of energy (emissions efficiency). How efficient are different regions with respect to their per-income emissions? Which regions are emitting pollution frugally vs. gratuitously?

Table 3.19 provides a snapshot of the 2022 data. Countries can produce a dollar's worth of GDP with about 0.1 KWh (the equivalent of about 1-2 cent in primary energy cost). Energy is an important ingredient into almost every economic activity, but it is a relatively cheap one.

The Europeans are most energy-efficient, followed by other Asian countries and Africa. China is again least efficient. This is not just an accident but

3. Greenhouse Gases

Table 3.19. Efficiency

	Energy	Emissions	Total
	GDP/Energy	Energy/CO_2	GDP/CO_2
	($ / KWh)	(KWh/Kg)	($/g)
OECD	8.39	5.88	4.94
USA	7.19	5.88	4.23
Europe	10.56	6.24	6.59
Not OECD	6.51	4.79	3.12
China	5.71	4.32	2.47
India	8.49	4.43	3.76
Other Asia	9.63	4.92	4.74
Africa	9.48	5.43	5.15
World	7.22	5.16	3.73
Forward-Looking Expected (Log) Growth			
2020–2050	+41%	+17%	+57%

Note: Higher numbers are better. OECD countries are more efficient than non-OECD countries per GDP. European countries are most efficient, China is least efficient.

Source: US EIA International Energy Outlook, Oct 2021. GDP in purchasing power parity, CO_2 Emissions, and Primary Energy.

policy-based. The Europeans have also been the most aggressive in taxing fossil fuels. For example, all European countries are levying gasoline taxes of about $2.50 per gallon (€0.55 per liter), compared to about $0.20 per gallon in the USA. Europeans drive smaller cars, live in smaller houses closer to their work, and have more energy-efficient industries than the United States.

The Europeans are also most emission-efficient. This is determined primarily by the source of energy. China uses mostly coal for its electricity, which is why its emission-efficiency is so poor. Putting energy and emission efficiency together, Europe is about 2–3 times more efficient in terms of GDP per gram of CO_2 emissions than China and about twice as efficient as non-OECD countries.

The least energy-efficient economies are generally in poorer regions and predominantly in Asia. As they become richer, they will also use energy more

7. The Kaya Components

efficiently. This seems to be the typical economic dynamic as countries move up the value chain.

Chinese state propaganda often likes to tout improving energy-efficiency per GDP as the country's green commitment to the world. Frankly, this improvement is natural and to be expected. It has been happening almost everywhere. Yet, our dismissal of Chinese propaganda may not be entirely fair to Asian countries and China in particular. Comparing CO_2 efficiency across countries is a little like comparing apples and oranges. For one, there is an unequal availability of local non-emitting sources of energy. Hydro-electric power can power Sweden, but it could not power China, India, or Saudi-Arabia. Moreover, if Sweden had to supply energy for 100 times as many people (1.4 billion Indians or Chinese instead of 10 million Swedes), Sweden would also pretty quickly run out of hydro-electric power and could be forced to resort to fossil fuels. For another, China does much of the world's manufacturing. Manufacturing requires relatively more power than services (such as banking or tourism).

Nevertheless, it seems plausible that other countries could have manufactured the same goods as China but at a higher price with lower CO_2 emissions — if only because China still relies on coal as its main source of energy. (There are of course other reasons for China's low manufacturing costs. Cheap energy is just one input. Cheap labor is even more important.) Yet imagining that higher-cost European countries could have produced the same *exportable* goods with a little higher cost is an unrealistic dream: industries that make goods that compete in world markets tend to move to where production is least costly. Local industries that do not move to cheaper locales tend to be eliminated by competition.

Thus, country-based emission controls will always be limited in their reach. When Western countries increase their CO_2 taxes or mandate zero net emissions, the unintended consequences are often counterproductive. Factories in Asia could appear and produce the same goods with even dirtier energy. This is not a minor theoretical nitpick, but supported by a lot of evidence. When Europe and the United States lost much of their manufacturing base to China over the last two decades, they reduced their own emissions but these losses did not curb global emissions.

The big takeaway in the efficiency data is that most countries and thus the world overall have become more energy-efficient per unit of real GDP

3. Greenhouse Gases

both over time as they have become wealthier. However, Figure 3.18 showed that richer countries still emit more per person. The improved efficiency from higher GDP was not enough to outweigh the effect of higher GDP on total emissions in non-OECD regions. China, India, and African countries will become more frugal per unit of GDP in the future, but they will still emit more net on net when their GDP grows.

Looking forward to 2050, all regions of the world are expected to be able to produce more GDP with fewer grams of CO_2. Production efficiency will improve even faster than emissions efficiency (clean energy). There are often easy fixes — like better insulation. Nevertheless, efficiency improvements will not be enough to cover the increasing energy needs of the world.

8 Reducing the World's Emissions

There are hard facts of life — facts that acolytes would prefer to ignore. Al Gore was insightful when he called them "inconvenient." The world is warming. The cause is almost surely fossil fuel emissions.

But there are also other hard facts of life that Al Gore and we need to face:

- We in the West are so ethnocentric that we have lost perspective about how much less important we have become — and thus also of what we can accomplish.

  Emissions are no longer primarily an OECD problem. They are primarily a non-OECD problem. By 2050, the OECD will no longer contribute only the minority *one-third* of world emissions that it does today, but only a minority *one-quarter*.

  Even if the OECD could halve its emissions — via efficiency gains or painful forced reductions in economic activity, an unattainable aspiration over the next 30 years absent major scientific breakthroughs — it would reduce global emissions by only 15%. A whole 85% of global emissions (and growing) would still remain.

  If we truly want to reduce (or merely arrest) global emissions, it makes no sense to try to accomplish this primarily via reductions in OECD countries — no matter how appealing this may be to fairness and climate activists. We can do so only via (shared) reductions in non-OECD countries.

  Fighting climate change just in rich countries is like fighting a fire only on one side of the house while letting it expand on the other side.

8. Reducing the World's Emissions

- CO_2 emissions are not only increasing but still accelerating in absolute terms. Non-OECD regions are responsible. The word "culprit" seems inappropriate, though. The obstacles to slowing down their emissions growths are
    1. widespread poverty (that energy-fueled economic growth can reduce); and
    2. the lack of access to technology that would allow them to leapfrog over fossil fuels.
- We need to find viable solutions to allow poor countries to grow their way out of poverty without increasing their emissions (as much). The United Nations donation box will almost surely not deliver such solutions. Neither does it appear that United Nations climate conferences will deliver global emission cuts.

    The only viable lever to slow worldwide emissions is to break the link between energy consumption and emissions. Poor countries will have to find it in their interests to grow out of poverty with clean energy rather than with fossil fuels. They will have to want to leapfrog over the fossil-fuel stage right into a clean-energy stage — much the same as many of them have leap-frogged over telephone landlines right into cellular mobile phones.

    Whether developing countries will want to go the clean-energy route will mostly depend on which technology will be less expensive on a large scale. Making clean and reliable energy a lot cheaper is the only viable solution.

Western climate activists have had little of real use to offer to the poor countries that are becoming the key to combating climate change. Despite all the publicity that climate activism in the West has been garnering, its results have been mostly mutual accusations and finger-pointing, squabbling among rich countries, and empty declarations of progress. Almost all actual progress in lower emissions has been due to the decline in the cost of clean energy — even in richer countries.

The world's playbook response to bad climate news year after year has always been to "lament and repeat." Then again, there is no world collective that could have had a playbook to begin with. Thinking there could be such a collective in the future is fantasy. Trusting in such a collective to save the world is deluded.

Don't blame the messenger.

3. READINGS

Further Readings

BOOKS

- Archer, David, 2009, The Long Thaw, Princeton University Press, Princeton, NJ. Explains the long-term history and effects of CO_2 (and global temperature).
- Gates, Bill, 2021, How to Avoid a Climate Disaster, Knopf, New York. Contains many useful emission estimates and calculations.
- Kaya, Yoichi; Yokobori, Keiichi (1997). Environment, energy, and economy : strategies for sustainability, United Nations Univ. Press. The Kaya identity expresses the total emissions of CO_2 as the product of four factors: human population (P), GDP per capita (GDP/P), energy intensity (per unit of GDP, E/GDP), and carbon intensity (emissions per unit of energy consumed, $CO2/E$). Our own flavor is a closely related variation. In the previous chapter, we used $E = P \times (E/P) = P \times (E/GDP * GDP/P)$. In this chapter, we used $CO_2 = (CO2/E) \times E$.

ARTICLES

- Colt, Stephen G and , 2016, Economic Effects of an Ocean Acidification Catastrophe argues that the economic harm would be very low — perhaps too low.
- Qiu, Chunjing, et al., 2021, Large historical carbon emissions from cultivated northern peatlands, Science Advances. Northern peatlands converted to croplands from 850 to 1750—i.e., long before the industrial revolution—can account for 120 $GtCO_2$.
- Mooney, Chris, 2021, An enormous missing contribution to global warming may have been right under our feet, Washington Post.
- Union of Concerned Scientists, 2021, Peatlands and Climate Change, Issue Brief. (Peatlands include permafrost and store more CO_2 than forests or the atmosphere.)

REPORTS

- US EIA International Energy Outlook, Oct 2021, our primary data source.
- OECD, 2017, Green Growth Indicators, Country-Based.

SHORTER NEWSPAPER, MAGAZINE ARTICLES, AND CLIPPINGS

- O'Hara, Fred (ed), 03/2018, Carbon Dioxide Information Analysis Center — Conversion Tables, Carbon Dioxide Information Analysis Center.
- Curtis, Tom, 07/25/2012, Climate Change Cluedo: Anthropogenic CO2 (Attributing atmospheric CO_2 to human emissions), Skeptical Science.
- Friedrich, Johannes, et al., 08/10/2017, 8 Charts to Understand US State Greenhouse Gas Emissions, World Resources Institute.

- Johnson, Scott K., 11/05/2021, Recent CO_2 emissions flattened out by revised forest data, Ars Technica.
- Mider, Zachary, 08/20/2021, The Methane Hunters, Bloomberg.
- Painting, Rob, 11/12/2015, Why were the ancient oceans favorable to marine life when atmospheric carbon dioxide was higher than today?, Skeptical Science.
- Unnamed, 10/24/2021, The Chinese Companies Polluting the World More Than Entire Nations, Bloomberg.

Websites

- https://www.eia.gov/outlooks/ieo/ is the main source of our information.
- https://ourworldindata.org/ curates data on important phenomena.
- https://skepticalscience.com/ debunks many climate-skeptics' claims.
- Global Carbon Atlas, based on Andrew and Peters, 2021.
- https://www.rff.org/geo/, Resources for the Future, offers a meta outlook based on different projections from different sources.

Country Classifications

West, Europe: Austria, Belgium, Denmark, Finland, France, Germany, Greece, Ireland, Italy, Luxembourg, Netherlands, Norway, Portugal, Spain, Sweden, Switzerland, United Kingdom.

Anglo-American: Australia, Canada, United States.

Latin America: Argentina, Bolivia, Brazil, Chile, Colombia, Costa Rica, Cuba, Dominican Republic, Ecuador, El Salvador, Guatemala, Haiti, Honduras, Jamaica, Mexico, Nicaragua, Panama, Paraguay, Peru, Uruguay, Venezuela.

Indian Subcontinent: Afghanistan, Bangladesh, India, Pakistan, Sri Lanka.

Asia (Other): Cambodia, Hong, Kong, Indonesia, Japan, Laos, Mongolia, Myanmar, North, Korea, Philippines, Singapore, South, Korea, Taiwan, Thailand, Vietnam.

Sub-Saharan, Africa: Angola, Botswana, Burkina, Faso, Burundi, Cameroon, Chad, Congo, Ethiopia, Gabon, Gambia, Ghana, Kenya, Liberia, Madagascar, Malawi, Mali, Mozambique, Namibia, Niger, Nigeria, Rwanda, Senegal, Tanzania, Uganda, Zambia.

Middle East (Oil-Rich): Bahrain, Saudi Arabia, Iran, Iraq, Kuwait, Libya, Oman, Qatar, Syria, Tunisia, Turkey, United, Arab, Emirates.

Middle East (Not Oil-Rich): Egypt, Jordan, Lebanon, Mauritania, Morocco, Syria, Turkey, Tunisia, Yemen.

CIS: Armenia, Azerbaijan, Belarus, Georgia, Kazakhstan, Kyrgyzstan, Moldova, Russia, Tajikistan, Turkmenistan, Ukraine, Uzbekistan.

Posit that cornell-welch argue that "spend money now" vs "wait until certain" is irrelevant.

Chapter 4

Climate Science

Solar radiation and greenhouse gases undoubtedly determine the planetary climate. And humans are undeniably altering the greenhouse gas concentration in the atmosphere. So what is the human influence on earth's climate balance and to what extent has it caused warming of the planet?

If you follow the media and read books on climate, especially those that set forth only one or the other vantage point for the story's sake, then you may think you already know everything there is to know. We thought we knew it all, too. We didn't. There is a lot more here than meets the eye. Allow yourself to be surprised. We were.

Ironically, it wasn't just the sceptics or the anti-climate movement that obstructed climate change research...

4. Climate Science

1 Climate Versus Weather

A good way to think of climate is that it is weather averaged over many decades and centuries. These averages remove both seasonal weather variation and interseasonal weather variation (most importantly, the El Niño recurrent multi-year patterns in the Pacific). Climate is then the remaining very slow long-term trend in weather, and conversely weather is the short-term variation around climate.

An intuitive way to distinguish between climate and weather is that humans can far more easily perceive weather changes than climate changes. Short-run weather changes are much larger than climate changes.[1] Weather swamps most people's perceptions. The best way to obtain trustworthy information about climate change is via accurate scientific instruments that can measure averages 24/7, 365 days a year, over many decades. Scientific instruments can then calculate long-term averages precisely enough to smooth out short-term weather variations. Data less than a few decades and the averaged multi-year weather records could be influenced by complicated and random variations unrelated to underlying long-run changes in climate — such as the occasional volcanic eruption or longer weather phenomena that even scientists do not understand.

For those of us in our fifties and beyond, maybe — just maybe — we can get some feeling for our *local* climate changes by remembering how weather seemed cooler when we were young. But like many human memories, such perceptions can also be mistaken.

There is another important complication if we want to assess the *global* trends are not necessarily representative of global trends. Over ochs (a few million years), there is good evidence that climate

---
se the term climate primarily for temperature, but it really includes many ıl parameters (especially humidity), as well.

1. Climate Versus Weather

trends were distinct in different parts of the globe. The arctic may have gotten colder, while the subarctic may have gotten warmer (or vice-versa).

Figure 4.1. Temperature Anomaly, 2020-2021

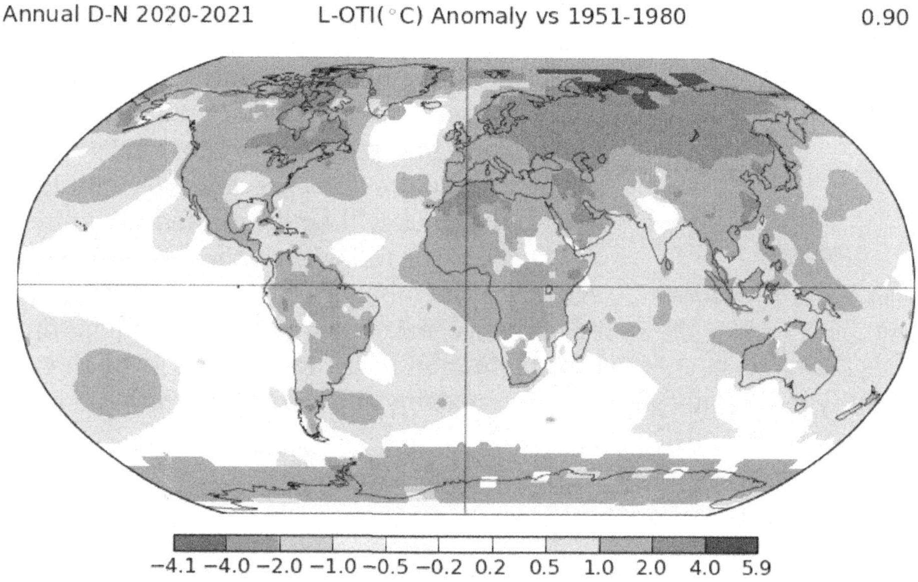

Annual D-N 2020-2021   L-OTI(°C) Anomaly vs 1951-1980   0.90

−4.1 −4.0 −2.0 −1.0 −0.5 −0.2 0.2 0.5 1.0 2.0 4.0 5.9

Source: NASA GISS Surface Temperature Analysis (v4). The base years are 1951–1980. Units are in °C. Another version of this map appears in Figure 5.3.

Even over the last few years, different parts of the globe have warmed at different rates. Warming has been more severe in Russia, North-East Canada, and the Arctic compared to, say, India, the South-Eastern USA, and the Antarctic. To assess *global* climate change, scientists need multi-decade measurements of many aspects of weather not just in a few spots but in many spots all over the globe. The best data is typically from satellites, even though some data sets go back to the 19th century.

As far as human impact is concerned, there is yet another related problem. As we explained in the previous chapter, human emissions have been accumulating slowly. And many planetary responses to those emissions would have been even slower. For instance, it is taking decades or centuries for oceans to warm and for Arctic ice to melt. This "glacial" speed makes our human impact more difficult to gauge. And it makes it more difficult to educate the

4. CLIMATE SCIENCE

public about climate change. Our human lives may be short, but our attention spans are even shorter. It is easy to lose a sense of urgency, given all the other pressing problems in our 24-hour news cycle.

Don't worry...The warmer it gets, the less we have to worry about the cold.

When it comes to climate change, humans are like frogs in very slowly warming water. On the plus side, the glacial pace of climate change gives civilization time to react and to adapt. For example, if the sea level rises slowly, our children can move inland towards the new shoreline and build stronger structures, so that damages and deaths from hurricanes will be lower (and they already are *much* lower today than they were a century ago). On the minus side, the glacial pace of climate change makes it an especially insidious threat. Procrastination is just too tempting. By the time humanity may finally get around to reacting appropriately, it may already be too late.

Activist Versus Scientific Views of Extreme Weather Events

What are climate activists to do? How can they catch the public's attention? Some of them try to take advantage of dramatic weather events. Weather changes can attract attention in ways that climate change cannot. This approach of blaming climate for all extreme weather may be well-intended, but it is not entirely honest. And, more importantly, it is not the scientific approach.

➤ Hurricanes (Tropical Cyclones)

Let us give a prominent example. When scientists want to explain that the analysis of climate change is not that simple and push back on over-active imagination, they sometimes discuss hurricanes. Most headlines in the popular media proclaim that hurricanes have been increasing. Indeed, there was a record number of 30 named storms in 2020. Hurricane season also now seems to start about a month earlier than just a few decades ago. Yet, the scientists themselves remain more circumspect. Unlike activists, they do not consider the past hurricane incidence data to be the unconditional smoking gun for

1. Climate Versus Weather

global climate change. (There are smoking guns, but they are elsewhere.) The scientists prefer to stress that the evidence is more nuanced.

Kossin et al. (2021) explain that tropical cyclones form not just when (global) air temperature is high, but when many regional influences come together. Local sea temperature — often linked to ocean circulation — is important, but other factors come into to play as well. Temperature gradients (differences) play a role. Dust from volcanic eruptions can play a role. And, in the Atlantic, fossil-fuel aerosol particulates and Saharan dust play a role. Global warming can influence both these factors and hurricanes, but not in straightforward monotonic direction. Their effects can come together regionally in different constellations. Based on the clearly increasing earth temperature (and the clearly but more modestly increasing global ocean temperature), climate scientists are now predicting more hurricanes of higher intensity and perhaps with different paths. However, this does not mean that they are predicting the number of hurricanes to increase strongly — they do not know. In any case, they will tell you that a few decades of hurricane data are *not* a good measure of global warming. (Scientists have much better ones.) Put differently, although it is not expected (global warming does not predict it one way or another), we could even see a few decades of reduced hurricane activity. If this were to occur, we should not conclude that global warming has subsided.

Figure 4.2 shows the state-of-the-art in tropical cyclone research, using data from six basins in which they occur. The left plot shows that the number worldwide increased from 1980 to 1995 and decreased since. This is sometimes touted as evidence against global warming by skeptics—incorrectly, of course. The models do however predict that when cyclones do form, they will be more intense. The right plot shows that the evidence for this prediction has been slowly accumulating. But our point is not that cyclones do this or that — it is that it requires scientific evidence to draw conclusions, not publicists and news.

▶ Sea Level Temperature and Rise

The evidence of consequences of global warming is stronger in the ocean data. However, some mysteries remain. As the temperature rises, sea-levels have and will continue to rise — it is a simple fact of physics. Figure 4.3 shows the actual evidence. Most activists would just extrapolate the sea-level rise

4. CLIMATE SCIENCE

Figure 4.2. Tropical Cyclones (aka Hurricanes)

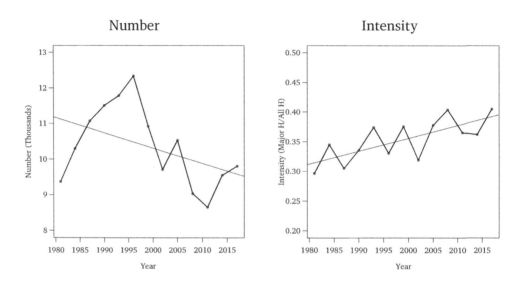

Source: Data were aggregated over all worldwide basins from Kossin et al., PNAS 2020. See also Environmental Protection Agency (EPA), Climate Change Indicators: Tropical Cyclone Activity via the National Oceanic and Atmospheric Administration (NOAA), 2016, and Vecchi and Knutson, 2011. See also Knutson et al., 2019, for a scientific survey.

exponentially and call it day. Scientists agonize *both* about the early evidence (that shows a mismatch in temperature and sea-level rise), *and* about whether they should extrapolate past trends linearly or exponentially.

The physicist Steven Koonin has openly questioned how much certainty there is about the IPCC's extrapolative predictions of impending *dramatic sea-level rise*. When scientists disagree, they blame each other for cherry-picking of evidence. Earth is a tough spot to do research in — but the process of science demands exactly such skepticism and debate. (We just wish it were less personal.)

➤ **Heat and Cold Waves**

Conversely, there was an epic cold-wave in the continental United States in February 2021; and Antarctica's 2021 polar winter was the coldest on record.

1. Climate Versus Weather

Figure 4.3. Global Sea-Level Temperature and Rise

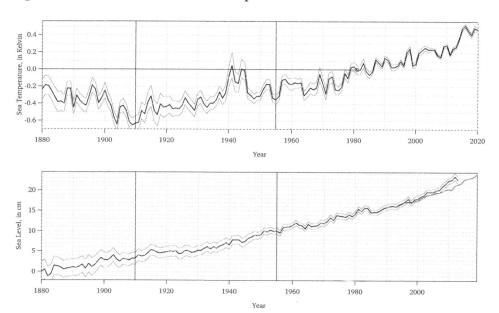

Source: The source is the U.S. EPA, both for Sea Temperature and Sea Level. This was the full data set, as available in March 2022. The blue line in the lower graph are based on satellite measurements; the black line on tidal gauges. There were three temperature patterns: 1880–1910, 1910–1955, and 1955–2020. Sea temperature rose most after 1955. Sea level has been steadily increasing (3mm per year), and perhaps slowly accelerating.

Should this make you think that the climate skeptics may have a point, that the data are ambiguous, or that the world could even be getting colder again? No! Neither a few cold waves nor a few heat waves nor a few hurricanes prove much about global warming.

▶ **Don't Misunderstand Us!**

To avoid any misunderstanding, our examples do not mean that most other climate-change-blamed phenomena in the news are based merely on click-bait, biased reporting, and incorrect human perception. Furthermore, make no mistake: **All serious scientists agree that earth's temperature has been rising and at an accelerating rate over the last 50 to 100 years.** Climate change is real and it will have stark consequences.

Yet, it is difficult to connect any single specific heat-related event (like Europe's hottest summer on record, 2021) to global warming. This does not imply that heat-related events are necessarily unrelated to global warming, either. Many almost surely are.[2] Increasing temperature *must* eventually lead to increases in many heat-related phenomena — such as heat waves in Europe or Arctic melting.

Our point here is simply that meaningful analysis requires more than just an impression from the news. It requires detailed scientific observations collected over decades with care and appropriate caution in interpretation.

2 The Global Thermostat

A good starting point to understand global temperature is to ask: Why does Earth have the (average) temperature now that it does? Currently, the global mean temperature across day and night and across all latitudes is 14 degrees Celsius (14°C) or 57 degrees Fahrenheit (57°F).

There are two forces maintaining this temperature: solar radiation and greenhouse gases.

Equilibrium

The impact of solar radiation is described by the Stefan-Boltzmann law. When solar radiation increases, the Earth starts to warm. As the ambient temperature rises, the Earth sends more radiation back into space. Eventually, the temperature rises to the "equilibrium" point at which the outgoing radiation from the earth matches the incoming radiation from the sun. If the Earth were an ideal radiator with no greenhouse gases, the Stefan-Boltzmann law implies that the earth's average temperature would be about −18°C (0°F). Most of the world would be an uninhabitable snowball.

Fortunately, greenhouse gases absorb some of the outgoing radiation and re-radiate it back to Earth, thus preventing a "snowball Earth." The greenhouse effect works as follows. Gas molecules can be thought of as little oscillators. Each gas has specific frequencies at which it resonates. Any radiation that

---

[2] Scientists, like Daniel Swain, are beginning to quantify the impact of global warming on the probability of extreme weather events. See our references.

does not push it at a resonance frequency passes through with little interaction. However, if the radiation frequency matches that of the molecule, then the molecule absorbs the radiation briefly before sending it off again but now in a random direction — and, most importantly for our purposes, some of the radiation is directed back to earth.

Figure 4.4. Greenhouse Effect

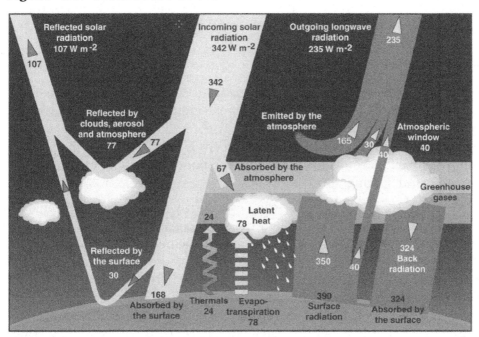

Source: Courtesy of Gregory Bothun based on Trenberth et al, 2016. See also Bothun's more detailed coverage. Yellow is visible light, red is infrared (re-radiated) light.

Figure 4.4 shows that when visible sunlight hits the surface of the earth, it is absorbed and then re-emitted as lower-frequency infrared light (thermal radiation). The now-infrared light is reflected back into space. Greenhouse gases — that happen to be transparent in the visual spectrum — resonate in the frequency of infrared light, which makes them absorb and re-radiate some of this infrared light back down to earth. Just as a greenhouse traps the sun's infrared light with glass to warm the plants inside, so do greenhouse gases trap the sun's energy higher up to warm the earth. This process continues until earth reaches a new equilibrium at a higher temperature. That is why earth's global temperature is +14°C (57°F) and not −18°C (0°F).

4. Climate Science

Despite the beneficial effects of greenhouse gases, there can be too much of a good thing. A horrific example is Venus. Without CO_2, its average temperature would be a reasonable 28°C (82°F). Instead, Venus' actual temperature is 460°C (870°F) — hot enough to melt aluminum and rain sulfuric acid. If you are now concerned that humans could push Earth into a Venus equilibrium, don't worry. Recall from the last chapter that earth's atmosphere is only 0.04% carbon dioxide, possibly reaching as high as 0.1% at the high end of future estimates. Venus's atmosphere is 97% carbon dioxide!³

Near-Perfect Prediction

Remarkably, it is not only possible to measure Earth's temperature, but it is even possible to measure whether it is already in thermal balance. If you can measure the heat your stove sends to your pot and how much heat your pot emits in turn into your kitchen, the difference is only zero when the pot is in equilibrium. If more heat is going into the pot than coming out, you know the pot is still heating up.

Analogously, NASA satellites can now directly measure both the amount of incoming solar radiation (called the solar constant) and the amount of outgoing thermal radiation. The difference is the heat uptake *dis*equilibrium of Earth (also called radiative forcing). For this reason, scientists know that earth is yet not in an equilibrium. They expect Earth to continue to warm until it reaches its new equilibrium, where incoming and outgoing radiation will again be balanced.. Let us repeat this: **Earth is currently absorbing more energy that it is sending back into space, so it will soon be warmer than it is today.**

How much warmer? In 2005, the planet absorbed a net influx of about 0.5 Watts per square meter. More energy was coming in than going out. Ergo, Earth was heating up. By 2020, the difference had doubled to about 1.0 Watts per square meter. Ergo, Earth is in the process of warming at a rate that is twice as fast as it was 15 years ago.

Remember that this is a measurement that is independent of whatever Earth's actual temperature is, whatever emissions humans may be releasing,

---

³Most likely, Venus first ran out of water, because it does not have a magnetic field that would have shielded any water in its atmosphere from the solar wind. Once the atmosphere had run out of water, not only had Venus lost its CO_2 water sinks, but it had also lost its terrestrial sinks because rocks cannot absorb CO_2 through weathering in the absence of water.

whatever solar or volcanoes may have been doing, etc. It is a direct measure of the rate of temperature change that is currently occurring.

3 The Temperature Record

Earth has never been and will never be in an entirely stable equilibrium. When climate activists state that we live in an era with unprecedented higher CO_2 and temperatures, skeptics counter that the Earth used to have both far more CO_2 in its atmosphere and far higher temperatures than it does today. And both are correct!

They want to know whose original climate we're restoring it to.

To understand what this is all about we begin with a brief look at our planet's long-run climate history. (Again, David Archer's The Long Thaw has more detail.) However — *and this is also important to understand* — most of this history is no longer relevant to today's situation. It is remarkably unimportant to the issues facing humanity now.

The scientists' problem is academic. Scientists do not understand all the feedback loops of relevance in the distant past. (In plain English, they are not sure whether the chicken or the egg came first, though they do know that each causes the other.) In contrast, scientists do know that humans have injected significant amounts of CO_2 into the atmosphere over the last century — it was not caused by warming itself or some other unknown influence. Therefore, much of their debate about how to interpret ancient paleo-history (as chicken or egg) is more of academic than of pragmatic interest.

4. Climate Science

Deep Time: 500 Million Years

Figure 4.5. Global CO_2 Estimates in Deep Time

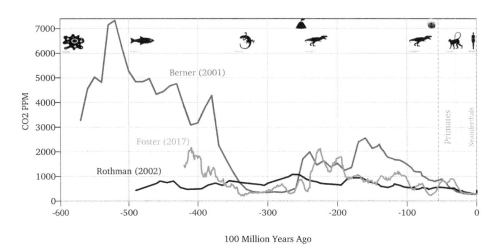

Source: Berner-Kothavala (AJS 2001), Foster et al. (Nature, 2017), Rothman (PNAS 2002). (According to Gavin Foster, recent revisions in the revisions in the "Paleosol d13c proxy" may have obsoleted the Berner estimates.)

Scientists have pieced together estimates over about six hundred million years. It seems miraculous that scientists can deduce anything about the continents and climate from hundreds of millions of years ago. But they can. Of course, as should be expected, the farther they go back, the more uncertain the data become.

Such *deep time* seems unimaginably long. Five hundred million years ago, fish were the pinnacle of vertebrate evolution. The first multi-cellular organisms had appeared "only" 50 million years earlier (in the Cambrian era). Not just the current continents, but even the ancestors of our current continents had not yet formed. Modern mammals took over after a measly 6-mile asteroid finished off most dinosaurs on a Tuesday[4] just 66 million years ago.

Let's start with scientists' estimates of deep-time carbon dioxide. Figure 4.5 presents estimated reconstructions of its atmospheric concentrations,

---

[4]Yes, scientists recently discovered evidence from that one very bad day!

3. The Temperature Record

measured in parts per million (ppm). The blue line in Figure 4.5 represents the most prominent estimate of CO_2 levels over this time span, although this study may be a little outdated by now. Science has progressed. But comparing the CO_2 estimates from three different studies shows how even the best scientific estimates can disagree.

Not even the most alarmed climate scientists believe that human emissions will push CO_2 concentrations beyond 1,000 ppm — although it is not completely impossible that unknown feedback effects could push the CO_2 concentrations higher for a while. (This would indeed be scary!) Let's say that 800–1,000 ppm is possible *if* humanity burns most available fossil fuels. Despite their large discrepancies, all series in Figure 4.5 agree that 800–1,000 ppm of CO_2 was not that unusual in deep time. In this comparison, Earth has been in a CO_2 drought for many millions of years. However, towards the right end of the graph, the lines also shows that 800 ppm is very high by human standards — *Homo Sapiens* appeared only about 200,000–300,000 years ago.[5] Thus, depending on the narrator's intent, the increase from 300 ppm to 400 ppm (and soon beyond to 600 or 800 ppm) can be proclaimed as earth returning to normal (by geological standards) or as being unprecedented (by genus-primate standards).

Temperature is even more difficult to reconstruct than CO_2 levels. Unlike CO_2, which is effectively a global gas, temperatures are largely local. If a researcher 500 million years in the future found a temperature record only from the Sahara or only from Mount Kilimanjaro or only from the Antarctic, even the best science could not rescue her from a wrong inference about earth's prevailing climate. Moreover, scientists know that there are not only periods in which all of earth's temperature moved up or down together, but also periods in which earth's temperature gradient changed — it became simultaneously hotter on the equator and colder on the poles or vice-versa.

Here we present today's best scientific estimates of deep history temperature, but do not consider the numbers to be definitive. Scientists may learn more and change them again.

About 600–700 million years ago and lasting for about 20-80 million years, the planet was (for a second time) in a state called "snowball earth." In this

---

[5] At 800 ppm, as occurs in fully occupied lecture halls, many of us begin to suffer modest adverse health effects — we tend to lose mental acuity and fall asleep. Most of us were not designed for 800 ppm — though some are and/or will be.

so-called Cryogenian period, the entire planet was a frozen wasteland. Ice reflected most sunlight, thereby keeping earth cold. Life was likely limited to a few hearty microbes. After this last snowball earth ended about 540 million years ago, the Cambrian explosion of complex multicellular life began.

Starting around 500 million years ago, and for about 85% of the time since, the earth was in a state called "greenhouse earth" (sometimes also called "hothouse earth"). The average global temperatures exceeded 70°F (21°C) or possibly even 80°F (27°C) at times. Recall that it is 57°F (14°C) today.

For the remaining 15% of the time, the earth was in yet another state that geologists call "ice ages." Formally, an ice age is an era in which there is year-round ice on the polar caps. We are still living in an ice age that started about 50 million years ago. Large primates first evolved during this our current ice age. The first ape appeared about 20 million years ago. Ice ages may be geologically unusual, but they are all that many mammalian orders alive today (including our own) have ever known.[6]

Figure 4.6 plots estimates of deep-time temperatures over the last 500 million years. The figure shows two different estimates — and, again, clearly, the estimates differ. Wing and Huber place more emphasis on temperature closer to the poles, Verard and Veizer on temperature in oceans closer to the tropics. (And they could both be correct! It is difficult to know.) Our current ice age is at the far right end. The darker purple line suggests that earth's current temperature is near the threshold between an ice-age and a greenhouse earth.

Climate-change skeptics often point out that the connection between planetary greenhouse gas levels and planetary temperature over 500 million years seems weak. And they are right again. However, this observation is not relevant. Why not?

First, over such long time spans, our record of earth's history becomes so uncertain that all estimates must be viewed with a healthy dose of skepticism. Second, the sun was about 5% cooler 500 million years ago. Third, the earth's orbit changes over time and it could have been a little further away from the sun — we will likely never know. Fourth, earth underwent massive geological changes, such as the formation and breakup of continents and mountains.

---

[6]There is some disagreement among scientists here, too. See Figure 4.6.

3. The Temperature Record

Figure 4.6. Global *Temperature* Estimates in Deep Time

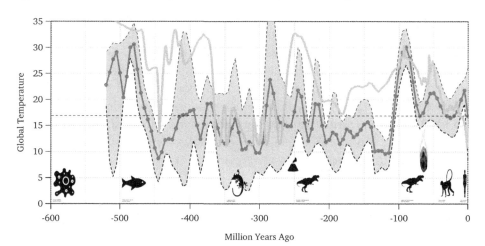

Source: The graph combines two different estimates of deep-time temperature. The orange line is a preliminary version from a Smithsonian Institution project led by Scott Wing and Brian Huber. It measures closer to the poles. The darker purple curve is from Vérard and Veizer (Geology 2019, Fig 3), with uncertainty in gray. It measures closer to the tropics. The horizontal line is roughly where ice ages might break. The icons denote the appearance of vertebrate orders.

This changed the exposure of different kinds of rocks with different abilities to absorb atmospheric carbon-dioxide. Fifth, not only would 1,000 ppm today have a markedly different impact on temperature than it had 300 million years ago, humans are creatures that evolved in 300 ppm conditions and not in 1,000 ppm conditions. Dragonflies 30 inches long would probably enjoy 1,000 ppm more than humans. And sixth, while scientists do not know what natural forces pushed CO_2 around in deep time, which makes interpreting causality difficult, we know exactly what has pushed CO_2 up in the last 200 years — our human emissions!

4. CLIMATE SCIENCE

Human Time: 500,000 Years

Within ice ages, there are further divisions. There are "glacial periods" and "interglacial periods." During glacial periods, Earth is cooling, and glaciers and ice sheets are advancing. During interglacial periods, Earth is warming, and glaciers and ice sheets are receding. Glacial periods thus end with the coldest interludes within cold ice-age periods. For the last 10,000 years or so — i.e., roughly the time span within which modern civilizations developed — earth has been in a warmer interglacial and also unusually stable period. The fact that glaciers have been receding is not new — they have been doing so for the last few thousand years.

In sum, our ancestors and we have been living near a traditional glacial minimum — a (shorter) interglacial warm era within a (longer) cold ice-age era. This positioning is fortunate. During the last major glacial maximum within our current ice age, conditions were far less hospitable. For example, just 15,000 years ago (well within human existence), the global temperature was 6°C colder and New York City was under a glacier 300 feet thick!

In rebuttal to the skeptics pointing erroneously to the deep-time graph, some climate activists then get to show off their own graph (in Figure 4.7). They point to the last 400 thousand years, a tiny blip at the end of the graph in Figure 4.6. This is roughly the time when Homo Sapiens first evolved. The close association between CO_2 and temperature is striking. In some quarters, Figure 4.7 has obtained a cult-like status as the iconic "smoking gun" — proof that CO_2 drives climate change.

But this simple interpretation is misleading. Although the data correlation patterns are literally correct, they don't mean what the presenter wants to imply. Just as we rejected the lack of co-movement of CO_2 and temperature over the last 500 million years as evidence of absence of a driving role for CO_2 on temperature, so too do we have to reject the co-movement over the last 500 thousand years as evidence of its presence.

The association in Figure 4.7 does not show (much less prove) that CO_2 drove temperature. Instead, it shows only that CO_2 and temperature moved together — correlation. Whereas correlation means only that there is a mutual relationship between two variables, causation is a much stronger concept. It means that one variable influences another in a *cause-and-effect* relationship. Figure 4.7 does not show such a cause-and-effect relationship. A deeper analysis of timing suggests that the comovement of the two series reflects feedback effects in both directions. Moreover, some other unknown variable could have driven both CO_2 and temperature.

3. The Temperature Record

Figure 4.7. CO_2 and Temperature over 0.5 Million Years

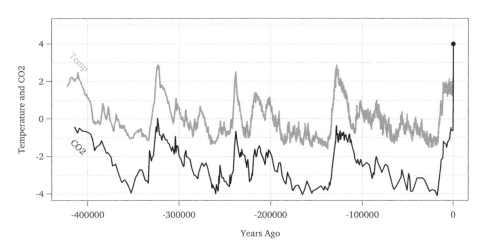

Note: This figure is often _falsely interpreted_ as showing that CO_2 concentration changes have caused global temperature changes. Instead, data analysis of changes in the two series suggests that it is more likely that some other unknown factor has caused the close association between CO_2 and temperature changes (or even that global temperature changes have caused more CO_2 changes than the other way around).

Source: The figure is based on the famous Vostok ice core data, discussed, e.g. in Petite et al. (Nature 1999).

The strongest empirical regularity of this data (again, not easily visible in the figure) is that earth seemed to have had a built-in regulator for these 400,000 years. When temperature was high and had recently increased, then it tended to fall again. When temperature was low and had recently decreased, it tended to increase again.

We do object to one common practice. When Figure 4.7 is presented to the public, it is usually with a purpose to mislead, suggesting that it is this figure

4. CLIMATE SCIENCE

that "proves" that CO_2 strongly drove temperatures over the last 400,000 years. It does not. Better evidence for (and a source of concern regarding) the role of greenhouse gas emissions in causing global warming is elsewhere. It is in the theory of physics and the calculations of radiative forcing.[7] And it is in the empirical evidence of the most recent 100–200 years (covered next).

Historical Time: 1,000 Years

This brings us to today's most "controversial" evidence (according to climate-change skeptics): The famous Hockey Stick Graph by Michael E. Mann (and coauthors) in Figure 4.8. Think of it as a "zoom" into the last 1,500 years. It shows that the global temperature has been on a sharp upswing beginning around 1800 and accelerating since (especially after 1970) — a hockey-stick-like pattern.

Ironically, this hockey-stick evidence is least controversial among scientists. It is here that the data are most precise. Over the last 100 years, scientists have real-time measurements of temperatures from all over the globe and of deglaciation and sea-level change. There is no longer any reasonable scientific disagreement about Mann's essential temperature observations. It's solid science.

The global temperature has been increasing and indeed accelerating for at least 100 to 200 years. This means that global warming is now faster than it has been for a thousand years, and scientists can observe it in daily satellite measurements. The last 50 to 100 years is, of course, also the time during which humans could have plausibly been influencing the planetary climate with their slowly accumulating GHG emissions. Before 1900, civilization's

---

[7]There are some sharp but isolated episodes in Figure 4.7 for which the culprit can be identified as a CO_2 shock. For these, it is possible to tease out a causal relationship. However, these episodes cannot be generalized to the longer 400,000 year range.

3. The Temperature Record

accumulated atmospheric emissions were simply too small to matter much. Global warming has accelerated so much that the last 20 years, 2000–2020, alone account for about half of human-induced warming.

The evidence is so clear and uncontroversial, and sometimes so distorted in the press, that it deserves reiterating a second time:

> **Climate-change deniers are simply wrong. Over the last 50 to 100 years, there is no question [1] that the earth has been warming at an accelerating rate and [2] that it has not yet reached a new stable equilibrium. Earth is continuing to heat up.**

Of even greater concern, Figure 4.9 shows that the temperature rise is still accelerating, in line with the satellite observation that more heat is still coming in than going out. About half of human global warming has occurred in the last 20 years. The single most important point is that all serious scientists agree that earth is now warming at an alarming rate.

Simultaneously, it is undisputed that it was human activity that has dramatically raised the CO_2 concentration in the atmosphere. CO_2 has been on an analogous increasing and accelerating trajectory. Of course, so have many other observed variables. Correlation is easy to come by. However, there is more than just correlation. The physics of greenhouse gases can explain what causal effects anthropogenic CO_2 should have played in the increasing global temperature.

Yet there are still a few (modest) mysteries. Figure 4.9 shows that temperature seems to have dropped by about 0.3–0.4°C around the time of the Renaissance (the onset of the "Little Ice Age"[8]). Scientists have some educated guesses as to why, but they do not know for sure.[9] Furthermore, this cold period was also likely *not* caused by a drop in CO_2 levels. Indeed, the temperature first dropped (around 1500) and the CO_2 level fell only later (around 1600). From about 1800 to about 1900, temperatures rebounded

---

[8]The Little Ice Age was not at all an ice age in the geological sense. We were in an ice age before the Little Ice Age and are still in it.

[9]Interestingly, it is not known whether it could have merely been cooler in the Northern Hemisphere. A new hypothesis links cooling to previous warming, that collapsed ocean circulation. Whether true or not, it illustrates the complexity of the climate.

4. Climate Science

Figure 4.8. CO_2 and Temperature over 1,500 Years

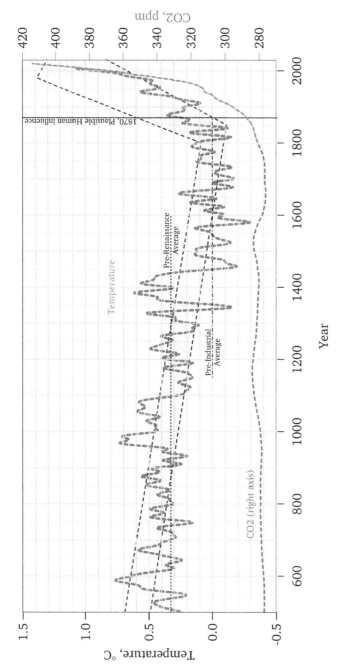

Note: The darker purple curve (mostly inside the "hockey stick") is Michael E. Mann et al.'s global temperature data. The 0°C mark is 1980. The lighter red line is the ETH Zürich CO_2 data. The vertical line shows when humans could have begun to meaningfully contribute to the global CO_2 concentration rise (around 1870). The figure also sketches the famous hockey stick.

The next figure zooms further into the final 500 years.

Source: Mann et al., 2008, "Proxy-Based Reconstructions of Hemispheric and Global Surface Temperature Variations over the Past Two Millennia," PNAS. Volcanoes, VEI 5 or greater, Wikipedia.

Figure 4.9. CO_2 and Temperature over 500 Years

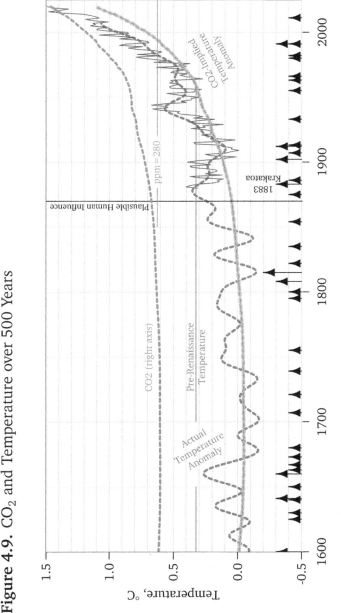

Note: See previous graph for more explanations. The thin blue line towards the right beginning in 1880 (and continuing Mann's curve) are in-time measured estimates provided by NASA. In addition, this zoom figure adds volcano eruptions at the bottom, with VEIs of 5, 6, and 7 (by arrow length). It also adds a line that sketches what temperature change the CO_2 increase would have predicted if a doubling of CO_2 in the atmosphere increased temperature by about **3°C** (50% immediately, 75% within 30 years, and 100% within 100 years). From 2000 to 2020, it seems too steep. Other reasonable assumptions don't fit slopes both early and late much better.

4. CLIMATE SCIENCE

from the unusually cold climate of the post-Renaissance. Almost all of this Renaissance temperature decrease and later recovery occurred *before* human emissions could have made much difference. Civilization just had not yet emitted enough GHGs to influence the climate this much.

Finally, a short sidenote: By necessity, all discussions of global warming have to be relative to a benchmark, and the choice of benchmark can make for a meaningful difference in the number of degrees of warming quoted. Earth has warmed about 1.4°C since preindustrial times (say, 1500–1800), but "only" about 1.0°C relative to the middle ages (say, 500–1500). And of the 1.4°C rise in temperature since preindustrial times, only about the last 1.0°C increase could reasonably be due to human-GHG emissions. Therefore, whenever the pundits discuss a (rounded) "1.5°C temperature increase since preindustrial times," you, the audience, have to keep in mind that it would be misleading to connect the full temperature increase to the arrival of industry or human activity. The more plausible human-emission caused abnormal increase *to date* (2020) is 1.0°C. *That's bad enough*, especially in places where temperature changes have stronger localized effects. And remember: human emissions will cause global warming of about 3°C. So far, only about 1°C has occurred. Another 2°C is still heading for us.

Strong Industrial-Age Trend Evidence

To summarize, what makes the recent evidence so much more powerful than evidence from paleo-history is not only that scientists now have satellite measurements, but also that they know that the CO_2 increase was not caused by some unknown phenomenon (or temperature changes themselves, the chicken and egg problem). Instead, the CO_2 increase was caused by human civilization. As we explained in the previous chapter, scientists know this (a) because they can count up how much CO_2 humans have emitted net of how much earth could have scrubbed, and (b) because they can assess the carbon source independently based on the scientific measurement of carbon isotopes. CO_2 from ancient fossil fuel burnt by humans has a different fingerprint than that of recent natural CO_2. Science offers up a unanimous verdict: About 130 ppm of CO_2 out of the total of 410 ppm in the atmosphere today is due to human activity.

We stated earlier that the source of more distant paleo-historic changes in temperatures is largely irrelevant. Even if CO_2 in the past had been caused by temperature changes rather than the other way around (itself an iffy proposition), this is no longer the situation today. Scientists know that human civilization has caused *our* current CO_2 increase. Whatever the causes of the CO_2 and climate-change dynamics may have been a few million years ago, industrial civilization has pumped enough greenhouse gases into the atmosphere in the last 100 years that scientists know it must have had temperature consequences. The basic physics of the greenhouse gas effect demands it.

It's almost as if humanity has been running an experiment to see what an increase in CO_2 would do, and the temperature has duly responded. And both the atmospheric CO_2 concentration and the global temperature are continuing to accelerate — of course, not each and every year, but in a reasonably consistent trending pattern. And neither the temperature rise nor humanity's experiment is done yet.

4 Greenhouse Gases and Temperature

Apologies — we have to circle back briefly to atmospheric gases in order to explain in more detail how they influence temperature.

Greenhouse gases can be grouped into two types. The first type are chemically stable greenhouse gases that stay in the atmosphere for a long time. This first type includes primarily what can be called fossil-fuel GHGs including CO_2 and Methane. The second type is more ordinary — water vapor. Think humidity.

Fossil-Fuel GHGs

We have already explained that carbon dioxide is the most abundant and important fossil-fuel GHG. We have also noted that human activity emits Methane, Nitrous Oxides, and F-gases, too.

➤ Global Warming Potential (GWP)

The three other GHGs are present in much smaller concentrations in the atmosphere than CO_2. However, pound-for-pound, the non-CO_2 gases are much more effective in absorbing and re-emitting infrared energy than CO_2.

To compare the impact of different greenhouse gases, scientists have developed a measure known as the Global Warming Potential (GWP), usually

4. CLIMATE SCIENCE

stated in terms of CO_2 equivalents (**CO_2e**). CO_2, by definition, has a GWP CO_2e of 1. We already used the CO_2e measure in the previous chapter, but had not explained it.

The GWP of a gas depends on two factors: (1) how efficiently it absorbs and re-radiates infrared radiation (i.e., how opaque it is to infrared), and (2) how long it stays in the atmosphere. For instance, Methane (CH4) is about 20–100 times more effective in absorbing and re-radiating infrared radiation than CO_2, but it disintegrates with a half life of 9 years. (It then decomposes into trace amounts of CO_2.) The widely accepted GWP figure for Methane is thus 30, meaning that each kg of methane emitted has 30 times the lifetime warming effect of a kg of CO_2.

Despite their higher GWPs, methane, nitrous oxides, and F-gases are so much rarer than CO_2 that CO_2 remains responsible for about 85% of human-caused global warming. CH4 is responsible for about 10%, nitrous oxide for 4%, and the remaining chemical GHGs for 1%. Humans emit about 40 GtCO_2, but the effective emissions rise to 51 GtCO_2e when we take account of the other GHGs (plus another 4–5 GtCO_2e for the land charge). The short-term total temperature effect of human GHG emissions is thus better measured by the 55 GtCO_2e per year; the long-term temperature effect is better measured by the 45 GtCO_2 (emissions plus land charge).

Doubling GHG

Physics can explain how much a specific increase in atmospheric GHG concentrations should raise the global temperature. A typical way to calibrate the temperature effect of a model is to ask how much the global temperature should ultimately rise for every doubling of CO_2 in the atmosphere. In the simplest canonical greenhouse model, doubling CO_2 raises the temperature by **1.2°C**. In a more elaborate model based on the sun's entire absorption spectrum, it is a little lower, **0.8°C**. Thus, roughly speaking, the direct long-term effect of CO_2 is about 1°C for every doubling of the CO_2 atmospheric concentration.

So far, humans have not yet doubled the CO_2 concentration but raised it by about 50% (from 280 ppm to 410 ppm). This implies a direct increase in the long-term equilibrium temperature of Earth of about **0.5°C**. Not all of it can have occurred yet, because the heating process takes a lot of time. More

4. Greenhouse Gases and Temperature

plausibly, the 50% increase in CO_2 has directly raised temperatures so far only by about **0.3°C**, with another 0.2°C on the way.

You should notice that something must be missing. The emissions-caused temperature change of 0.3°C is clearly insufficient to explain the already-observed 1.5°C global temperature change since 1800 (or the 1.0°C increase since 1500). Scientists need to reconcile the larger observed global temperature increase with the smaller theoretical CO_2-predicted temperature increase.

There is widespread agreement that the "missing link" is a second type of greenhouse gas: water vapor.

Water Vapor and Clouds

Think of water vapor as humidity in the air. Unlike CO_2 or the aforementioned GHGs, water is not long-lived in the atmosphere but circulates constantly. It evaporates and rains back down. This process is called the "water cycle."

Nonetheless, water vapor is very important because it is ten times more abundant than CO_2 in the atmosphere (0.4 percent compared to 0.04 percent). Some scientists estimate that, at any given moment, water vapor could have the potential to be responsible for about 85 percent of the atmosphere's ability to block outgoing infrared radiation. (CO_2 blocks "only" 7 percent, but does so for a much longer time. Water vapor also captures and moves heat around, heating the arctic and cooling the tropics.) Civilization has not directly pushed more water into the atmosphere. However, it has done so indirectly. The CO_2 has raised the global temperature, and warmer air holds (almost mechanically) more water vapor.

But the role of water vapor is not that simple. Water is also the essential ingredient in (white) clouds,[10] which reflect incoming solar radiation even before this radiation can reach the ground. Thus, on net, water vapor accounts for much less than 90 percent of global warming — perhaps only 65-85 percent. This range is so wide for two reasons: (1) Some uncertainty arises because the relationship between water vapor and clouds is not one-to-one. Cloud formation also requires seeding with tiny particles. (2) More uncertainty arises

---

[10] NASA/NOAA report that from 2005 to 2019, the planetary albedo (white cloud layer and sea ice) declined, partly due to the natural Pacific Decadal Oscillation, i.e., El Nino and La Nina.

4. Climate Science

because the effect of clouds on temperature is also still not fully understood. It appears that clouds sometimes have a warming effect on the local climate and sometimes a cooling effect — it seems to depend on the type of cloud, the local climate and a variety of other conditions.

NASA has only been measuring and recording global water vapor and clouds across different latitudes for a few decades.[11] Scientists have no direct observational record of clouds over the last few centuries, much less over the last few hundred-thousand years. And local observations are not enough — if it rains more over Illinois, it could easily rain less over New York.

5 Scientific Agreement and Disagreement

Let us summarize what we have covered so far. Over the last 50 to 100 years:

1. The global temperature has been sharply increasing at an accelerating rate.
2. The CO_2 concentration in the atmosphere has been sharply increasing at an accelerating rate.
3. This atmospheric CO_2 increase has been overwhelmingly man-made.
4. Some of the global temperature increase has been due to the direct CO_2 (and related chemical gases) greenhouse effect.
5. More of the temperature increase must have been due to water-vapor greenhouse effect, itself caused by rising temperatures, which were in turn caused by the direct GHG heating.

We will now present a version of remaining scientific disagreements as they make sense to us (as scientists but outsiders to the field) without endorsing or denying any specific views. (We cannot be referees.) Here are the two reasonable perspectives: The majority of scientists believes that human GHGs can already account for the heating that we have observed and that they have been, are, and will continue to be the sole driver of global warming over the last and next century. The minority view wonders whether enough "omitted factors" remain that could render human GHG's not entirely responsible.

In addition, both views allow for uncertainty. For example, solar activity could increase or decrease, a large supervolcano could erupt, etc.

---

[11]Dessler et al have now confirmed *with data* that an increase of 1°C seems to trap an additional 2 W/m^2.

5. Scientific Agreement and Disagreement

The Majority View

The mainstream model is that human GHG emissions have been and will continue to be entirely responsible for forcing the increase in earth's temperature. The long-lived fossil-fuel GHGs do so partly themselves but also (and more importantly) by priming the water cycle. GHGs increase the global temperature, which evaporates more water, which raises the humidity in the air, which is a potent GHG, which again raises the global temperature further. This makes sense: higher temperatures cause more water to evaporate and allow the atmosphere to hold more water. There is no disagreement among majority and minority here.

In the majority view, clouds play a mostly passive role. They are reactive, not proactive. Thus water vapor is a simple temperature multiplier for CO_2 (and then for itself). And the multiplier is not small. Water vapor amplifies the direct CO_2 effect on temperature by a factor of about two to three. This estimate is based on a model that best fits the historical CO_2-temperature data. Recall that CO_2 alone could explain only about 1.0°C for a doubling of CO_2. Calibrated from short-term physical observation of local responses, the models can explain the doubling of the direct temperature effect of CO_2. There is still quite a bit of uncertainty and unexplained variation here, though. Earth is a complex system.

A different approach — perfect if the mainstream is correct — is to take it as given that clouds would have behaved in a way that creates the best fit between (a) measured CO_2 increases in the atmosphere and (b) measured temperature increases on the planet. In this case, the revised predictions for the effect of a doubling of CO_2 is no longer just the direct CO_2-effect of **1°C**, but (including water vapor) the so-called **climate sensitivity**:

- **2.4°C** for the mainstream climate-science model.
- **3.0°C** for simulation models, tended by armies of scientists and running on super-computers. 3°C also the IPCC's preferred number.

These two- to three-fold calibrated amplification factors best reconcile the historically observed CO_2 concentration and temperature data. But climate scientists are not sure. Reasonable climate-sensitivity numbers can range from about 1.5°C to about 4.5°C— an uncomfortable wide range.

Importantly, one scientific drawback is that these amplification factors are not fully empirical. They are *not* based on two centuries of historical cloud records on Planet Earth. They are fitted, assuming the model is "as assumed." However, they are not arbitrary, either — there are many short-term associations that confirm the predicted local effects. Scientists are hard at work trying to measure the associations better on a global basis.

The Minority Dissent

The minority agrees that humans have caused a sharp increase in CO_2 accumulation in the atmosphere, that there is accelerating global warming, that CO_2 alone can explain at least one-third of global warming, and that water vapor feedback effects can amplify it.

The main disagreements center around the amplification factor of water vapor. The minority argues that aspects of clouds (and perhaps some other aspects of climate change) remain more of an enigma. Such skepticism is the bread and butter of the scientific process. Just like the critiqued model, the skepticism can be wrong. Scientists should remain skeptic about skepticism, too!

The minority notes a standard problem *in almost all fields of science*: the fact that CO_2 can explain most puzzles does not mean that there could not also be some other important omitted influences. What if what mainstream scientists call "natural random background fluctuations" happened not to have been so random over the last 200 years and thus distorted the inference by coinciding with the stark human GHG emission increase?

The minority also argues that the mainstream does not have enough empirical evidence to conclude that *only* CO_2 could have primed the initial temperature increase.[12] This is not an absurd hypothesis. Earth has experienced large and not-fully-understood temperature changes many times over the last 400,000 years even before the advent of large human CO_2 emissions. (Not all are attributable to other factors, like astronomical and solar cycles.)

---

[12]For example, geophysicist Jan Veizer writes that "I will argue that it is the other way around, with the tiny carbon cycle piggybacking on the huge water cycle, and the models are therefore reversing the cause and effect relationship." Veizer also suggests that solar activity (more specifically, ionospheric cloud nucleation pathways) could explain some of the changes in the water cycle over the last 500 years. See also Kirkby 2008 and Svensmark et al (2017). Ganopolski et al. suggests that Earth barely escaped a drop from the interglacial maximum around 1900 when solar radiation reached its minimum. It has been increasing since.

5. Scientific Agreement and Disagreement

If the minority view is correct, the carbon-cycle impact on the water cycle could be not the entire story. Even if CO_2-driven temperature increase drives most of the observed climate change, the omitted variables could mean that the cloud-modulated amplification factor could be smaller than three, perhaps even as low as two. If this is so, then harsh action to restrict fossil-fuel emissions would be somewhat less urgent.

The majority points out that it is difficult to conclusively reject the minority view, i.e., to measure the causal effect of the carbon cycle on the water cycle, because it is so broad and unspecific. There is not even one specific alternative mechanism widely agreed upon by critics. However, that is not proof that the minority views are necessarily wrong. Our interpretation is that scientists have the data to confirm that the majority theory could be true, but they lack the data (for now) to test whether the theory could be false: For example, to reject just one specific alternative about cloud formation, scientists would need data regarding whether there have been unusual spikes and systematic changes in cloud cover *unrelated* to CO_2-primed temperature changes that contributed to the planetary temperature response over the last 200 years.

By definition, a minority view is always controversial and not widely accepted. (And what are the consequences if the minority is wrong and civilization fails to act now?) From the perspective of the majority, the minority view has a big hurdle to climb, in that the majority view already has a coherent link from CO_2 to temperature change to humidity change to further temperature change — and with a good amount of evidence. The minority has little evidence to match this. From the perspective of the minority view, the burden of proof is on the majority and the case is not yet closed. Earth is a complex and chaotic enough system that, even lacking a specific alternative mechanism, the majority view could still be wrong. What if some other factor had primed the initial 0.3°C temperature rise?

One important meta-problem is that scientists' motives are difficult to judge. Has some minority dissent been promulgated by the fossil-fuel lobby to mask their own financial motives? The majority views some minority dissent as such (and rightfully so). Will engagement further fan the flames? The minority view wonders whether the majority view has become an echo chamber, with allegiance dictated by ideology and grant money, and with little tolerance for normal scientific skepticism. Scientists are just humans, too.

4. Climate Science

Widespread distrust has also made it surprisingly difficult for us to ask questions — scientists' first reaction when we question evidence is whether we do so because we are trolls coming to pick bones or whether we do so because we are genuinely curious and apply the same skepticism to their work as we do to our own. At times, this has sadly made it more difficult for us to learn more.

Making Sense of Data

Although there is no century-long global data on the role of clouds, the reader can puzzle over some of the temperature data in Figure 4.9. Roughly speaking, the overall hockey-stick graphs in CO_2 and temperature are well aligned, both in trend and acceleration. The alignment in trends favors the mainstream view. However, it is possible that the recent acceleration in global temperature could be due to the post-1980 reduction in anthropogenic SO_2 emissions from cleaner coal.

In addition, puzzling observations remain. They can be summarized by the statement that *if CO_2 is such a slowly increasing global gas, then why does global temperature not follow the same smoothly increasing path*, of course with suitable allowances for known solar and volcanic events? A year here or there may be chaotic noise that does not need to be understood, but on a global basis over decades, scientists should be able to explain the big deviations.

For example, from 1810 to 1850, there was an 0.4°C increase (ending the Little Ice Age) without a great change in CO_2 concentration — or for that matter, any other good explanation. Clearly, something other than human CO_2 emissions (which were still negligible) must have been responsible. What was it (and could something similar also be happening now)?

For example, the large Mount Tambora volcano eruption[13] caused the "winter without summer" of 1816, visible in Figure 4.9 — but what caused the large oscillations in temperature over the following two decades?

From 1860 to 1910, the temperature was stable or mildly declining. Volcanic activity probably contributed, but was it strong enough to nullify the steady increase in CO_2?

---

[13]Not all volcanoes may have emitted similar amounts of SO_2, so our description is not exact. Solar activity variation also does not seem to explain the observed patterns.

5. Scientific Agreement and Disagreement

From 1910 to 1945, global warming was strong with a sharp 0.5°C temperature increase. This is a large part of what is attributed to increasing GHGs. However, this also coincides with an increase in solar activity. Should we discount this warming?

However, just as it looked like the scientists had picked up a pattern, the temperature increase went on hiatus from 1940–1970 despite only modest volcanic activity. Why?

The hiatus was clearly over by 1970. For that time forward, there was now a sharp 1.2°C accelerating increase. And this time, solar activity could not have been the culprit because it had been on the decline since 1960! Volcanic activity also was not particularly low. Thus, the recent temperature acceleration seems generally a little too sudden to be attributable solely to the smooth atmospheric CO_2 increases, even giving CO_2 the 3°C power attributed to it by the mainstream models. (The imputed power is plotted in Figure 4.9.) Was warming delayed to 1970 by some global buffer that had filled up? Or was it delayed by reflective sulfur-dioxide particles from dirty coal that had peaked in 1980 and then declined sharply after 2000?

The majority points to the overall trends of CO_2 and temperature. The "signal" ($CO_2 \rightarrow$Warming) is strongly there and the physics are solid. The satellites' measurement of thermal disequilibrium further tell us more warming lies ahead.

The minority points to some deviations from the trend that are not fully understood. The majority might call this "natural background variation" — but calling it noise does not explain it. Some forces caused these large variations. Climate scientists are not 100% sure what it was. They are collecting more data, trying to find causes for each episode — but this could lead to overfitting the evidence. (Looking harder for confirmation than for rejection of theories also often tends to lead, not surprisingly, to overconfident confirmations.)

Perhaps, stating the argument as a match between "majority" vs "minority" is itself misleading. It could be that the truth lies somewhere in the range. What if mainstream climate scientists are 95% right and 5% wrong? What if something else that we do not yet fully understand still plays an important role?

As economists, we are not in a position to referee debates among the climate-science experts. We can merely present the agreements and disagreements to the best of our abilities. Interested readers can venture out to learn more.

Yet More Puzzles and Chaos

The role of clouds is not the only issue for which scientists desperately need more data. Most of the surface heat on the planet is not stored in the atmosphere, but in the oceans. Unfortunately, scientists have few direct measurements of how global ocean heat has varied over the centuries, especially with respect to the temperature *deep* in the ocean. Scientists can infer a little about deep water temperature indirectly by being clever. For example, with some extra assumptions, the observed increase in sea levels can be used to back out how much warmer the water has become. But with entire continents rising and falling, direct deep-water temperature measurements would be far better.

Here is another puzzle. Scientists do not know why it is primarily the global temperature lows that have risen, not the highs. Put differently, worldwide, climate change has *so far* brought primarily milder winters rather than hotter summers. From the perspective of melting a glacier or the permafrost, it matters less whether the average temperature increase is caused by lows or highs. From the perspective of habitability (and from the perspective of a scientist who really wants to know how the processes work), it does matter. Without understanding the past, scientists are not good at predicting how the highs and lows will react in the future.

Finally, there is a completely different point to consider: the complexity of the large chaotic system that is Planet Earth. Yes, it would be easier to understand Earth if scientists had long histories of all the data they want (which they do not have). But it would still not be easy. The relationships between solar activity, the atmosphere and especially water vapor and clouds, and climate are complex — and scientists do not have any other planets or small-scale systems on our planet that they can experiment with in order to obtain better guidance for the big-scale complex system that is Earth. Computer simulations are no substitutes for experimental evidence. Simulations reflect the assumptions that one puts into them.

5. Scientific Agreement and Disagreement

What is the Optimal Temperature and Change?

There is no disagreement that earth is warming and increasingly so. Our book is not about refereeing the scientists' modest disagreements. Instead, it is about pragmatic economic responses to climate change. Even so, we still have to grapple with further difficult questions. Here is a short preview:

How should one weigh the costs and benefits of global warming? For example, it is very likely that <u>heat-waves will kill more people</u> in the future — despite migration that will reduce the problem. Realistic reductions of emissions and warming cannot eliminate most of these deaths, but they can (modestly) reduce them. However, how should we count the fact that fewer cold-waves will save lives in the future? Is it appropriate to net one against the other?

Here is an even more basic question that sounds ridiculous at first but is not: What is the earth's optimal temperature? Was the cooler earth temperature 50 years ago better than the temperature today? How much better? What about the much colder temperature 12,000 years ago? If today's 6–7°C warmer temperature is better, how certain are we that another 2–3°C — after an appropriate adaptation interval — would be worse?

Are the costs of climate change so high that the optimal temperature is whatever it is at the moment? Is temperature variation and volatility the problem? In this case, any change would be undesirable. If this is so, then slowing the rate of increase would almost surely be beneficial, although it still would have to be weighed against the cost of doing so.

Our Perspective

Fortunately, the answers to many of the scientists' and economists' questions are not of as great an importance to our book as they are to other books about climate change. Our book is not primarily about how to eliminate *all* fossil fuels or *all* global warming. Instead, our book is primarily about pragmatic and affordable steps that can be taken to reduce reliance on fossil fuels and slow global warming as soon as possible. It is about the social blockages that have impeded moving the needle and how to get it moving now. We are not writing about planning for policies in 50–100 years; we are writing writing about policies this decade.

4. Climate Science

Ergo, for us, even in the unlikely case that the majority of scientists are wrong about global warming aspects and the optimal temperature is not today's temperature but 1°C more or less, we would still see no reason not to recommend that civilization curb fossil fuels a lot more (and more urgently) than it has done so far.

Our views may be less aggressive than those of the many mainstream earth scientists, but this is unimportant. The world is so far away from the optimal reduction of fossil fuels that our disagreements are small. Thus, we do not need to forecast whether aggressive action should ultimately reduce global warming by 0.3°C or 0.6°C (from 3.0°C to 2.7°C or to 2.4°C) in order to recommend curbing fossil fuels. The solutions that we will recommend in the rest of our book will largely remain the same, either way. They are limited not by the optimal climate path that a non-existing world collective order should follow, but limited by the hard social, political, and economic realities that actual individual decision makers will face.

6 Were the Models Wrong in the Past?

New York Times, 2012/11/24 — perhaps a little over the top?

At the start of our chapter, we asked the rhetorical question "What is a climate activist to do?" when climate change is so slow. But we can also ask the rhetorical question "What is a climate-change denier going to do?" when the evidence of global warming is so strong.

Climate-change deniers can cherry-pick past statements, often of hysterical public pronouncements by <u>some climate alarmists</u>, that have failed to come true — from <u>predictions of an ice age</u> (in the 1960s, long before humans had pumped up their emissions) to <u>predictions of Manhattan's west-side being underwater by now</u>, to imminent <u>predictions of "Peak Oil"</u>. (The figure of Lady Liberty is from the New York Times in November 2012.) Yes, these quotes exist; and yes, some scientists held these opinions. But by-and-large this is a misrepresentation of the scientific consensus. The less vocal majority of scientists are not primarily activists. If anything, they typically try to measure better and moderate and report conservative non-extreme estimates. But most new boring findings rarely receive the same attention that more alarming new findings do.

Figure 4.10. Performance of Temperature Models

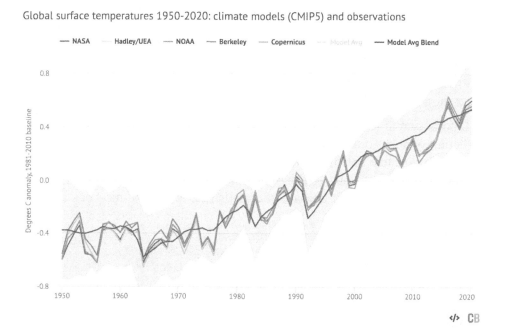

Note: Pre-2005 was fitted, post-2005 was predicted.
Source: Hausfather et al., 2019 and Gavin Schmidt, NASA.

But was it true that past models were bad? When skeptics repeat their claims often enough, some audiences become convinced that where there's smoke, there must be fire. But this is false. There is no fire. Figure 4.10 shows that some skeptics' broad claims are mostly an urban myth. The earlier-generation models were not perfect, but they were pretty good at least since the 1970s.[14] *On the whole* (not each and every one), past models were not hysterical, alarmist, or later contradicted by facts.

What about the models today? Of course, past performance is no guarantee of future performance. Even with much more knowledge, today's models could be wrong. And specific models disagree on precise numbers. However, most agree not only that human emissions have raised atmospheric GHGs and global temperature, but that both will continue (see our next chapter). We have already stated repeatedly that the planet is not yet back in temperature equilibrium. Any disagreement over the precise details should not be viewed as evidence that scientists don't know what they are talking about. They do know, and their disagreement is just the process of good science when the questions are difficult, the system is complex, and not all useful data are available. And, of course, all models contain errors. That is why they are called models.

If someone aims a rifle at you, our best models say you should duck. A prediction of a "zero model" (that the shooter will miss you) is also a model. Current models use the best scientific evidence there is — much better than the zero model that you would reject only after the bullet kills you. The scientists' models say that climate change is real and upon us. Let's not wait until you are dead.

---

[14]Earlier models were not only less sophisticated, but they also had good reason to predict global cooling. The world may have been on a path towards the next glacial period.

Further Readings

BOOKS

- Archer, David, 2016, The Long Thaw, Princeton University Press, Princeton, NJ. Explains the long-term history and effects of CO_2 and global temperature.
- Dressler, Andrew and Edward Parson, 2019, The Science and Politics of Global Climate Change: A Guide to the Debate, 3rd ed. Cambridge University Press.
- Kenny, Charles, 2012, Getting Better: Why Global Development Is Succeeding — And How We Can Improve the World Even More.
- Koonin, Steven I., 2021, Unsettled, BenBella Books, Dallas, TX, 2021. An opinionated analysis of the science and uncertainty of climate change and the human influence. Discusses much water vapor and sea-level evidence. Gary Yohe has published a critique in Scientific American on May 13, 2021. Yohe objects to cherry-picking, which Koonin similarly objects to in the IPCC. From our perspective, there is more agreement than disagreement on a second factor. Yohe agrees that scientists are still living "in a moving picture." Yohe objects not so much to the climate facts in Unsettled, but to the characterizations of the scientific discourse.[15] A more detailed pushback to Koonin appears in Climatefeedback.org.
- Pinker, Steven, 2018, Enlightenment Now: The Case for Reason, Science, Humanism, and Progress offers a good perspective regarding how people often underestimate human progress and adaptation.

ACADEMIC ARTICLES AND REPORTS

- Cruz, Jose-Luis and Esteban Rossi-Hansberg, 2021, The Economic Geography of Global Warming models the effects of climate change on different regions. Because warmer areas are more populated today, the net worldwide human costs can be quite high.
- Foster, Gavin L., et al., 2017, Future climate forcing potentially without precedent in the last 420 million years, Nature Communications.
- Ganopolski et al., 2016, Critical insolation–CO2 relation for diagnosing past and future glacial inception, Nature.
- Knutson et al., 2020, Tropical Cyclones and Climate Change Assessment: Part I: Detection and Attribution, Bulletin of the American Meteorological Society.
- Lenssen, N., et al., 2019: Improvements in the GISTEMP uncertainty model. Journal of Geophysical Research: Atmospheres, NASA GISS Temperature Analysis.

---

[15] In an interview with Tucker Carlson, Koonin's statements were badly edited and distorted. For example, Koonin *never* disputed the reality of climate change. We also consider both Koonin and his critics to be too harsh. No book and no critique can get everything right. In our opinion, Koonin lays out an important perspective. To the extent that the water cycle is not yet fully understood, one should view his perspective that "the science is not settled" as reasonable — or, in Yohe's words, that scientists are living "in a moving picture."

4. Readings

- Loeb, Norman G. et al., 2021, Satellite and Ocean Data Reveal Marked Increase in Earth's Heating Rate, Geophysical Research Letters. Summarized in The Washington Post.
- Medhaug, Iselin, et al., 2017, Reconciling controversies about the 'global warming hiatus', Nature.
- Sixth IPCC Assessment Report Summary for Policymakers, 2021. A comprehensive and readable summary of the underlying 1,500 AR6 page report.
- AR6 Synthesis Report: Climate Change 2022 The most recent IPCC assessment of the impact of climate change.
- Swain, Daniel et al., June 2020, Attributing Extreme Events to Climate Change: A New Frontier in a Warming World, *One Earth*.
- Trenberth, Kevin E. et al, 2016, Insights into Earth's Energy Imbalance from Multiple Sources, AMS Journal of Climate.
- Vecchi, Gabriel A. and Thomas R. Knutson, 2011, Estimating Annual Numbers of Atlantic Hurricanes Missing from the HURDAT Database (1878–1965) Using Ship Track Density. See also the survey in Knutson et al., 2019, some recent evidence of increasing tropical cyclone strengths in Emmanuel, PNAS, May 2020, and Kossin et al. (2021) from which our hurricane figure was plotted.
- Veizer, Jan, 2011, The role of water in the fate of carbon: Implications for the climate system, 43rd Int. Seminar on Nuclear War and Planetary Emergencies, 313–327. It discusses the perspective that the water cycle is primarily driving the CO_2 cycle rather than the other way around.
- van Wijngaarden and William Happer, 2020, Dependence of Earth's Thermal Radiation on Five Most Abundant Greenhouse Gases. This paper calculates "...from [at least] about 100 ppm to about 800 ppm, the solar energy increase that ends up affecting the troposphere increases roughly the same with each doubling of GHGs (especially CO_2)." We are not going to "enjoy" a situation where further CO_2 suddenly (or more increasingly) stops mattering at some level, because the atmosphere would have already become opaque to the relevant wavelengths.

Shorter Newspaper, Magazine Articles, and Clippings

- Chandler, David R., 2007, Climate myths: Carbon dioxide isn't the most important greenhouse gas, New Scientist.
- Chinchar, Allison, 2021, Antarctica's last 6 months were the coldest on record, CNN.
- Golden Gate Weather Services, 2021, El Niño and La Niña Years and Intensities.
- Lee, Howard, How will our warming climate stabilize? Scientists look to the distant past, 2021/12/20, Ars Technica.
- Lee, Howard, Scientists extend and straighten iconic climate "hockey stick", 2021/11/10.
- Kossin, James et al. Should the official Atlantic hurricane season be lengthened?, RealClimate.org, Apr 2, 2021.

- Lee, Howard, 2021, Scientists extend and straighten iconic climate "hockey stick", Ars Technica.
- Lindsey, Rebecca, updated Oct 7, 2021, Climate Change: Incoming Sunlight, NOAA, Climate.gov.
- McCully, Betsy, 2020, Ice Age New York
- Open Source Systems, Science, Solutions. Global Warming Natural Cycle.

Websites

- `https://www.ipcc.ch/`: The *International Panel on Climate Change*. It contains a wealth of information and data. It is also the best expositor of the majority view. More recent reports have ventured more into the social sciences and policy making.
- `https://earth.google.com`: shows Google Earth Climate Change animated over time.
- `https://www.realclimate.org`: climate scientists background for interested non-scientists.
- `https://skepticalscience.com/`: debunks many ill-informed climate skepticism claims.
- `https://www.noaa.gov/`: The National Oceanic and Atmospheric Administration. Hosts the Atlantic Hurricane Data discussed in Section I.
- `https://nsidc.org/`: National Snow & Ice Data Center.

Video / Audio

- Alley, Richard B., 2005, American Geophysical Union Bjerknes Lecture.
- Sobel, Adam, 2021, How Climate Change Fuels Extreme Weather, a podcast analyzing the link between climate change and extreme weather events.

4. READINGS

> *request for select reader feedback*
>
> Request for Reader Feedback and Corrections: We have one goal: to present the fairest description of climate change possible. We are *not* wedded to any views. We are not trying to cherry-pick. Where we perceived a *valid* scientific controversy, we have attempted to present both sides. If you believe that we have been misrepresenting the science, please let us know.

> *sidenote*
>
> Skepticism not only from inside the climate-science community but also from beyond is important — as it is in *any* field of science. As outsiders, we want to allow ourselves some liberty commenting on the state of climate science itself.
>
> Like many non-climate researchers (including many physicists), we have often found it difficult to ask probing question. We understand the hesitation of expert climate scientists when dealing with us. Not only have they been the subjects of personal attacks by large fossil-fuel companies (as if they were politicians), they also have had to deal with "trolls" (often paid) whose motives are not to understand the evidence, but to speak to a political audience.
>
> Nevertheless, it seems to us that the emotions have become too high. Although science is (and should be) adversarial by nature, the tone of the debate and mutual (sometimes personal) attacks eevn among the scientists have become excessive to the point of being counterproductive. It seems to us that one source of (remaining) disagreement among good climate scientists has arisen not because climate science is shoddy, or because the scientists are conflicted, evil or stupid — or that those questioning existing explanations are evil. (Yes, there are many charlatans, shills, and trolls, too, but this is not who we are writing about.) Instead, the principal source seems to be that it is difficult to attribute causality to slowly moving variables in an environment as complex as Earth and with the naturally limited data at hand.
>
> The public in particular has difficulties understanding the natural process of science. Science is never certain — but climate science is (probably) as good as it gets. Even Newtonian mechanics, Einstein's relativity, and Darwin's evolution are not "proven" in the mathematical sense. Instead, it is "just" that the scientific evidence is overwhelming.

Chapter 5

A Warmer Future

The previous chapter explained the science of climate change and some remaining scientific disagreements. This chapter explains what scenarios humanity should prepare for. There is no scientific disagreement that the global temperature is rising and at an increasing rate. NASA satellites tell us that the world is not even close to a new temperature equilibrium yet. Global warming is about to accelerate — and it will do so for a long time even if humanity managed to greatly reduce its emissions.

In the last chapter, we explained that a minority of scientists are wondering whether anthropogenic greenhouse gases are solely or just largely responsible for global warming. *From our book's perspective*, this does not matter much. The economic theory, models and evidence (explained in detail in Chapter 7) strongly suggest that humanity would collectively be better off if it greatly reduced its fossil-fuel emissions. And when one country reduces its emissions it has a positive effect on the others. The world is better off not just for global greenhouse gas reasons, but also for local particle pollution reasons.

Our perspective is that the real problem is not that the world has not yet gone far enough in its assessment of the needed cuts, but that individual decision-makers — countries and people — have their own and different incentives. It is in their self-interest to pollute when they don't bear the entire cost of polluting — and we see no realistic scenario in which this will change.

The world as a whole is currently so far from where it should optimally be that environmentalists' penultimate goals seem irrelevant to us. It is not important how far fossil-fuel reductions should go. It is important to get viable

5. A Warmer Future

cost-effective reductions jump-started. For now, although climate science and activism have successfully saturated the media, they have not meaningfully reduced worldwide emissions.

To us, the world seems like a team that is behind 0:5 and its players and coaches are now arguing about whether they will need to score 5 goals or more than 5 goals (because the other team could score again). Meanwhile, ten of the team's players have already sat down on the field or gone to the locker room, unwilling to suffer more personal exhaustion and injury, while a minority of fans (activists) are impotently yelling at their TV screens that the coaches should get the players running again.

It seems to us that environmentalists should care less about Utopian changes, and more about getting the world going *now* on realistic and cost-effective reductions in fossil fuel consumption that policy makers and the general public *world-wide* (and not just in the US and Europe) can sign onto.

This chapter also covers a related and perhaps even more important issue. In our view, the world has to be prepared for the unexpected. There is tremendous risk and uncertainty in climate forecasts. What could plausibly happen but is unlikely to happen is potentially far worse than what scientists *expect* to happen. Worst-case scenarios are important and have to enter the analysis — but they must also be kept in perspective. Ships are "overengineered" with safety features and even carry life boats, but they are not built and operated for worst-case scenarios. That is, they still set sail despite the fact that there is a danger of sinking. Climate policy cannot be based exclusively on the worst possible outcomes — though worst cases must still be evaluated. Moreover, there are also more than a handful of other existential risks for our civilizations. If we live to ensure against all of them, the only thing that will be ensured is that our civilizations will slowly disappear. An analogy is a soldier in a battle who knows that death could come from many places. It is impossible to avoid them all. It's not a situation one wants to be in, but it is what it is. Earth is not and never has been a safe place.

This chapter will first look at the most likely scenario — accelerating global warming — and then pivot to further risks, such as feedback loops and tipping points.

joke

> According to a new U.N. report, the global warming outlook is much worse than originally predicted. Which is pretty bad when they originally predicted it would destroy the planet. — Jay Leno, comedian

1 The Expected Warming Path

Adjustment Speed

For a comprehensive analysis of Earth's expected warming path, we recommend Archer's book. This section is our quick-take.

Assuming that civilization were to stop all emissions today and the CO_2 concentration remained at 410 ppm, how long would it take for the planet to settle down to a new equilibrium temperature? The best-case answer is...*centuries*. The planet has many heat buffers. The most important ones are again the oceans. The oceans are large and warm only slowly. And glaciers, polar caps, and Siberian permafrost are melting into the oceans, spreading the rate of temperature increases further over longer time spans.

What do the mainstream physics models say? As noted in the previous chapter, they predict roughly an ultimate increase of 1.5°C for the 50% increase in CO_2 that human fossil-fuel use has caused.[1] Only about half (0.8°C) of this increase has occurred to date. If CO_2 concentration stabilized at today's 410 ppm level (not a chance!), the planet would reach the three-quarters point of adjustment to final equilibrium by about 2100 — in other words, there would be a further increase of 0.3–0.5°C. The remaining 0.3–0.5°C warming would then take centuries.

This description is a good first take, but it is also an impossibly static view. It ignores the fact that civilization continues to pump ever more GHGs into the atmosphere. It ignores further amplifying or mitigating effects. To prepare for climate change, the world needs to work with more detailed projections.

Common Representative Scenarios (RCPs)

Historically, one of the many problems of planning for global temperature change on human time scales was that different scientists had used different models that had resulted in different predictions. And before they could even discuss any new findings, they always first had to synchronize their jargon, data, models, and backgrounds.

In response to this cacophony, the World Meteorological Association decided in 1988 that it needed more standardized assessments. Thus, it founded

---

[1] The increase from 280 ppm to 410 ppm is about 50%. Applying this 50% increase to the aforementioned effect of doubling (3°C) gives this 1.5°C.

5. A Warmer Future

the <u>Intergovernmental Panel on Climate Change (IPCC)</u>. The IPCC has developed a set of scenarios that are designed to summarize broadly the impact of rising greenhouse gases from many different models.[2]

With our new super-fast computers, we can get the forecast wrong TWICE as fast as we used to!

The most useful IPCC scenarios have the "memorable" name of <u>Representative Concentration Pathways</u>, more commonly abbreviated RCP. The name designates the extra net solar radiation in Watts per Square-meter that will heat up Earth. There are many paths that remain on the same pathway. (For example, the world could emit more this year and less next year.) We will use reasonably representative numbers for emissions, CO_2 in the atmosphere, and temperature based on pathways that we will be discussing.

The RCPs are firmly based on the mainstream view that water vapor and clouds produce an amplification factor of approximately 2–3 in order to best reconcile the historical global trends in CO_2 and temperature. The RCP model scenarios primarily consider how much pollution humans emit; how long these emissions will stay in the atmosphere; and how much solar radiation the atmospheric gases, water vapor, and clouds will trap. The RCPs are a group of reasonable scenarios that describe the paths of future warming under different assumptions.

The RCPs have also played important roles in all international climate negotiations. Their key advantage over the even simpler "ultimate-equilibrium-outcome temperature" models is that they provide dynamic time paths, rather than just the static equilibrium end-points with which we started this section.

---

[2]The IPCC has also developed social, political, and economic analysis, called SSPs. As economists with knowledge of the performance of macroeconomic models, we find these to be less convincing. Our disagreements are not important here. We just advise our readers to keep an open but skeptical mind when reading IPCC reports. Importantly, SSP1-2.6, SSP2-4.5, SSP4-6.0, SSP4-7.0, and SSP5-8.5 are SSP scenarios that follow RCP 2.6, RCP 4.5, RCP 6, RCP 7.0, and RCP 8.5. The SSPs can be thought of as potential narratives that feed into the RCPs. (We sometimes abbreviate RCP 4.5 as RCP 4. Scientists do not have enough data to reliably distinguish between the two, because the differences are minor.)

1. The Expected Warming Path

Figure 5.1. IPCC Emission Pathways With Uncertainty

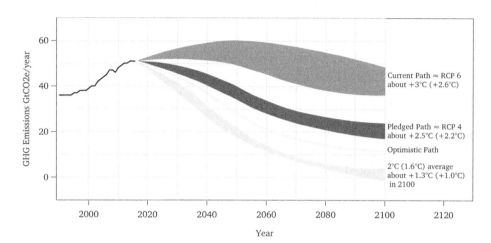

Note: The first temperature increase quoted on the right is relative to pre-industrial times; the second (in parentheses) is reasonably human-caused and thus about 0.3–0.4°C lower.

Source: ClimateActionTracker, May 2021 update.

Figure 5.1 includes reasonable time paths of future emissions and temperatures for the two most important and widely-referenced RCPs.

RCP 6 (in red) is a realistic "modest effort" scenario. As of 2020, think of it as a "minimal-intervention business as usual" scenario. Under RCP 6 (i.e., net warming of 6 W/m^2) global emissions would peak in about 30 years,[3] and temperatures would rise by about 3°C by 2100 relative to pre-industrial times. We consider it the most realistic scenario looking forward 50 years. Few analysts predict much higher emissions paths (such as RCP 8.5).

Beginning with the 2021 IPCC report, RCP 7 has become a newly prominent scenario—with modestly more warming than RCP 6, of course.

RCP 4 (or its near twin RCP 4.5, in blue) is an "active intervention" scenario, based on current government pledges. World emissions would peak this decade. Global temperature would rise by about 2.5°C by 2100. This

---

[3]This is in line with current expectations by the U.S. Energy Information Association.

135

5. A Warmer Future

scenario seems unrealistically optimistic. The world is already falling short of RCP 4. Few believe that it could be rescued.

Just like simpler models, the RCP estimates are expected values with a lot of uncertainty around them (shown as bands in the figure, which become larger further into the future). Think of the RCPs as discussion scenarios, not as famous last words.

Look closely at Figure 5.1: The *expected* differences between the RCP 4 and RCP 6 scenarios is 0.4–0.5°C, the difference between 2.5°C (requiring a lot of effort in RCP 4) and 3.0°C (requiring little effort in RCP 6). Recent estimates of the global damage are that the economic effect of the extra 0.4–0.5°C warming will be a lowering of GDP by about 0.1%, with perhaps half coming from agricultural damage. (We have never seen GDP damage increase estimates above 0.3% when comparing RCP 4 and RCP 6.) 0.1% of world GDP is a lot of money — a few hundred billion dollars — but these are damages that are eminently survivable for human civilization.

If you suffer from existential climate-change anxiety, you need perspective. The World Health Organization (WHO) predicts about 250,000 additional deaths per year from climate change up until 2050. These extra deaths will occur almost exclusively in poorer countries and arise primarily from malnutrition and diseases. (Yet, famine mortality has been declining — largely because fossil-fueled ships now provide large amounts of food to famine-stricken areas — and the specifically mentioned diseases may even be curable in 30 years.) 250,000 people are a lot of people, but this is "only" 0.003% of humanity. Far more people die from poverty. It's cruel, but economists have to think in millions and billions of people. The latest (2020 academic research predicts significantly more deaths after 2050 (mostly in Africa) if RCP 8.5 were to come about — although this research also states that its estimates are so uncertain that it cannot even reject the null hypothesis that there will be zero excess deaths due to climate change. It's hard to predict 80 years into the future. How would you have predicted 2020 in 1940?

Under RCP 6, high emissions will continue well into the next century (Fossil fuels will become more difficult to find at that point, too.) Thus, the RCP 6 harm will be more persistent than the RCP 4 harm, too. The temperature under RCP 6 would stabilize only in 2200-2300, while under RCP 4 it would stabilize around 2100-2200.

1. The Expected Warming Path

Any damages beyond the year 2100 will be caused more by future generations than by us (except in the sense that we are responsible for getting them accustomed to more wealth and higher standards of living). And those future generations will likely be richer than us. With better technologies, they may find it easier to get off fossil fuels. Furthermore, it seems unlikely that today's generation cares more about the world three generations into the future than they care about themselves.

The modest *percentage expected* temperature and GDP differences should not be interpreted to imply efforts to reduce emissions are unimportant, but they put the *expected* outcomes into perspective:

1. Even with great effort and activism success in stemming climate change, Earth will still be warming. Realistically, activist effort is only about slowing down warming by a modest amount. No one believes that it is possible to reduce warming to less than 2.5°C (1.5°C above today).

2. Even without special efforts, i.e., if the world ignores climate activists altogether, Earth will be warming by a similar magnitude, maybe 3.0°C instead of the just-mentioned 2.5°C. As long as "only" the *expected* bad outcomes occur, humanity will end up worse off, but by no more than, say, about 0.1–0.3% in welfare. This may be bad, but it is far from the end of the world.

However, from our perspective, the precise temperature change is not what really matters. We leave it to others to debate whether the ultimate global temperature increase will be 2°C or 4°C. It doesn't even matter whether the optimal intervention from a global perspective is to keep the temperature increase under 2°C or under 4°C.

Our book's "shtick" (colloquial for "theme") is that the best global choice is ultimately irrelevant. There is "no one home" who can pursue the optimal choice for the global collective. Individual decision-makers can only decide for themselves. Because the decision-makers do not receive all the benefits, they don't lower their emissions enough. Whenever they do lower their emissions, it has further positive effects on everyone else.

Thus, in our view, activists should care about how they can realistically nudge decision-makers towards *moving the needle* **now**. Lower fossil fuel consumption may not help one's own country, but the effects on other countries will be positive not negative... And, as we previewed in Chapter 2, there are good choices (and bad ones). We will return to them many times in later chapters.

At the time of the 2016 Paris International Climate Accords, the IPCC had also laid out the already briefly mentioned RCP 2.6, roughly the same as the green band in Figure 5.1. This scenario could have been met only under unrealistically aggressive policies. This scenario is basically obsolete. That train has already left the station. RCP 8.5 was the opposite — a pessimistic scenario based on the assumption that humanity would take no steps to abate emissions. This would have resulted in about 4.5°C of warming. Technological advances have thankfully rendered RCP 8.5 obsolete as well. (However, the ghosts of RCP 8.5 still haunts many analyses on the Internet and pieces in the popular press.)

2 The Expected Warming Harm

Planet Earth has been warming at an accelerating rate, and we have seen only a small part of it so far. How bad should we expect the harmful economic consequences of global warming to be? Of course, that depends on which RCP we are talking about, but we can sketch the general mechanisms.

RCP-Based Temperature Changes

To review, the world currently emits about 30 $GtCO_2$ per year, with about equal shares of 10 $GtCO_2$ each by the OECD (6 $GtCO_2$ from the USA alone), by China, and by everyone else. Adding other greenhouse gases and a charge for land use (deforestation) brings this up by another 20–25 $GtCO_2$e. In cumulative total, humans have emitted about 1,700 $GtCO_2$ since about 1800, of which about 1,000 $GtCO_2$ have remained in the atmosphere. The temperature is now about 1.4°C higher than it was before industrial times and 1°C higher than it was before the Renaissance. (Table 5.7 on pg. 149 provides a handy summary.) An increase of at least 0.7°C has already been caused by human greenhouse gas emissions, perhaps more, but only half of this effect has occurred so far.

Figure 5.2 plots the expected temperatures over the next few hundred years under four RCP scenarios. Let us start with the two implausible extreme scenarios. To limit Earth to the current temperature would require us to stay

2. The Expected Warming Harm

Figure 5.2. Climate Change Under RCP Scenarios

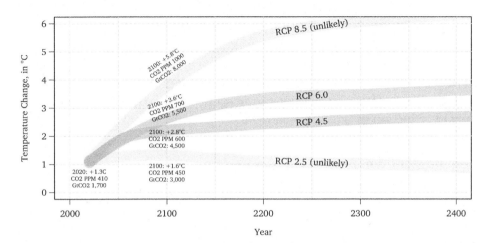

Source: IPCC, Nazarenko et al. and Meinhausen et al.. The base year is 1980, suggesting a world temperature increase of about 1°C by 2020. The more important aspects are the relative temperature paths.

on RCP 2.5. Total emissions would have to be no more than about 3,000 GtCO$_2$ by 2100. With 1,700 GtCO$_2$ already emitted, the remaining 1,300 GtCO$_2$ budget allows for about 20 GtCO$_2$ per year *on average*. This scenario has become so exceedingly unlikely that it is now only an interesting hypothetical. Emissions are already running closer to twice this 20 GtCO$_2$ and they have not yet peaked. The opposite extreme is RCP 8.5. It assumes continued fossil-fuel use at peak levels without much regard for climate change. It is outdated largely because clean energy technology has been progressing at a rapid rate.

This leaves the two plausible scenarios. RCP 4 limits humanity to about 45 GtCO$_2$ per annum, while RCP 6 limits it to about 55–60 GtCO$_2$ per year by mid-century. We deem RCP 6 to be more likely than RCP 4, because many poorer nations still want to develop and this requires more energy. Roughly speaking, the goal of climate conferences is to push civilization from RCP 6 to RCP 4. By 2050, the differences will still be quite small, perhaps too small (0.1–0.2°C?) even to distinguish between the temperatures in the two RCP scenarios. The main difference is what will happen after 2050. By 2100, the two RCPs are expected to diverge by about 0.3–0.5°C.

5. A Warmer Future

Interestingly, as of 2021, the MIT Technology Review reports that some trackers are beginning to revise their expected temperature estimates *downwards*. Both the climate action tracker and the United Nations now forecast a best-estimate scenario squarely between the two RCPs. The former expects a temperature increase of about **2.7°C** this century, on top of the 0.7°C last century. Perhaps even more importantly, their estimates of the extent to which activist policies could reduce global warming reductions have also moved towards the lower end. Their activist intervention scenario may only offer an expected reduction of 0.3°C, down from 0.5°C.

Terrestrial Changes

What will be the impact of climate change? We start with the direct impact of warming on land. For example, how bad would a temperature increase of, say, 2°C be in the United States? 2°C is roughly the difference in mean temperatures between Boston, New York, Washington DC, and Raleigh (each). Frankly, a 2°C increase in annual temperature won't mean the end of habitability for any of them. Even 3°C would qualify as more of a major nuisance than an outright catastrophe to most Carolinians. And so what if New York's climate would become more like Raleigh's?

Have you noticed global warming?

It is also not the case that 2–3°C would suddenly render much of Arizona or the Sahara uninhabitable. For practical purposes, they have already largely been uninhabitable for millennia. Parts of the adjacent Sahel would, however, likely become newly uninhabitable. This would heavily impact its 135 million inhabitants — but there is a lot of uncertainty even here. It could be that changing weather patterns could turn the Sahel wetter again, which could offset the harmful increase in temperature in terms of human habitability. Yet the change in precipitation patterns could then devastate altogether different areas. (Active human replanting, a form of geoengineering, could probably help in the Sahel, too.)

Global warming has also not been uniform. Figure 5.3 shows the changes in temperature across different regions of the globe so far. The warming

2. The Expected Warming Harm

Figure 5.3. Map of Planetary Global Temperature Anomaly

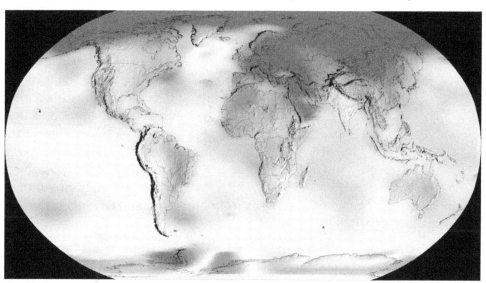

Source: NASA Scientific Visualization Studio. The increased temperature in 2021 is measured relative to 1951-1980. Except a small portion of Antarctica and south of Greenland, all temperatures have increased. The strongest increase has occurred over Eurasia. Similar to Figure 4.1.

has been most dramatic around the poles on the Eurasian continent and in Western Australia. In the tropics, where relatively more poorer people live, the increase has so far been more modest. Much of the United States has not been greatly affected by climate change yet. This could change.

5. A WARMER FUTURE

Sea Level Rise

The second important consequence of global warming is rising ocean temperatures and sea levels.[4] (Ocean acidification, discussed in Chapter 3, is not primarily a warming consequence.) The sea level rises primarily because warmer water expands in volume. Glacial meltwater from land (primarily from Antarctica and Greenland) raises global sea levels further.

The oceans are an effective heat buffer, because water holds much more heat than air. Whereas land temperatures have risen by about 1°C over the last 100 years, even the upper layers of the ocean have warmed "only" by about 0.13°C. The lower layers have probably warmed much less, but scientists have no long-term measurements of how far down global warming has reached and how unusual any warming would have been.

For now, the sea level rise seems to have been a slow and steady process. (Maybe it is very mildly accelerating.) Figure 5.4 shows that the rate was about 2 mm per year for the last 150 years or so.[5] Not shown in this graph, it also seems that this 2 mm was much faster than it was for millennia before. The increase almost surely indicates a warming ocean.

However, there is a small mystery here. Recall from Figure 4.9 on Page 111 that Earth has warmed much more since about 1950. The temperature anomaly was barely 0.3°C by 1950. The warming rate was much slower before then. (And warming likely wasn't due to the still small human CO_2 accumulations in the atmosphere.) Yet, why was the sea level rising at almost the same rate around 1900 — before most of the temperature anomaly took hold — as it has been rising after 1950?

---

[4]Koonin, Unsettled, Chapter 8 offers a more detailed but controversial discussion of sea level rise predictions.

[5]Interestingly, it is difficult just to measure what the global sea level actually is, in part because water and ice themselves have enough weight to press down entire continents.

2. The Expected Warming Harm

Figure 5.4. Global Mean Sea Level Rise Since 1880

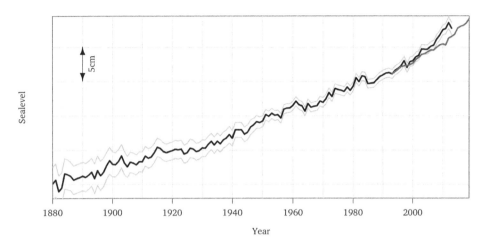

Source: Environmental Protection Agency (EPA). The blue line shows satellite data; the other data are from tide gauges. For a longer-term perspective and predictions, see Realclimate.org.

As of 2022, the IPCC projects for 2021 a global sea water rise of 55 cm (22 in), plus or minus 20 cm, under RCP 4 and about 65 cm (25 in), plus or minus 22 cm, under RCP 6. Some critics (specifically Koonin, Chapter 8) have argued that the IPCC has predicted a sea-level rise that is too high.

Other considerations, however, suggest that there is a risk that the sea level rise could turn much higher. Accelerated melting of ice sheets could increase the rate to as much as 180 mm/year (about 3.4 meters per century) in a near-worst-case scenario. An increase of 60 cm is bad but not catastrophic; 340 cm is on a totally different level!

There is good reason to be more concerned about long-run future sea levels (though humans can move over time spans of millennia). Figure 5.5 shows that just 20,000 years ago — about the time when modern humans spread across the continents — ocean levels were likely 120 meters lower than they are today. *This 120 meter figure is not a typo!* Some forecasts 10,000 years into the future even predict sea levels that will be 35 meter higher than they are today.

Figure 5.5. Long-Term Global Mean Sea Level Rise

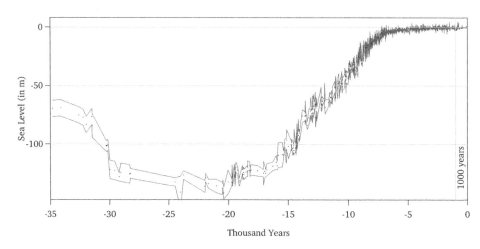

Source: Clark et al., 2016, Consequences of twenty-first-century policy for multi-millennial climate and sea-level change, Nature.

> sidenote
>
> There are many lost lands — perhaps not Atlantis, but surely a lot more than just Doggerland, a large swatch of land connecting Britain to the continent that was settled by ancient and now displaced tribes.

The harmful consequences of sea-level rise are easier to predict than the harmful consequences of terrestrial weather changes. This is because annual water temperature variation is smaller and satellites provide near perfect elevation data. Shore dwellers can look up online whether they will be harmed by future sea-level rise. The most vulnerable region in the United States is Florida. Worldwide, the most exposed regions include Bangladesh, Indonesia, the Netherlands, Venice, and parts of England. Some islands will disappear. But for most of the rest of the world, sea-level rise will be a yawn. Don't expect residents of Nebraska or Russia to be greatly concerned.

3   Inevitable Change and Associated Harm

Earth has always been changing. Maybe it was an accident or maybe it was not, but civilization has evolved in the most pleasantly warm and least volatile 4,000 years of the last few hundred thousand years. We cannot expect this unusual low environmental variability to continue forever — and even the best geoengineering will not be able to stave off large climate changes at some point in our (hopefully more distant) future.

Civilization will have to learn how to deal with more climate change — if not now, then in the future. For readers prone to fear and anxiety, we can provide some comfort (although there are better other books dealing with climate anxiety). If the IPCC's most likely outlook is correct, climate change will be modestly harmful, but it will be no threat to human civilization. The majority of humans (and especially those in rich countries) will come out just fine. They will barely be affected. A minority of humans (most of them in poor countries) may not.

All change creates winners and losers. And change is itself costly. On average, any change creates harm, because adapting is costly (just as fighting to prevent it is). For this reason, any change typically imposes suffering disproportionately on poorer people.

Nevertheless, there is an important point to keep in mind. Environmentalists and activists often paint a distorted picture, because they count up only the losses of climate change without netting these losses against the gains that will accrue to others. (Yes, climate change will create winners, too!)

Where?

Unfortunately, the temperature map in Figure 5.3 does not show what economists are most interested in: where do we expect the most extreme economic consequences of warming?

Many terrestrial impacts are difficult to judge; some are easy. The worst harm to humanity will clearly not occur in the Sahara or in the Antarctic. They are already uninhabited. Instead, the worst harm will occur somewhere in between.

However, the economic effects are also not always intuitive. If Earth warms by, say, 3°C on average, it could be that it warms 6°C in parts of Canada and

5. A Warmer Future

Siberia. If you now rejoice that this is great, because the Northern regions would become more pleasant, this conclusion may well be wrong. Although warmer temperatures will lengthen the Northern growing seasons, make their climates more temperate, and raise Russian and Canadian GDP in some areas first and the rest in the very long run, the melting permafrost could initially turn large parts of these regions into uninhabitable and stinking slush for decades, if not centuries.

To date, future terrestrial impacts due to global warming are difficult to judge. Scientists neither fully understand where most future global warming will occur nor how bad the consequences will be. (And they understand even less when they try to take into account changes in future weather patterns.) At best, scientists can offer only rough overall global guesswork with few confident specifics.

Sea-level impacts near shores are much easier to judge than terrestrial impacts. For one, we can comfortably predict that inland residents of continents will suffer little harm. Residents of Denver, La Paz, Addis Ababa, and Kabul have nothing to fear. Many million people who have settled near shorelines do — about 770 million people live at elevations lower than 5 meters. For them, sea level rise could cause a lot more misery.

The first harm will be that local drinking water becomes salt-water contaminated and useless. This is already a concern in Miami. Bangladesh could lose 10% of its land over the next century. As many as 400 million people could be forced to move, primarily in Bangladesh and Indonesia. It is cold comfort to those affected, but sea-rise caused and climate-change-caused migrations have happened many times before in human history — sometimes accompanied by warfare.[6]

In rare cases, communities have built sea barriers. The Dutch have shown the rest of the world how to survive below the sea-level by building dykes (which they built when they were still quite poor). Manhattan will almost surely not recede underwater for centuries, although it will have to pay dearly for better flood control measures. Some valuable land and beach houses will be lost to erosion (expensive when summed up over all the world's coastlines), but there will still be plenty of newly habitable oceanfront, too.

---

[6]Besides, the central problem in those countries may not be sea-rise over the next 200 years, but widespread poverty. If they were richer, they could also deal better with rising sea-levels.

3. Inevitable Change and Associated Harm

Who?

Many of the remaining chapters of our book will focus on how civilization should respond to impending warming. But one thing is clear: environmental changes will always harm the poor more than the rich. Wealth makes adaptation to change easier. At an individual level, rich people can buy air conditioning and move away more easily. At the country level, rich countries can build better sea barriers. It has simply always been better to be rich and healthy than poor and sick.

Realistically, it seems unlikely that the disproportionate climate suffering of the poor will sway anyone (other than a few activists who will dedicate much of their lives to helping them). If we rich cared more for humanity's poor, we would not have to wait for climate change. We could alleviate plenty of human suffering, malnutrition, homelessness, sickness, and death *today*, e.g., by donating money to UNICEF.[7] The fact that the rich collectively and individually (including most "salon activists") do not send significant fractions of their incomes to the poor speaks volumes about how much voluntary altruism and poverty concern we can expect to see from humanity in the future. It's a lot cheaper to lament poverty than to alleviate it.

---

[7] However, no country has ever escaped poverty through global donation. Countries escape poverty through economic growth.

5. A Warmer Future

4 Temperature Change Summary

Figure 5.6. Climate Action Tracker Summary, Nov 2021

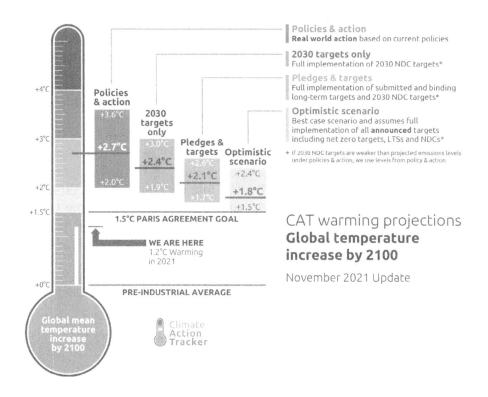

Source: Climate Action Tracker: Glasgow's 2030 credibility gap: net zero's lip service to climate action for more detail.

The climate action tracker (CAT) is a non-profit collaboration of scientific institutions. It tracks forecasts of global warming on a monthly basis.

Figure 5.6 shows their thermometer as of November 2021. We recommend reading the CAT source web-page more carefully. It offers more insight into their analysis, including further details on countries and policies. If the 2030 targets of Glasgow are fully implemented, CAT forecast a reduced temperature increase of about 0.3°C by 2100 relative to a scenario of current policies. Full target implementation is already a big and optimistic *if*, as we will explain in

4. Temperature Change Summary

Chapter 9. We consider pledges and optimistic scenarios not to be realistic—best-case and beyond best-case scenarios. We suspect that the CAT scientists would largely agree with our assessment, some exceptions notwithstanding.

Table 5.7. Summary of Mainstream Temperature Estimates

Renaissance Decrease (1450–1800)	−0.4°C	
Increase, 19th Century (1800–1900)	+0.2°C	
Increase, 20th Century (1900–2000)	+0.7°C	
Increase, Recent (2000–2020)	+0.6°C	(included below)
Exp. Increase, 21st Century (2000-2100)	+2.7°C	(IPCC: 2.1–3.4°C)
Optimistic Activist Scenario (2000-2100)	+2.4°C	(COP26, Glasgow 2021)
Activist Curtailment		
Global, Effect by 2100	0.3–0.5°C	
All-OECD, Effect by 2100	0.1–0.2°C	(pro-rated)

Source: These are rough estimates. Note that differences of 0.1–0.2°C are within margins of error and quoted differently by different sources (e.g., relative to slightly different benchmarks or updated over time). The top part of this table is based on Mann et al., as in Figure 4.9. The 21st Century estimates in the bottom part were summarized by the MIT Technology Review, which in turn based its estimates on the IPCC and Climate Action Tracker assessments.

For easy reference and discussion in later chapters, we summarize the (majority) assessment in Table 5.7. For perspective, 2–3°C is bad and almost surely unavoidable — but remember that Earth has already warmed by a much starker 6°C since the last glacial maximum just 15,000 years ago. The current predicted increase of 2–3°C is so remarkable only because (a) the change is very rapid; (b) Earth has not seen such large temperature changes, up or down, since the advent of advanced human civilization about 5,000 years ago; and (c) Earth is already at the high end of its temperature range at least since the appearance of the human species.

To make smart decisions, it is important to understand that climate-change activism is *not* about avoiding expected global warming of about 2–3°C. Instead, it is about pushing Earth from about RCP 6 to RCP 4, i.e., from 2.7°C to 2.4°C. Consequently, activist intervention could at best push down warming by about 0.3–0.5°C. Reducing future emissions in all OECD countries to become consistent with RCP 4 could reduce global warming by only 0.1–0.2°C, because the OECD is only responsible for about one-third of the world's emission. We will pick up the subject of economic *analysis on the margin* in Chapter 6.

Don't be too comforted by this relative comparison of 3°C now vs. 6°C a few millennia ago, though. The 3°C is based on expectations — and what if those expectations are off?

5 The Well-Known Unknowns

We have focused thus far on the *most likely* changes in planetary temperature associated with human emissions of GHGs. But our discussion of the most likely scenario has hidden the risks associated with deviations from expectations. Fortunately, the scientific models make it possible to assess not only the most likely outcomes but also *some* of the modeling uncertainty.

Given that there are still many active debates in the climate sciences even in the best-understood and most-likely scenarios (that is, on a global basis over long time horizons), it is not surprising that it is even more difficult to quantify the uncertainty. Thus, without trying to referee the science we discuss the evidence primarily from the perspective of the mainstream IPCC viewpoint. As a handy approximation, imagine that the minority view predicts only about one-half the temperature increase that the mainstream predicts.

With this caveat, Table 5.8 shows how higher future CO_2 concentrations will likely affect future average temperatures, including an estimate of the uncertainty around the prediction. At the current state (410 ppm) and rate of net increase in atmospheric CO_2 of about 2.5 ppm per year, scientists expect an eventual GHG-caused annual temperature increase of about 0.025°C per year. This does not sound like much, until you think in decades. It amounts to just under 1°C for every human generation (35 years) — and that *is* a lot.

We can summarize mainstream scientists' current understanding of the influence of the link between CO_2 and temperature *with uncertainty* as follows: The predicted long-term global temperature increase for a doubling of CO_2 in the atmosphere ranges from 1.5°C to 4.0°C, with **2.5°C** a good working middle. By this metric, civilization's 50% increase from 280 ppm to 410 ppm has pushed up the new and not yet fully realized long-term equilibrium temperature by about **1–2°C** (rather than a precise 1.5°C), with ongoing

Table 5.8. A Map from CO_2 Concentration to Global Temperature in Long-Run Equilibrium, Mainstream View

| Estimated 66% Range | | | Temperature |
Low ⟵	Mid ⟶	High	Increase
	270		0
320 ⟵	340 ⟶	380	+1°C
370 ⟵	430 ⟶	540	+2°C
440 ⟵	540 ⟶	760	+3°C
530 ⟵	670 ⟶	1,060	+4°C
620 ⟵	840 ⟶	1,490	+5°C

Note: The range is the 66% confidence range of atmospheric concentrations associated with warming *above pre-industrial levels*. In other sciences, it is often more common to cite 95% confidence levels, which would be bands that are roughly twice as wide as those quoted in this table.

Source: Azimuth and the Climate Modelling Intercomparison Project.

emissions now pushing the planet towards **2–4°C**. (The minority view would lower this to "only" 0.5–1°C and 1–2°C, respectively. Of course, even this would still be a stunning increase over such a short time frame.)

6 The Less-Known Unknowns

A 2–3°C increase is bad enough, but we are far more worried about unknown worst-case scenarios than we are about the expected scenario or even an expected worse-case (but not worst-case) scenario. Because humanity is conducting an unprecedented giant experiment in climate engineering by raising CO_2 concentrations in the atmosphere, it is possible that scientists have not recognized all that could happen.

5. A Warmer Future

The Unprecedented Speed of the Human Impact

Our first worry is not the equilibrium temperature at which Earth will settle in 100 years, 500 years, or 1,000 years, but the sheer speed with which the planet is being pushed. Because scientists know of no recent historical analog for such a fast rate of change, they find it hard to predict whether the impact of such a rapid change will be worse and cause new problems (such as temporary exhaustion of buffers).

The fossil record tells us that 75% of all species vanished in response to the great asteroid strike of 65 million years ago. Only a <u>few creatures weighing more than 25 kg</u> survived. Of course, this comparison is a bit over the top. Human-induced climate change is neither as fast nor as dramatic as this asteroid strike.

However, human changes are still lightning-fast and dramatic by geologic standards. They are so fast that our scientific methods are not yet precise enough to allow us to detect whether there have been short-term isolated CO_2 shocks in the last 50 million years comparable to those created by civilization over the last 50–100 years. Even the largest events, like the volcanic eruptions 250 million years ago that created the Siberian traps (roughly the size of the United States) and wiped out 80% of all species, took more than a million years. Although humanity will never match the total CO_2 released by the Siberian eruptions, the average speed with which CO_2 in the atmosphere is rising is more than ten times faster than it was then.

How many species will be unable to cope? We do not know. How will this impact the biosphere and food chain? We do not know.

6. The Less-Known Unknowns

> sidenote
>
> Climate change is a contributor to the biodiversity crisis, but it is not the only one. Human population growth, with accompanying appetites and habitat reductions, is probably more central. We agree with scientists and environmentalists that ecosystem collapse is another existential crisis for humanity — perhaps even more than climate change. It's good to see the United Nations raise the issue and see the media report on it. There is a danger that the attention on climate change could distract from ecosystem collapse. Humanity needs to urgently address both.
>
> We see technology as a good non-collaborative universal approach to reducing emissions and slowing climate change. We don't see good universal approaches to extinction crises and ecosystem preservation. However, we do know that workable solutions will have to make it in the self-interests of local populations to protect ecosystems. (And unfortunately, local populations do not own and thus cannot cover ocean habitat protection.) We hope that younger generations will do a better job of protecting the world's ecosystems than our generation has.

Harmful Feedback Loops

Our second worry is feedback loops. A perfect example is the impact of thawing permafrost already mentioned in Chapter 3. The frozen ground, which covers large areas of the near-arctic north, holds 3,000 $GtCO_2e$ — potentially 300–600 ppm worth of atmospheric CO_2. Worse yet, it could be released in the form of methane, which is even more potent than CO_2. The rising temperatures would release yet more methane from the permafrost. And so on.

It is unknown how realistic and damaging the permafrost feedback loop is, but the signs are not good. Scientific studies have been revising upward their estimates of thawing speed almost as fast as the thawing itself is happening.

Table 5.9 shows a few more possible harmful feedback examples. The most important one is the (also aforementioned and central) water-vapor feedback loop. As the atmosphere starts to warm, the amount of water vapor in the atmosphere tends to rise. As a result of the increased water vapor, the atmosphere warms further, enabling more water vapor to be held in the atmosphere, and so on.

Even clouds could constitute a harmful feedback loop. Depending on where the clouds form, they can either warm or cool the atmosphere. High-level clouds tend to keep heat in. Low-level clouds tend to reflect heat. Unfortunately, it seems now that both a warming atmosphere and increasing

5. A Warmer Future

Table 5.9. Examples of Feedback Loops

Harmful, Amplifying	Helpful, Reducing
• Permafrost melt (methane) • Less ice albedo • Ocean circulation disruptions • Sea level rise to glacier melt • Rainforest drought and loss • Wetland methane release • Forest fires • Shallow gas hydrates	• More cloud reflectiveness • Higher rainfall • Photosynthesis • Chemical weathering • Meltwater CO_2 absorption. • Altitude temperature

Source: Earthhow Climate Feedback Loops.

water vapor favor the formation of high-level clouds — but scientists are not yet certain which way this will go.

Melting ice can also produce a feedback loop. Sea-ice and ice sheets provide large white surfaces, reflecting solar radiation. As long as melting ice does not expose darker soil or ocean below, the melting process is slow and steady. However, once ice has melted enough (usually first near the edges), the darker surface below absorbs more solar radiation rather than reflecting it. This solar absorption in turn speeds up the heating and melting process. Warming climate and associated fires, along with rising human population, are also leading to an increase in deforestation. (This feedback loop could have created the Sahara in the first place.) Deforestation is currently particularly pronounced in tropical forests such as the Amazon. The deforestation reduces the uptake of CO_2, which in turn leads to more warming, which in turn leads to more forest fires, and so on.

Not all feedback loops are harmful. Some feedback loops are beneficial, partially offsetting of the impact of warming. For instance, a rise in CO_2 speeds up the growth of plants that remove CO_2 from the atmosphere. This beneficial feedback helps stabilize the climate. The oceans and terrestrial sinks have always been working against global warming, stabilizing atmospheric GHGs, absorbing relatively more CO_2 when the atmospheric CO_2 concentration is higher. (However, higher temperatures can be destabilizing, because they can cause the bubbling release of CO_2 from the oceans.)

6. The Less-Known Unknowns

Scientists do not understand all possible feedback loops, if only because they have never observed in real time the large temperature increases and high temperatures that may be required to initiate many of these loops. The uncertainties regarding these feedback loops are also partly why the bands associated with the RCP scenarios in Figure 5.1 are so wide. But what we don't know yet should scare us, not comfort us.

Catastrophic Tipping Points

Our third worry is tipping points. Think of a glass of water on a table. With small nudges, the glass shakes and the water sloshes but soon everything returns to normal. However, if the initial shock is big enough, a sudden harmful feedback effect appears: the glass tips over and the water pours out. Such thresholds are called "tipping points."

An example of a smaller tipping point being triggered occurred in 2012. It is ordinarily unimportant to New Yorkers whether the sea-water level is a little lower or higher. Even normal storm surges don't matter much. Yet Hurricane Sandy exposed a tipping point. It caused little damage — up until the moment when its storm surge reached the level that it could enter the subway and basements — and then it caused $70 billion of damage!

Humanity has never lived through a period of sharply rising greenhouse gases above 400ppm. All that humanity has lived through have been small nudges. The tipping points could be a "killer."

Table 5.10. Predicted or Plausible Tipping Points

1–2°C	Arctic and Greenland Melt; Indian Monsoon disappears
2–4°C	West-Antarctic and Amazon
3–5°C	West-African Monsoon, Boreal Forest Disappears
3–6°C	El Nino ceases
4–5°C	Atlantic Circulation ceases
5 °C+	East-Antarctic Melts

Source: Bild der Wissenschaft, July 2019, p22f.

Table 5.10 lists a number of tipping points that could be triggered by climate changes. (No one knows for sure.) Many scientists suspect that we are just about to cross the first one. Our children may get to observe the melting of the Arctic and Greenlandic ice sheets and perhaps also the disappearance of the Indian Monsoon.

The faster the climate changes, the more harmful its effects are likely to be. This is because species and civilization will have less time to adapt. If the Indian monsoon were to stop and/or the Himalayan glaciers were to melt, it could destroy the livelihoods of more than a billion people. If the Atlantic circulation were to change and the jet stream moved north, Great Britain could turn into the climatic equivalent of Iceland. If India and Great Britain had centuries, they could slowly plan and adapt. But if these changes occur too fast, say, within one generation, the result is likely to be much greater human misery.

Previously, we described the IPCC estimate of up to 45 cm of sea level rise by the end of the century. It's harmful, but not catastrophic. However, the IPCC suggests that the sea level rise could increase to 90 cm under a higher RCP — and this would matter more. OK, but even this is not our main worry. We fear that it is possible that sea level rise could suddenly become far more abrupt. Global warming could quickly melt Greenland's and part of Antarctica's ice sheets. (We may already be beyond the point of no return.) In fact, paleo-historic records show that both the Greenland and the West-Antarctic ice sheets have melted and collapsed in the last 125,000 years. They may well do so again, this time perhaps within our lifetimes. David Archer warns that ice sheet collapses could conceivably increase the global sea level by — get this — 3,000 cm. Obviously, unlike the 1 meter increase in the IPCC scenario, such a 30 meter increase would be catastrophic. Are such multi-meter-level changes in sea levels so high that their predictions can be dismissed as absurd alarmism? No. Just 20,000 years ago, the sea level was 120 meters lower, so another 30 meters is not inconceivable. Again, we are not suggesting that 5, 10, or 20 meters of sea level rise is likely. We are only suggesting that it is not impossible.

Given our three grave concerns (speed, feedback loops, tipping points), basing climate policy decisions on the middle of the band seems dangerous. What if the outcome turned out to be on or beyond the extreme of the band — and not on the good side? Heaven (won't) help us.

7 Appropriate Perspectives

We started this chapter with a description of our planetary temperature history. The civilizations of the last 5,000 years have evolved in a temperature interval of about plus or minus 1–2°C. Such climate stability has been unusual. Our current stable temperature is *not* Earth's "standard normal." Earth's climate is variable and it will change again.

> 'All conditioned things are impermanent' – when one sees this with wisdom, one turns away from suffering. — The Buddha, Dhammapada v 277.

Normal Variation

Earth was approximately 8°C *colder* than it is today just about 20,000 years ago in the most recent glacial maximum. As we have stated repeatedly, we are currently enjoying as high and stable a temperature as the world has seen for millions of years. Primates have lived in worlds colder than it is today, but never in one 2–3°C *warmer*.

The large and long planetary cycles in Earth's deep geological past (with Earth remaining in a greenhouse or icehouse for hundreds of millions of years) raise the specter that there could be stark hidden positive feedback loops, invoked by tipping points, perhaps even Domino-like, just waiting for us around the next corner.

There is a realistic chance that by pushing Earth's temperature to levels not seen in millions of years, civilization may wake long-dormant processes and feedback loops that could push us much further — perhaps out of the planet's current ice age altogether. We wish scientists could run a quick "high-temperature trigger test" to learn what is lurking behind the corner without triggering it permanently. This is wishful thinking. Our scientists and policy makers are essentially driving blind.

As if global warming wasn't bad enough, there is even more to worry about. Although it seems far less likely, there has been enough natural variation on Earth to remain concerned about the opposite, too. Scientists are not sure what drove the onsets of glacial periods. They do know that glacial periods have tended to occur every 100,000 years or so, perhaps related to astronomical phenomena. Many scientists think that humanity's CO_2 emissions may have nipped the next glacial maximum in the bud — but they again are not sure.

5. A Warmer Future

The error terms in climate models are about the sudden and unexpected. These include possible unforeseeable random disasters. In less than a decade, on April 13, 2029, the 200 meter <u>Asteroid Apophis</u> — with an energy potential roughly equivalent to all human nuclear arsenals together — will whiz by Earth *inside* the orbits of our geostationary satellites. An asteroid of this size hits Earth on average <u>every 40,000 years.</u> And asteroids with ten times more energy hit Earth about every 100,000 years.

I am an optimist. I have every confidence that global warming will be nullified by nuclear winter.

In addition to asteroids, there are about 20 supervolcanoes, and one or the other has erupted every 50,000 to 100,000 years or so. If the <u>Lake Toba Supervolcano</u> were to erupt, Earth's temperature could fall by 5–15°C for a decade or more and even trigger another <u>glacial period</u>. An even bigger <u>supervolcano eruption</u> could lower Earth's temperature by as much as 15°C.

Either a large asteroid or a supervolcano could wipe out most human crops for a few years and dramatically reduce agricultural output for decades. The fatalities would be much worse than those from the expected 2–3°C human-made global warming (that we should rightly be more worried about today). Fortunately, this is exceedingly unlikely within the next thousand years. Unfortunately, this is not impossible. In fact, over long enough a time span, it is a near certainty.

Our point here is not that a change of 2–3°C is little to be concerned about, much less to expect global cooling. Our point is that large temperature surprises are nothing unusual. Yes, if Earth experiences 3°C warming, it can cause a lot of human misery — but it's nothing compared to the environmental changes that bigger 5–10°C changes have brought in the past and could bring again. Should we be ready for either?

7. Appropriate Perspectives

The Power To Alter Our Fate

Human civilization developed in a 5,000 year period of remarkably stable conditions and largely has come to rely on them. It may not last.

But there is something new. Technology has now advanced so far that, for the first time, humanity has the power to influence the planet via actively designed geoengineering — directed human intervention in climate on a world-wide scale.

We think it is important that humanity develops the means to react — if need be, to stabilize the planet's temperature. Naturalists will of course recoil. But hear us out first. Simpler geoengineering is nothing new. Humanity has been actively geoengineering for millennia, possibly since it had mastered fire. Agriculture is geoengineering at a gigantic scale.

But humanity now has something more powerful at its disposal — scalpels rather than a stone knife. Scientific advances have brought more powerful technologies within civilization's grasp. More are being developed all the time.

Among the better geoengineering ideas are technologies to remove CO_2, including reforesting and accelerated stone weathering, and solar radiation management, including cloud seeding and rain-making (without the prayers). More radical and controversial interventions could send reflective sulfur dioxide (SO_2) particles into the upper atmosphere (as explained in Chapter 3), where they will last for a few years (and before they will eventually come down again in the form of mild acid rain). Unlike CO_2 reductions, whose temperature effects will take many decades to start working, geoengineering can reduce temperature almost immediately. And deflecting an asteroid is similarly human tampering with natural processes.

Geoengineering may turn out to be a terrible idea. But we believe that if planetary warming awakens catastrophic feedback loops, scientists should understand all choices for last-resort interventions. We judge the "moral hazard" (of no longer worrying about climate change) that this would engender as relatively modest. Humans are already polluting without much abandon. We view geoengineering as "in case of emergency, break glass." We will return to geoengineering in Chapter 14.

5. A Warmer Future

Natural vs. Man-Made?

This brings us to our final point. When should humans intervene? Does it matter that (or even whether) humans are the primary cause of global warming?

The answer should be an emphatic no *No*!

Even if the current temperature increases had come entirely from natural sources, civilization should still want to learn what the effects are and how to manage them.

Mass extinctions are entirely natural, too — but we wouldn't want to live through one if scientists could do something about it. Climate change and CO_2 changes may have caused or contributed to the five major mass extinctions in the last 500 million years. The volcanic release of 36,000 $GtCO_2$ over 15,000 years is a prime suspect for the Permian mass extinction about 250 million years ago. It wiped out 90% of all species. In the Devonian extinction, about 370 million years ago, an estimated 96% of all species disappeared.

Stop being a naturalist! Just because something is "natural" does not mean that it is good and that it is in the interest of humanity not to interfere with it. The Black Death was natural! Even if the sun were the cause of all global warming, if its activity were increasing, the planet would still suffer the same temperature consequences. Scientists should still research how humanity could respond.

8 Planetary Roulette, Anyone?

Knowledgeable climate skeptics no longer deny global warming. Instead, they point out that the *expected* harm from slow planetary warming may not be that damaging. They may even be right (depending on how bad is bad). But for a long time, many of these critics were missing an important point. (Although most no longer do today.)

The impact of the expected warming scenarios pale in comparison to the unlikely, but not impossible, worst-case scenarios. Many of us pay for home insurance, not because we believe that our houses will burn down, but because we do not want to take the small risk of a really bad outcome. Most of us also pay for the opposite, water and flood insurance. Of course, buying insurance makes sense only if the insurance is less expensive than the house. On a planetary level, humanity should be prepared for global warming, but being prepared for (far less likely) unexpected cooling is also not a bad idea.

At the same time, humanity cannot make decisions for the absolute worst-case scenario. Yes, an asteroid could hit us. (Or, more realistically, nuclear weapons could destroy our cities.) However, we cannot move our civilizations underground in order to avoid such worst-case scenarios. In fact, there are so many worst-case scenarios, we don't have the resources to avoid them all. The universe has never been risk-free and never will be risk-free. What are prudent risks to take, what are not?

Perhaps the best way to describe the global climate situation is that civilization is playing Russian roulette. What would you pay for not having to participate in one round of Russian roulette? What should civilization be willing to pay for not having to participate in one round of climate-catastrophe roulette? How does it depend on the number of loaded and unloaded slots in the magazine?

5. READINGS

Further Readings

Please see also the references in Chapter 4. In addition:

BOOKS

- Lomborg, Bjorn, 2020, False Alarm, Hachette Book Group, New York. An opinionated view of the costs and benefits of climate change.
- Lomborg, Bjorn, 2001, The Skeptical Environmentalist, Cambridge University Press. Well worth considering, despite receiving scathing critiques upon publication. Provocative. Deserved more sober disagreement and critique.
- Mann, Michael E., 2021, The New Climate War, Hachette Book Group, New York, NY. Describes the misinformation and delay campaign of the fossil-fuel lobby.
- Robert Pindyck, 2022, Climate Future: Averting and Adapting to Climate Change covers uncertainty and unknowable much better than we do.
- Wagner, Gernot and Martin L. Weitzman, 2016, Climate Shock: The Economic Consequences of a Hotter Planet, Princeton University Press, Princeton, NJ. Focuses on risks off the "expected" path.
- Wray, Britt. Generation Dread: Finding Purpose in an Age of Climate Crisis, Knopf 2022, discusses how to handle climate anxiety (not climate change).

REPORTS AND ACADEMIC ARTICLES

- IPCC, Climate Change 2014 Synthesis Report Summary for Policymakers (AR5); IPCC Synthesis Report: Future Climate Changes, Risks and Impacts (AR5); and Climate Change 2021 The Physical Basis: Summary for Policy Makers (AR6).
- Lenton, Timothy M., et al., 2019, "Climate tipping points — too risky to bet against." Nature.
- Martin, Ian, and Robert S. Pindyck, 2015. Averting Catastrophes: The Strange Economics of Scylla and Charybdis, American Economic Review.
- Nazarenko, L. et al., 2015, Future climate change under RCP emission scenarios with GISS ModelE2, Journal of Advances in Modeling Earth Systems.
- Wayne, G.P., 2021, The Beginner's Guide to Representative Concentration Pathways, Skeptical Science.
- Sayedi, Sara Sayedah, et al., 2020, Subsea permafrost carbon stocks and climate change sensitivity estimated by expert assessment, Environmental Research Letters.
- Zhao, Zi-Jian et al. 2020, Global climate damage in 2°C and 1.5°C scenarios based on BCC_SESM model in IAM framework, Advances in Climate Change Research. Suggests most warming damage will be in agriculture.

Shorter Newspaper, Magazine Articles, and Clippings

- Dietz, Simon, et al., 2021, Economic impacts of tipping points in the climate system, PNAS. Estimates a 25% increase in the SCC due to tipping point—including a 10% chance that it could double the SCC.
- Hausfather, Zeke, 2021, Explainer: The high-emissions 'RCP 8.5' global warming scenario. Views RCP 8.5 as an unlikely but not impossible scenario.
- Herr, Alexandria et al., Points of No Return, Grist.
- Johnson, Scott K., Study finds we're already committed to more global warming—sort of, Ars Technica, 2021/01/06.
- Johnson, Doug, 36,000 gigatons of carbon heralded history's biggest mass extinction, Ars Technica, 2021/09/17
- Lavelle, Marianne, 2016, Crocodiles and Palm Trees in the Arctic at 8°C, National Geographic.
- Lee, Howard, 2021, How will our warming climate stabilize? Scientists look to the distant past. 2012/12/20. The Eocene was dramatically hotter than today. The article also describes different proxies for warming and uncertainty estimates.
- Lindsey, Rebecca, 2021, Sea Level Rise. NOAA.
- Moor, Rose, 2019, 15 USA Cities That Will Be Underwater By 2050 (10 Already On The Ocean Floor), The Travel.
- Nuccitelli, Dana, 2013, A Glimpse at Our Possible Future Climate, Best to Worst Case Scenarios, Skeptical Science.
- Rahmstorf, Stefan, 2016, Millenia of Sea Level Change, 2016, and Sea level in the IPCC 6th assessment report (AR6), 2021, RealClimate.org.
- Rutgers University, 2020, New Research Affirms Modern Sea-Level Rise Linked to Human Activities, Not to Changes in Earth's Orbit.
- Sobel, Adam, 2021, How Climate Change Fuels Extreme Weather, a podcast analyzing the link between climate change and extreme weather events.
- Stockholm University, 2021, Sea Water Predictions Are Tough.
- Temple, James, 2021, The rare spot of good news on climate change, MIT Technology Review.
- Various, 2020, Explainer: Nine 'tipping points' that could be triggered by climate change, Carbonbrief.
- Wikipedia, List of cities by average temperature.

Websites

- Munich Re's Climate Change Edition: An insurance-based risk assessment site for climate change. (From an insurance perspective, it can make sense to plan for a low-probability RCP 8 scenario.)
- NOAA Sea Level Rise Viewer shows whether you should consider moving.
- RCP Database.

5. READINGS

FILM AND TELEVISION

- An Inconvenient Truth, 2006, Al Gore's path-breaking documentary raising public awareness.
- PBS, 2020, Climate Change — The Facts.

---

request for select reader feedback

Request for Reader Feedback: We were surprised not to be able to identify more catastrophic global harms in the *expected* IPCC scenarios of RCP 4–6 (in the first half of this chapter). Of course, on a worldwide basis and in specific locales, the absolute harm can be large. Yet the % total harm seemed not as bad as we thought, and the latest 2021 IPCC report was as vague as alarming. (We could identify enumerated sets of damages in here, plus some proprietary insurance estimates.) Please email us if we have forgotten to describe other major (in %) detrimental consequences that a 2–3°C expected rise in global temperature would bring (or any other relevant omissions and errors for this matter). In particular, please email us links to damage estimates.

(We do discuss the more serious consequences of worse-case scenarios, such as a small probability of a catastrophic rise of 6–10°C, e.g., through sudden permafrost melts with feedback loops, in the second half of this chapter.)

Part

The Social Problem

Chapter 6

A Crash Course in Economics

We now move from the natural environment to the economic environment — a transition that requires an understanding of some key economic principles. The first set of questions that we want to answer are when society can reasonably rely on individual self-interest and when not; and how society can and should bend individual self-interest towards the common good. The second set of questions are about marginal thinking and tradeoffs (including issues such as best delay). The third set of questions regard innovation, research, and development. It includes how costs and economic decisions can change with scale and time.

1 Human Self-Interest and Free Riding

The basic premise of economic science is that individuals tend to act in their own self-interests. This premise is, of course, never perfectly true. There are situations in which we are more likely to make voluntary sacrifices, such as when we are close to the recipient and when it costs us less. Most of us will gladly buy dinner for our families or friends.

But beyond sacrificing modest amounts for these chosen few, the evidence indicates that most of us humans put our own interests far above those of others. For example, children in poor countries could be pulled out of privation for the price of one cappuccino a day — and yet the donations from rich nations remain by-and-large pitifully small. (And it is not a valid objection that some of it is wasted in the transfer. The remaining portion would still go far.)

It would be nice if we could assume that people would put the climate problems of the *other 8 billion* high up on their priority lists. Unfortunately, this is not a realistic way to approach the problem. The evidence speaks against it.

It follows that for a voluntary-participation climate-change plan to work, it must be in the self-interest of most people to participate. Each one of us 8 billion people won't make much difference. The world outcome will ultimately remain the same no matter what our friends and we do. (Whatever other people do, most will continue to do so, regardless of what we as individuals will do.)

For example, if avoiding polluting costs you \$1 and this pollution imposes costs on the world's 8 billion people of \$0.0000000125 each, the total world pollution cost would sum up to \$100. For the social good of the world, you should not pollute. Yet for your own good, why spend your \$1 for a return of \$0.0000000125? If you care about 100,000 other people as much as you care about yourself, or if your example somehow convinces 100,000 others also not to pollute, your return increases to \$0.00125. Let us ask again: Do you really want to spend your \$1 for a return of \$0.00125? Your answer matters little, though. What do you think the majority of people will do?

1. Human Self-Interest and Free Riding

Economists call this free-riding — a situation in which you can consume a common good but without having to pay for it. If you are the free rider, it may be a great deal for you, but it is not a great deal to those who have to pay for you. If they could, they would want to exclude you (to make you choose to pay!)  — but this is precisely the problem when exclusion is not possible (as in many cases). There are many examples in which groups cannot start profitable businesses, because they realize that they cannot prevent subsequent shirking by team free-riders. (The most famous example may well be the prisoner's dilemma.)

Economists believe that it is a primary role of governments to ensure that their populations do not become too selfish. Societies cannot function if there is only individual self-interest. Thus, they need governments that can coerce their citizens.

For example, every sensible person would agree that any civilized society needs public services (streets, schools, police, courts, military, social security, welfare, hospitals). Only failed states, such as Sudan and Libya, are so libertarian that they do not provide these services. (And only a few predators enjoy living in them.)

These services need to be paid for by taxes. And any tax system requires a collection agency and laws to punish tax evasion. Otherwise, who would pay taxes? If paying was a matter of individual choice, everyone — both poor and rich alike and even most idealists — would find good reasons why only others should be paying the taxes. Furthermore, economic evidence suggests that altruism is not contagious at scale. In small communities, there are often individuals with enough concern for the common good to volunteer paying up. Others may pay more when they see these peers doing it (and especially when other villagers notice whether or not they are participating). However, voluntary contagious altruism almost never works in large societies. How many other Americans would be more likely to pay more taxes voluntarily just

6. A Crash Course in Economics

because *you* decided to set a good example and pay more taxes voluntarily? Not a meaningfully large fraction.

Nevertheless, although we need forced taxation and large governments, they are less than ideal: it costs money to collect taxes; it wastes money, because tax-rich governments can be twisted by lobbies, voters, demagogues, and disinformation; and it reduces work incentives, because individuals prefer to work for themselves instead of for the government. But taxes are an unavoidable and necessary social arrangement.

The best level of taxation is determined by what economists call a tradeoff — sacrificing one thing to obtain another. In our tax example, true libertarianism makes as little sense as true totalitarianism. Governments should not be too big or too small. (They are also never entirely benevolent and rarely entirely malevolent.) The right amount of government taxing and spending is a difficult and constant balancing act. If it weren't, it would have been a solved problem long ago and concerned citizens would not have been discussing the best scope and scale of governments at least since Plato's times.

The best solution with respect to the tax example is never "no taxes" or "all taxes." The best solution is somewhere in the middle. It is an eternal question where in the middle it should be relative to where we are at the moment: have we gone too far or not far enough?[1]

Unfortunately, institutions and communities are also almost always less efficient the larger they are. This also applies to governments. A village can usually administer local public spending more frugally and efficiently than Washington D.C. can administer national spending or the United Nations can administer international spending. This is why small collective problems are easier to solve than big ones. And climate change is the biggest collective problem of them all.

---

[1]Despite self-denial, the United States is already more "socialist" than many countries in Europe, except at the tiniest 0.01% sliver of its top wealth distribution. Social Security and Medicare, two hugely popular programs, are the largest socialist redistribution programs in the world.

2 The Tragedy of the Commons

What should government take care of, and what should individuals take care of?

In some cases, the answer is obvious. What would happen if traffic laws were voluntary? Considerate drivers would stop at red lights, but less considerate drivers would just take advantage of them and cross first. Yet at some point, even the most inconsiderate drivers become worse off — once there are too many other inconsiderate drivers on the road. By setting reasonable traffic rules and fairly enforcing them, governments can greatly increase economic well-being for all. Good traffic rules speed up the flow of traffic and reduce accidents. No sane economist thinks it would be a good idea to leave it to drivers to do whatever they please. In this context, "free driving choice" is not in the collective interest.

In other cases, the answer is not so obvious. The best choice is often for government not to intervene. This is so counterintuitive that social philosophers before Adam Smith did not understand it (to the detriment of their states), and many intellectuals still bristle when they are told to keep their hands off because they cannot design a mechanism to allocate resources better than letting individuals make their own decisions. Free-market capitalism can coordinate prices and activities in a way that is not just in the individual ("private") but also in the collective ("social") best interest.

Suppose, for example, that you are considering taking a business trip to New York. The cost of a round-trip ticket is $1,500. All you have to decide is whether the benefits of your trip, net of your other costs, exceed $1,500. If they do, your private welfare increases when you take the trip. More importantly, if no one else is harmed, total social welfare also increases. If someone else benefits from your trip, e.g., the airline and its employees, all the better! Social welfare increases even more.

However, your purchase could lower the social benefit if all costs of your air travel are not reflected in the price of your ticket. The airline is not likely to be the source of such a problem. Presumably, it would not sell you a ticket for $1,500 unless it was in its own interest, too, covering the cost of the fuel, personnel, wear and tear, and a fair return for the airline's investors.

Yet there is a problem with the free-market solution. Airplanes also emit pollution, and neither you nor the airline pay for the damages the pollution

6. A Crash Course in Economics

causes. These are called "external costs" — for short, "*externalities.*" In this case, they are negative externalities — because they "spill over" negatively onto others beyond the parties to the transaction. Because your ticket cost does not reflect these externalities, your trip may be beneficial to you and the airline, but not necessarily to society. For example, if your emissions cost others $500 in health costs, then you should only make the trip in the social interest if the benefits exceed $2,000.

Figure 6.1. Los Angeles Before the E.P.A.

Source: Sierra Club.

Even the most ardent free-market economists agree that government intervention may be needed when large externalities are involved, because private transactions fail to take into account all effects on third parties. On the other hand, the opposite alternative is also not desirable. Governments should not be in the business of making airline booking decisions for individuals. They are not as good at knowing what you need as you are.

So how can governments arrange it so that the participating parties will also take the social costs into account? To do so, the government must find a way to include the $500 external cost in the price of the ticket, so that you and the airline make the best decision in the social interest. Most of this chapter is about explaining, in detail, how and why this is best done with a tax on jet fuel emissions.

2. The Tragedy of the Commons

If you are still not convinced that we need a government to regulate emission externalities, just learn from history. Prior to the formation of the *Environmental Protection Agency* in 1970, there were cities in the United States that looked just like Beijing, Lagos, or New Delhi today. Figure 6.1 shows off Los Angeles. Smog was often so thick that visibility was limited to one mile. Oil slicks on rivers were widespread. Arsenic, cyanide, and mercury poisoning of groundwater were common. Lead in paints and gasoline poisoned our children. Asbestos killed thousands. Thank God those worst of times are over — and this is due to the intervention of government curbs on self-interested individuals.

Economists also call such situations "The Tragedy of the Commons," based on an academic study by Garret Hardin. He looked at examples in nineteenth century England, in which farmers let their sheep overgraze the "commons" (a publicly-shared area). It was so bad that many commons became completely run down. Because each farmer had acted in his self-interest, they collectively lost all benefits of the shared resource in the end. In our case, the analogy is that pollution emitters have been overgrazing the common good of clean air.

Conflicts over shared resources are difficult to resolve.[2] Even state governments often fail to handle state instances well (and there are good reasons why, which we will explain soon) — much less do so when the problems cross state borders and require coordination among multiple states or even nations.

For example, water in Southern California (where we live) is a scarce resource. Los Angeles could not exist without managed water access. Yet until very recently, many Californians considered it their civil rights to drill water wells on their lands. Not surprisingly, some aquifers ran dry and the rest may do so soon. There have been many water wars fought in the Western U.S. states over the last two centuries. These states are now struggling hard to find a collaborative solution that makes everyone better off than the free-for-all. It is not certain that they will succeed.

Yet it will always be a valid question as to where we are versus where we should be: has government intervention gone too far or not far enough? We study economics to answer thorny questions like this.

---

[2] Elinor Ostrom won a Nobel Prize for studying cases where individuals and communities did manage to avoid the tragedy. However, these were *local* situations, not *global* ones.

6. A Crash Course in Economics

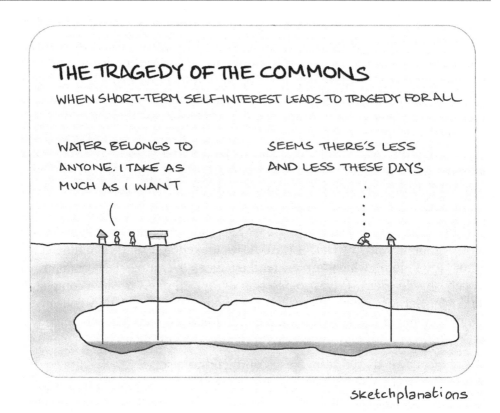

3 When Should Governments Intervene?

So when should society leave it to free parties to make decisions, and when and how should it involve government coordination and coercion? The answer is easy if an activity is infinitely bad (ban it!) or completely harmless and without externalities (allow it!). But as with all things economic, the tradeoff is interesting and nuanced when it is in the intermediate gray zone.

Greenhouse gases (GHGs) fall into this gray zone. They are not infinitely bad, and they are not totally harmless. We must evaluate how bad the externalities associated with GHGs are. But how? And should it be different if we know that any government we might appoint to work on the problem will be inefficient or even corrupt? And what if some parties lose (like coal miners)? And what if GHGs could later be sequestered again from the atmosphere for a price? Read on.

3. When Should Governments Intervene?

A Numerical Game-Theory Example

We are now going to try to make a real economist of you with a small dose of game theory. Game theory studies situations in which the payoffs of "players" depend not just on their own actions, but also on the actions of others. If you've ever played poker, you already understand this dynamic.

The concepts are so important that they are worth illustrating with a numerical example. However, if you are truly allergic to numbers of any kind, it is possible to skip the numerical example and still understand most, but not all, of what we will discuss later.

Our kinds of "games" always include three elements: (1) players, (2) choices (strategies), and (3) payoffs. The players in our example are three companies, X, Y, and Z. (You can also think of them as three sectors of the economy. The payoffs can also go to both owners and employees.) Our choices allow each company to pick one of two different technologies, $\mathscr{C}$ or $\mathscr{D}$. Our payoffs need to be shown in a two-dimensional table, because they depend on all three players (X, Y, Z, in columns) and both choices ($\mathscr{C}$ and $\mathscr{D}$, in rows):

	Players: Company			Total
	X	Y	Z	Social Welfare
w/ Tech $\mathscr{C}$	**$300**	$400	$500	$1,321
w/ Tech $\mathscr{D}$	$290	**$421**	**$600**	

The boldface in each column highlights which of the two technologies yields bigger payoffs for each player. The table shows that player X prefers $\mathscr{C}$, and that X and Y prefer $\mathscr{D}$. If each firm acts in its own self-interest, the total social welfare (the sum of the three firms' payoffs) is maximized at $1,321 (billion, of course). There is no better collective choice, even if firms could be forced to do whatever society wants them to do. In this game instance, firms can be trusted to make the best social choices.

Now we introduce the next step: pollution. Let's say that $\mathscr{C}$ is a "clean" technology, and $\mathscr{D}$ is a "dirty" technology. $\mathscr{D}$ leaks paint all over the machine and requires regular cleanup, which costs $60. Taking pollution cleanup costs into account, here is how each firm will select its strategy:

6. A Crash Course in Economics

	Players: Company			Total
	X	Y	Z	Social Welfare
w/ Tech 𝒞	**$300**	**$400**	$500	$1,240
w/ Tech 𝒟 – pollution	$230	$361	**$540**	

Now that 𝒟 is polluting, it is less profitable and so there is only firm Z still using 𝒟. Once Y has to take care of its own cleanup costs of $60, it prefers the clean 𝒞 over the dirty 𝒟. The social welfare is lower at $1,240 because we have changed the technology. Now 𝒟 requires cleaning, which was not the case in the previous $1,321 example.

Most importantly, the fact that the technology is dirty makes no difference in one respect: We still do not need a government. X and Y are choosing the clean technology; Z is not. Government could not do better by telling any company to switch.

A Dirty Externality

Next, we introduce external effects into the example. To do so, we change the manner in which 𝒟 is dirty: rather than leaking into the company's own plant, the paint releases its polluting vapor beyond the firm. We assume that the total pollution of using 𝒟 is still the same $60, but the victims change: each of the three companies suffers $20 in required cleanup costs when any one of them chooses 𝒟. Of the $60 in pollution, the polluter bears $20 itself, and the externality imposed on the other two is $20 each for a total of $40.

Let us first confirm that if we have a Utopian dictatorial government that forces X and Y to use the clean technology and Z to use the dirty technology, social welfare is still $1,240. With one polluter, each firm will suffer $20 in pollution, caused by Z but imposed on each of the three.

	Players: Company			Total
	X	Y	Z	Welfare
w/ Tech 𝒞 – pollution	$280	$380	Verboten	$1,240
w/ Tech 𝒟 – pollution	Verboten	Verboten	**$580**	

It may be unfair to X and Y to suffer the consequences of Z's pollution, but our economic concern is not fairness. Our concern is to achieve the

3. When Should Governments Intervene?

highest social welfare achievable. If you want to worry about fairness across companies, just realize that nothing prevents the government from taxing any company (such as polluter Z) any amount it wishes and giving the tax revenues to others (such as X and Y).

The Free-Market Solution

We are now ready to explain the problem that arises under the "everyone for themselves" free-market solution. If left to their own devices, what would each company do? For now, assume that each company takes the behavior of others as given, and just focuses on its own choice and welfare. In this case, here is the solution:

Free-Choice Solution: X uses $\mathscr{C}$; Y and Z use $\mathscr{D}$.
(Each company suffers $40 in pollution [bold "in equilibrium" box].)

	Players: Company			Total
	X	Y	Z	Social Welfare
Proposed Solution	$\mathscr{C}$	$\mathscr{D}$	$\mathscr{D}$	
w/ Tech $\mathscr{C}$ – pollution	**$260**	$380	$480	1,201
w/ Tech $\mathscr{D}$ – pollution	$230	**$381**	**$560**	

Game theory solutions are like puzzles or brain-teasers. Let's <u>trust but verify</u> that the solution is correct:

Cmpny X: Because Y and Z use $\mathscr{D}$, X will suffer pollution of $40, over which X itself has no influence. If X chooses $\mathscr{C}$, X will earn $300 − $40 = $260. (The first term, here of $300, is from the first table on page 175.) If X chooses $\mathscr{D}$, X will suffer the $40 pollution from the others, plus an extra share of $20 from its own pollution. This would give it only $290 − $40 − $20 = $230. Thus, we have confirmed that X prefers $\mathscr{C}$ over $\mathscr{D}$.

Cmpny Y: Because Z uses $\mathscr{D}$, Y will likewise suffer pollution of $20, over which Y itself has no influence. However, Y has control over its own pollution. If Y chooses $\mathscr{C}$ (and thus does not pollute), Y will earn $400 − $20 = $380. If Y chooses $\mathscr{D}$, Y also suffers its own pollution and will earn $421 − $40 = $381. Thus, we have confirmed that Y chooses $\mathscr{D}$.

6. A Crash Course in Economics

Cmpny Z: Because Y uses 𝒟, Z will suffer pollution of $20, over which Z itself has no influence. However, Z has control over its own pollution. If Z chooses 𝒞, Z will earn $500 − $20 = $480. If Z chooses 𝒟, Z will earn $600 − $40 = $540. Thus, we have confirmed that Z chooses 𝒟.

Summing up all the boldfaced best choices gives the result that social welfare is now only $1,201. Although the technology still pollutes by the same $60 as it did in the previous $1,240 example (in which there were no externalities), the social welfare in the free-market solution is lower when there are these externalities than when there is coercion. This is because Y is not doing the right thing from a social perspective. Y ignores that its pollution has a $40 negative effect on two other parties.

This is the classic *public goods* problem — the tragedy of the commons in numbers. The reason why we wanted to explain it with a numeric example is that we can now use it to demonstrate what government can do to improve the social welfare.

Government Dictates

➤ Green Dictatorship

Our Utopian government is not a realistic possibility. So we need to look at other solutions. Our first is a "dictatorship of the environmentalist": Outlaw all pollution!

	Players: Company			Total Social Welfare
	X	Y	Z	
w/ Tech 𝒞	$300	$400	$500	$1,200
w/ Tech 𝒟 − pollution	..... Verboten			

Although there is now no pollution and thus no externalities, from a social welfare perspective, this turns out to be a bad solution, too. Government should let Z pollute. (In our example, the government can later clean up the pollution for $60.) Z's gain from 𝒟 is higher than the total externalities it imposes on X and Y. The problem with this dictatorial green solution is that it prohibits a socially valuable but polluting activity. (Yes, pollution can be in the social interest!)

3. When Should Governments Intervene?

Most modern activities create pollution. We would not want to outlaw them all. The key is intelligently trading off the benefits of the activities versus the costs of their pollution.

▶ Smart Dictatorship and Command-and-Control

As noted on page 176, the Utopian dictatorial solution, in which a benevolent government tells X and Y to use $\mathscr{C}$ and tells Z to use $\mathscr{D}$, is better. In fact, it yielded the best possible outcome, $1,240. This was the theory behind old-school Soviet-style planned command economies. If the commands are correct, the social outcome is best.

The problem with these command economies is not necessarily that the planners are evil (although they often are), but that even the best planners usually don't have all the necessary information, and their clever subjects will do all they can to exploit any system to their advantage. How would our dictator know whether to tell X and Y not to pollute, and allow Z to pollute? Y will make all sorts of claims why its payoffs with pollution are not just a little but a lot higher. What if the planner's cost and benefit estimates turned out to be wrong?

The economist James Gwartney tells a wonderful anecdote of how economic planners in Moscow tried to incentivize managers and employees in the Soviet Union to produce window sheet glass. The planners instituted a reward system based on the tons of glass produced. What could possibly go wrong? The glass plant managers realized that they could earn the rewards by producing sheet glass so thick that it was barely transparent. When the planners realized that they had set the incentives wrong, they changed the rules to reward square meters of glass produced. The results were similarly predictable. Under the new rules, the plants produced glass so thin that it broke all the time.

The Soviet planning system was an extreme example of directed production that was not based on the prices that free-market consumer demand would set. But even capitalist governments must institute incentives to accomplish social tasks. And they all make similar mistakes. For example, under French colonial rule in Hanoi, a government rodent extermination program paid bounties for each rat tail handed in. Instead of eradicating rats, many hunters caught rats and amputated their tails (so that the rats would continue to breed), while others started farming rats for profit.

6. A Crash Course in Economics

Allow us a cautionary tale from the domain of regulating global emissions. In the Montreal Protocol to protect the ozone layer, firms were awarded valuable credits for destroying the potent greenhouse on hydrofluorocarbon gas HFC-23. However, production of HFC-22 as feedstock ingredient for other products was allowed. The production of HFC-22 creates HFC-23 as a by-product, the subsequent destruction of which would now yield extra credits. Not surprisingly, HFC-22 production has been rising.

> sidenote
>
> One possibility for dealing with the public goods problem is for rich countries to pay poor countries for behaving in a more climate-friendly manner. For instance, rich countries could pay Brazil, Congo, and Indonesia to retain their rain forests. In effect these countries could probably be convinced with very modest subsidies to harvest the wood and replant it rather than burning it to clear agricultural land. (We will discuss international treaties in Chapter 9 in more detail.)
>
> Unfortunately, if one pays for reducing harmful actions, the incentives to commit those actions in the first place in order to obtain the benefits for the reduction also increase. If rich countries were to pay Brazil and Congo to stop burning down rain forests, would it be a reward for their current practices of aggressive burning? Furthermore, how many American voters would be willing to pay taxes so that the money can be transferred to Congolese warlords? (And what would they really do with the money if they got it?)

But don't laugh too hard at the follies of Soviet planners, French colonialists, and climate treaties. Firms are essentially small economies, in which management tries to coordinate economic activity without resorting to free-market price-based contracting. (Otherwise, why have a firm at all? Individuals could just transact in free markets.) The CEO's role is essentially that of a planner and even dictator — and many firms have failed for the same reasons of poor information, mistaken incentives, and clever employees as government programs have failed.

Information and incentive problems are at the heart of much modern economics. And these same problems also lie at the heart of solving the climate-change problem.

3. When Should Governments Intervene?

A Managed Market With a Pollution Tax

Returning to the numerical example, we still need a better solution in which firms do not pollute when it is not in the social interest but a dictatorial planner does not have to tell firms how to operate, either. After all, firms should know best about the relative costs and benefits of their ($\mathscr{C}$ and $\mathscr{D}$) production technology choices. We now explain how to do this. Government can set a tax equal to the externality (the "social harm"), and *then* let firms decide for themselves.

Recall that the externality we discussed above was $40. What would happen if government taxed the use of the dirty technology by exactly this amount? Again, assume that each company takes the behavior of others as given and focuses on its own behavior and welfare. With a $40 tax, the firms will choose the following:

Solution With-$40-Tax on $\mathscr{D}$: X and Y use $\mathscr{C}$; Z uses $\mathscr{D}$.
(Each company suffers $20 in pollution [in bold "in equilibrium" box].)

	Players: Company			Total Welfare
	X	Y	Z	
w/ Tech $\mathscr{C}$ – pollution	**$280**	**$380**	$500	$1,240
w/ Tech $\mathscr{D}$ – pollution – $40 tax	$210	$341	**$540**	

The $1,240 is the sum-total of $1,200 to the firms plus $40 in tax revenues, which belong to society collectively. Again, we need to check that we have solved the puzzle correctly, i.e., confirm that this solution is in every company's self-interest:

Cmpny X: Because Z uses $\mathscr{D}$, X will suffer Z's pollution of $20 over which X itself has no influence. If X chooses $\mathscr{C}$, X will earn $300 − $20 = $280. (The first term, here of $300, is from the first table on page 175.) If X chooses $\mathscr{D}$, X will still suffer Z's pollution of $20, plus another $20 of its own pollution, plus the tax of $40, for a private benefit of $290−$20−$20−$40 = $210. Thus, we have confirmed that X chooses $\mathscr{C}$.

Cmpny Y: Because Z uses $\mathscr{D}$, Y will suffer Z's pollution of $20 over which Y itself has no influence. If Y chooses $\mathscr{C}$, Y will earn $400−$20 = 380. If Y chooses $\mathscr{D}$, Y will still suffer Z's pollution of $20, plus another $20 of its own

Cmpny Z: pollution, plus the tax of $40. This gives Y $421−$20−$20−$40 = $341. Thus, we have confirmed that Y chooses $\mathscr{C}$.

Cmpny Z: Because X and Y use $\mathscr{C}$, Z only suffers pollution over which it has no influence. If Z chooses $\mathscr{C}$, Z will earn $500. If Z chooses $\mathscr{D}$, Z will earn $600 − $20 − $40 = $540. Thus, we have confirmed that Z chooses $\mathscr{D}$.

The social welfare is the same as that chosen by the Utopian benevolent dictator — except that all that this limited government planner had to measure was the social externality and set a tax equal to it. Government could then leave it to the companies to measure their relative benefits from the two technologies and make their own decisions. The pollution tax provided the proper incentives.

A subsidy to clean technology instead of taxes, or a combination of clean technology subsidies and dirty technology taxes, could also work. Again, the point of these policies is not for the government to collect or distribute funds; it is to alter the incentives of firms so that it is in their self-interests to do the right thing.

It is important to recognize that there is a distinction between pollution costs and taxes. Pollution is a negative, both from a private and a social perspective. Taxes are a negative only from a private perspective, not from a social perspective. The taxes paid to the government do not evaporate. Instead, they can be used for the common benefit — such as removing the paint vapor from the air or paying X and Y to remove the $20 in vapor damage caused by Z.

sidenote

> There is a close alternative to a pollution tax, called cap-and-trade. Emitters need permits (carbon credits) to pollute, but they can trade these permits. We will not go into further details on the advantages vs. disadvantages. There are many online explainers, such as one at Brookings or at the World Resources Institute.

4 Practical Problems of Pollution Taxes

A government tax on pollution equal to the externality is a great solution, because it puts each party in charge of what it knows best. Firms know their businesses, but they are often not aware of how much of an externality they impose on others. A firm should be in the business of its business. We cannot expect it to research the social good. Government should be in the business of finding out how its members work together and making them collectively better off. Thus, judging social harm, which affects non-contracting parties, is what governments should be in the business of.

Good economics suggests that governments should tax more when the pollution problem is worse. The challenge is finding a good balance. If you want certainty, stick to death and taxes — usually no one knows exactly. In the case of pollution externalities, even the best level of taxes are often uncertain. And the implementation problems go beyond just the uncertainties of estimation. The theory is easy, but the practice is hard. In the real world:

- What if it is not easy to judge harm? What if it costs $100 to investigate whether there is any externality at all?

- What if the government is inefficient? For example, what if administering a $40 tax costs $10 or $100? What if the government does not use the funds to clean up the pollution as it is supposed to do, and this allows the pollution to cause a lot more harm?

- What if the government is corruptible and untrustworthy? What if one of the three parties in our example has so much influence that it practically "owns" the government? In the non-legal sense, all governments are corrupt to some extent. We cannot trust governments to always do the right thing — yet we have no better alternative.

- What if voters clamor for subsidies and lower energy prices? This is not far-fetched. When energy prices experienced a small spike in 2021, even the voters in <u>Norway</u> suddenly wanted relief from high energy taxes.

Economics generally suggests that the worse governments are, the less they should be involved; and, in extremes, it is sometimes better simply to accept a bad free market-solution than to replace it with a worse government "solution."

But what if the pollution is seriously harmful? What if the pollutant is not just paint but cyanide? Would you want to allow the <u>Cyanide Manufacturers</u>

6. A Crash Course in Economics

Trade Association of America to provide the information to judge how harmful cyanide is to the people in the neighborhood just outside of their factories? Would you leave it to their profit margins and industry wages to decide how much cyanide they deem safe to emit? Would you want to allow them to fund the election campaigns of politicians? Would you want them to fund those scientists who (truly or ingeniously) happen to believe that cyanide is not that bad? Would you want to have an EPA administrator who was a lawyer lobbying for the cyanide manufacturers in a previous life...or will be again in a future life?

Your Majesty, according to our study, the shoe was lost for want of a nail, the horse was lost for want of a shoe, and the rider was lost for want of a horse, but the kingdom we lost because of overregulation.

In real life, both Congress and the administration (including the EPA) are often partly captured by industry and by single-issue or single-industry voter interests. The average top-level U.S. government agency administrator worked for a political or industry think-tank and will do so again after leaving the (low-paid) job. However, the alternative of letting every industry do as they please without any control in a free market could be even more disastrous. Ideally, administrators should be free of conflicts and well-informed — but this is simply not how the real world works.

4. Practical Problems of Pollution Taxes

Winners and Losers

But even with a perfectly informed and benevolent government, there is still another big problem. Higher social welfare does not mean that each and every party will be better off. There will be winners and losers. In our example, Z will be worse off, because it will have to pay taxes (which all parties, collectively and jointly, will get to keep and use productively). Y will be miffed, too. Y did not have to pay the $\mathcal{D}$ tax but could have made more profits without it. Similarly, we did not concern ourselves with how to compensate victims of pollution. Should the tax receipts flow to the pollution victims? Seems fair, but what if more powerful lobbies have better lawyers and are experienced in getting government grants?

The problem with dictatorships is that they rarely have the social interest at heart. Benevolent dictatorships are rare. The problem with democracies is that even a government that wants to have the social interest at heart is often ill-suited to solving the social pollution problems. We may chuckle that "poor" Z is now rightly forced to pay taxes — until Z starts a public relations campaign about how unfairly it has been treated, sues based on a claim that the EPA has overstepped its bounds, gets its employees to vote against the current administration, and funds the opposition presidential candidate. If Z employs more voters than X and Y, it could even win. Z may not even need a majority of voters; it may be enough if it has a majority in pivotal swing states.

In the United States, coal miners in Pennsylvania and West Virginia may be few, but there may be enough of them to tip the balance in favor of one or the other presidential or senatorial candidate. Perhaps the most prominent nonsensical political subsidy in the United States for decades is for <u>ethanol</u> (alcohol as fuel). It exists only because corn-growing farmers in Iowa are the earliest primary voters. Ethanol is not a green fuel, because it costs more diesel fuel to farm ethanol than it creates.

6. A Crash Course in Economics

Elasticity of Escape

But even these dire problems in choosing among lesser evils still paint too rosy a picture of real life. The most serious problem of taxes (other than the fact that they may be set wrong) is that payers can seek to escape them. If government taxes technology $\mathcal{D}$, what prevents Z from setting up shop in a neighboring country, right across the border (which keeps the pollution externalities we suffer the same), which happens to be a country that is looking for more employers and has thus not imposed any tax? (Subsidies for $\mathcal{C}$ don't solve the problem, either, because they will attract new "pretend-polluters" that will emerge primarily to collect subsidies on $\mathcal{C}$.) As far as CO_2 is concerned, it does not matter whether it is emitted in the US or in China.

And, for an extreme example of this problem, what if X, Y, and Z were not firms but countries? Would Z ever impose a pollution tax on itself? Once you answer "no," then you understand the single most important aspect of our global CO_2 tax problem. The fox will not protect the henhouse. When it is strongly against their self-interests, countries will not volunteer.

joke
> [Isn't instituting a pollution tax in just one region]...like making a peeing section in a swimming pool?" — George Carlin, comedian.

◇

Economists have earned a deserved bad rap for being more pro-capitalist and less interventionist than other social scientists. In defense of our discipline vis-a-vis the social science disciplines of our colleagues, most of us economists are not so much enamoured with the free market as we are more skeptical about real-world governments. Just because the free market has a problem does not mean that the government, especially over the long run, will solve it better.

5 Margins, Costs, and Scale

Understanding pollution taxes and the limits of both the free market and of governments was the most important point of our chapter's lesson in economics. But there are just a few more important concepts that play central roles in the economics of climate change. This section discusses margins (perhaps the central concept of economics for most students) and the economics of costs.

Marginal Thinking on Global Warming

As we explained in the previous chapter, climate change is real and it is big. The global change of 2-3°C will cause large changes in climate. Worse, this large a change could push Earth over tipping points and into feedback loops that scientists do not fully understand.

However, when climate activists advocate for reductions in CO_2, e.g., at UN COP conferences, many make a basic mistake. They compare apples with oranges. They fail to understand that they are not fighting against the terrible consequences of 2-3°C warming. They are fighting "only" for reducing global warming by 0.3-0.5°C. (It is realistically not possible to fight against 2-3°C warming any more than it is possible to fight against the sun going down at night.) It is only the *marginal* benefit of *lesser* harm that world leaders have to weigh against the *marginal* cost of GHG reduction measures.

Scientists are of course aware of this. In the next chapter, we explain that the *global marginal benefits* of reducing warming by "only" 0.3-0.5°C relative to the *global marginal cost* of curbing CO_2 are still large enough to warrant a global carbon tax of about $50/t$CO_2$. The result would be fewer GHG emissions and an increase in global social welfare of about 0.1-0.2% — a large sum, on the order of about $100 billion per year. However, although it would be collectively suboptimal for humanity to forego this 0.1-0.2% gain from a reduction of 0.3-0.5°C via CO_2 reductions, it would *not* be world-ending. In fact, in aggregate economic terms, it would make only a tiny difference.

6. A Crash Course in Economics

Diminishing Returns to Scale

The majority of dwellings in the United States house more cockroaches than people. To exterminate 90% of the roaches in a home may cost only $100 in spray cans. However, it could easily cost $1,000 to get rid of 99% and $100,000 to get rid of all 100%. (Roaches are good at hiding.) To get rid of all of them, the average cost is $1,000 per roach-percent. But after the first 99% have been wiped out, the last 1% cost an incremental $99,000. Economists call this $99,000 the *marginal cost* of the last 1%. (Most residents follow the sensible route and wipe out only the first 95% year after year.)

Here is another example. Let's say you are the typical American driving 15,000 miles per year. If we asked you to drive 100 fewer miles per year (e.g., to reduce pollution), we may not even have to pay you. You could easily find a few unnecessary trips that you could omit. If we asked you to drive 1,000 fewer miles per year, it could become uncomfortable. You might have to curtail some fun trips that you would otherwise have taken. If we asked you to curtail 10,000 miles, it would seriously affect your life and work. We would probably have to pay you a lot of money to get you to agree. Because the 10,000 miles is the sum of 100 trips of 100 miles each, you can think of the first 100 mile reduction as costing you almost nothing, the tenth reduction costing you more, and the one-hundredth reduction costing you a lot. Every additional cut is progressively more painful.

The marginal cost of the first trip reduction is lower than the average cost, which is lower than the marginal cost of the last trip reduction. You can think of the average of all earlier marginal cuts — from the first zero-cost reduction to the last most expensive marginal-cost reduction — as the average cost. Non-economists often confuse the two, and the most important lesson for new economics students is to get them thinking in terms of margins for decision-making, not averages.

When marginal thinking is applied to purchases, economists call the increasing cost phenomenon the *Law of Diminishing Returns* (LDR). You probably enjoy the first hot dog, but when forced to eat the 70th hot dog, you would probably no longer enjoy it as much as the first.

The LDR is of pervasive importance in economics. From our perspective, the most important consequence is that it is likely much cheaper to clean up 2% of each of 50 U.S. states (when each dollar of cleanup still goes far) than to clean up 100% in one state (where the last 1% of pollution cleanup could cost billions of dollars). Similarly, it is likely to be relatively cheap and easy to cut the first 1-10% of global fossil fuel use. It could even turn out to be profitable — we will discuss the options in the third part of our book. Entrepreneurs could plant trees and sell the lumber. But eventually it would become harder and harder to find good land for profitable tree farming. Similarly, governments could purchase the dirtiest least profitable coal plants very cheaply in order to retire them. But eventually they would run out of bad cheap coal plants. At some point, it would become more expensive to reduce CO_2.

The conclusion? It will be increasingly more expensive, per ton of CO_2, for the world to cut, say, 90% rather than 10-50% of global CO_2. And it will likely be almost impossibly expensive to cut 100%, at least for many decades.[3] Humanity should not be dogmatic. An ambitious goal to reduce as much CO_2 as is economically feasible is good. A stupid goal to eliminate every last bit of CO_2 is not.

The LDR also implies that the world should immediately take care of the "lowest-hanging fruit" (CO_2-emitting activities that can be cleaned up), because they are very cheap to take care of. And this is exactly our point— let's *immediately* take care of the first 10%. *Move the needle.* We know it's economically worthwhile. From a collective perspective, the world is recklessly delinquent in not moving this needle now.

Sunk Costs

The marginal concept has another startling consequence. After any generating plant has been built, its marginal cost of electricity production is far lower than the average cost (which does include plan construction cost). The extreme cases are nuclear power plants. It costs $5-10 billion to build one; but, once built, it can deliver power at a per-MWh cost lower than other technologies. (Uranium is dirt-cheap.) Consequently, closing already constructed nuclear power plants prematurely is wasteful. (We will explain nuclear technology and its alternatives in greater detail in Chapter 12.)

The average cost of electricity from many large fossil-fuel plants is also much higher than the marginal cost, although fossil fuels are more expensive

---

[3]For example, polar researchers should probably never build a clean power plant for their Antarctic winter research station. Diesel and diesel generators are just too efficient and cost-effective to be replaced by alternatives.

fuel than uranium. Coal plants have lifetimes of about 30-50 years. Once built, it becomes inefficient not to use them. Thus, the game is largely rigged at construction time.

It is ominous that China is building new coal plants this decade at a record pace. It is an easy prescription that the world should do everything in its power to convince China not to build them. If the world could only get China to replace its aging coal plants not with newer coal plants but with nuclear plants, it would make a huge difference in global CO_2 emissions. Unfortunately for the world, it may already be too late to convince China. China already suffered acute power outages in 2021. These outages make it even less inclined to put its current coal plans on hold. It may be too late for China, but it may not yet be too late for India, Africa, and other regions. They need to get to the point where it is in their economic interests not to build coal but clean electricity plants!

Learning Curves and Returns to Scale

Over small quantities and in the short run, most — but not all — activities follow the LDR, with its decreasing marginal benefit and increasing marginal cost. Over large quantities and in the long run, this is not so clear. Learning and mass production can often reduce marginal per-unit costs.

For example, when Thomas Thwaites decided in 2021 to find out how much it would take to build just one toaster, it cost him $1,792. If you add his time, it would probably be ten times this. Engineers call the first pilot product *FOAK (first of a kind)*. If Thwaites had to build a few more, the average cost per toaster would be much lower — in engineering lingo, these toasters would be *NOAK* (next of a kind). The cost decline would reflect his learning curve.

Furthermore, if Thwaites built 1 million toasters and appropriately farmed out tasks, he could probably do so for as little as $20/unit because of economies of scale. For example, mass producers do not hand-make single pieces of plastic. Instead, they make casts (that costs a few thousand dollars one time) and then pour thousands of plastic pieces at almost zero marginal cost.

One central question when it comes to re-engineering humanity's energy and pollution systems is: "When and how will per-unit-costs increase or decrease?"

5. Margins, Costs, and Scale

It is almost surely the case that some clean technologies will become cheaper with more experience and mass production and thus have increasing returns to scale. Other technologies will become more expensive with declining resources and thus have *decreasing returns to scale*. We don't yet know which are which. Will electric energy storage become more expensive as we need to mine increasing amounts of lithium, cobalt, etc., to build more batteries; or will we learn how to build better cheaper energy storage solutions from other ingredients altogether? (History suggests the latter.) Will CO_2 sequestration become more expensive as we run out of space to plant more trees or will we find cheaper ways, e.g., by breeding trees that are more efficient in converting CO_2 into wood?

The Economics of Delay

Virtually all instances of increasing returns to scale due to mass production share one feature: They require a high upfront fixed cost. The construction of large-scale factories is expensive and not undertaken lightly. In many cases, it is more efficient not first to build factories but to invest more in research and development to bring down the factory cost as much as possible, and only then to spend the much larger amounts necessary for deployment of better plants in large quantities.

Depending on the specific technologies, humanity today may not yet be well-served to pursue zero-CO_2 solutions. This does not mean it could not be well-served investing in researching how to build zero-CO_2 solutions in the near future. Research and development is often orders of magnitudes cheaper than deployment. By saving on deployment now, scientists might be able to spend ten times as much on research and development. Most scientists and engineers believe that investing in clean-tech R&D should be highest priority for humanity today. Have we already mentioned that we are strongly in favor of more clean-tech R&D?

When should entrepreneurs deploy a new technology? The right time is not obvious. If a technology costs decline by 10-15% every year (as battery costs have declined over the last 10 years), then it will cost only half as much to deploy the batteries five years later. Ergo, is it better to buy a battery that saves emissions of 1 $MtCO_2$ (over the next 30 years) today, or a battery that saves 2 $MtCO_2$ but beginning in 5 years? Or half-half? Applied economists build models to inform such decisions. Generally, when costs are coming down fast, it makes more sense to delay. However, delay is also dangerous. What if progress will be slower than anticipated? (For example, controlled nuclear fusion has been 30 years in the future seemingly forever.) And what if delay leads to procrastination?

6   The Economics of Innovation

Our final topic combines the previous two. It is about innovation, and how the world should foster and pay for it.

➤ The Research Externality

Whereas pollution is a classic negative externality, research is a classic positive externality. Few for-profit companies conduct basic fundamental research (not related to specific products), because the benefits would rarely accrue primarily to themselves even if the research succeeds. Instead, their fundamental-research insights will likely also benefit their competitors, other firms, and eventually all of us. We are in effect free-riders. Ergo, companies do not invest the much larger fundamental-research sums that would be in the social interest.

For argument's sake, let's say there is a good chance (say 10%) that chamomile tea would create free energy, and it would only cost $10 billion to do the basic research to find out. If $10 billion would possibly solve the world's energy problems, it would be a bargain for the world! Yet, no private company would spend the money. If the research succeeded, every company could just brew chamomile energy. Thus, each company would prefer to follow rather than lead.

It is because of positive externalities like this that many countries fund basic research at their universities rather than leave it to for-profit companies. Although funding for the *National Science Foundation* and some other agencies helps reduce the externality problem in the United States, it does not solve the problem fully. Beside the fact that science funding levels are generally quite low, why should the United States fully support basic research that will end up also helping China and Europe (and vice-versa)? Most silicon technology (including computers, displays, and solar cells) was invented in the United States, but nowadays are produced overwhelmingly in Taiwan,

6. The Economics of Innovation

Korea, and Japan. Vanadium flow batteries were invented with U.S. taxpayer money but are now manufactured in China. The Far East has been free-riding on American inventions for much of the 20th century — just as America was free-riding on the European inventions of the 18th and 19th century.

One solution is patent protection. Yet global patents are remarkably expensive to obtain and enforce. (Except for medical inventions, it is unclear if they have benefitted anyone except the legal industry.)

▶ The Sharing Externality

This leads us to the second related problem. Let's presume that chamomile energy and patent protection were perfect and some company had invested its $10 billion, solved the problem, and now owned the perfect chamomile energy patent. At this juncture, our firm would not want to share its technology (which would reduce the global pollution externality). Instead, the company would want to charge a few trillion dollars as a reward for its inventors and investors. After all, they had staked their fortunes and sometimes their livelihoods on exploring chamomile when no one else was chipping in. Why give it away? However, this also means that chamomile won't be used as widely as it should be to solve the planet's emission problems.

Bill Nordhaus calls the issues of insufficient research and insufficient sharing the *double externality* of clean energy innovation. It is a central problem in fighting climate change.

▶ Perspective

Lack of sharing of proprietary technology, often developed at great expense, and fear of expropriation has and will continue to hobble many otherwise planet-optimal solutions. China has a reputation for ruthless expropriation and reverse engineering of foreign technology in joint ventures in the past. Such behavior now keeps fearful Western companies from offering to replace China's planned coal plants with cheaper and better nuclear-power technology. This situation hurts not only the Chinese, but all of us when it comes to CO_2 emissions. But who can blame the nuclear engineering companies? Why should they commit economic suicide?

Like many others, we believe that accelerating innovation in clean technology ranks at the top of the best ways to stem global climate change. We are excited about Bill Gates' _Breakthrough Energy Fund_ (BEF), which invests with a part-philanthropic mandate to take the social benefits of inventions into account. (If we had enough money, we would have invested, too.) However, BEF also has a financial mandate. Its hardest decisions may yet to come: If BEF succeeds in developing breakthrough inventions, will it share the know-how with parties that cannot afford to pay — or simply will not want to pay? Will BEF share its inventions with Chinese companies, which will then likely copy the technology and sell it cheaply in mass all over the world? It would be good for the planet but not for BEF.

In general, economists know of no universal solutions that solve the double externalities. The socially best partial solutions so far have typically involved individual countries funding universities to conduct the basic research with their public money. This is obviously less than ideal.

In real life, this approach has run into many difficulties. Universities and countries focus on their own self-interests, too. Universities do not even want to allow other universities to exploit their research breakthroughs. They want to patent their inventions and earn licensing fees, often by partnering with just one company, which can then earn monopolistic profits by restricting access.

Finding good economic solutions is a constant struggle. If the problem were easy to solve, we would not have needed to talk about it for so long.

➤ A Pollution Tax or Clean-Technology Innovation Subsidies?

Although our book advocates increasing subsidies for research and development in green technology, it would be incorrect to view our endorsement as a substitute for a pollution tax. Pollution taxes help redirect innovation to where it is more useful. By reducing the pollution (and thus the incurred tax), pollution taxes increase the incentives for clean-energy research.

7 Moving the Needle Now

We chose the subtitle of our book, *moving the needle*, to signal that the *marginal* costs of reducing emissions and taking action right now are already more than low enough to act on many fronts. Our book's goal is to help nudge the needle in the right direction. We hope that the world will invest in more clean research, learn as it goes, and as it learns from research, decide where it should then push next.

We understand and sympathize with the concerns of and the disagreements among scientists about how far humanity will eventually have to go over the rest of the century. But this long-term debate is not as important as picking the low-hanging fruit over the next decade.

A journey of a thousand miles begins with a single step. The world needs to start the journey sooner rather than later, and the first steps are exceptionally easy because they are cheap enough so as to be in the interest even of individual countries. We will offer some specific recommendations in Part III. But first, we must explain how integrated models of economics and climate change work.

6. READINGS

Further Readings

BOOKS

- Cowen, Tyler, 1991, Public Goods and Market Failures, Routledge.
- Dixit, Avinash K. and Barry J. Nalebuff, Thinking Strategically: The Competitive Edge in Business, Politics, and Everyday Life, Norton.
- Ostrom, Elinor, 1990, Governing the Commons: The Evolution of Institutions for Collective Action, Cambridge University Press.
- Ridley, Matthew, 2020, How Innovation Works, Harper.
- Sowell, Thomas, 2015, Basic Economics, Basic Books, New York, NY. An intuitive book on the basic principles of economics.

REPORTS AND ACADEMIC ARTICLES

- Nordhaus, William D., 2019, Climate change: The ultimate challenge for economics, American Economic Review, 109 (6): 1991-2014. Includes an explanation of why climate change is the "mother of all externalities."
- Hsiang, Solomon and Robert E. Kopp, 2018, An Economist's Guide to Climate Change Science, Journal of Economic Perspectives.

Chapter 7

Modeling The World Economic Impact

The first part of our book explained energy needs and earth sciences from a largely non-interventionist perspective — an essentially do-nothing scenario. The previous chapter explained some key economics principles — for example, the role of taxes in the control of pollution. This chapter puts earth sciences and social sciences together: what exactly *should be* the right amount of taxation to control global warming?

The models discussed in this chapter are not just dry economics from the "Land of Ivory-Tower Theory." Instead, they form the bases of all climate negotiations, including, e.g., those in the Paris Accord of 2016. The formal names for models that integrate all scientific areas for the purpose of assessing potential policies are "integrated assessment models" (IAMs). The primary purpose of IAMs is to recommend the appropriate level of global tax on CO_2, not just today but also in the future.

On a historical note, these models were pioneered by William Nordhaus and justly earned him a Nobel Prize.

7. Modeling The World Economic Impact

1 An Economic Sketch of Earth

Figure 7.1. A Simple Schematic for a Typical Integrated Assessment Model

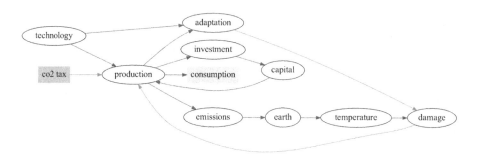

Note: Blue arrows are positive feedback; red arrows are negative feedback. Gray nodes are earth-science-based. The modeler tweaks the CO_2 tax controls (in orange) to maximize consumption (in light blue).

Source: Inspired by a DICE schema by David Garcia-León.

Figure 7.1 shows a basic diagram of an integrated assessment model, largely based on Nordhaus' prominent Dynamic Integrated Climate-Economy (DICE) model. DICE is a relatively simple model — there are far more intricate ones. The diagram, though *not* complete, retains the most important blocks and links. Each of the blocks contains a lot more detail in the actual model and relies on parameters that have to be set by the modeler.

On the left are the two external inputs into the model. The first input is in the orange box, labeled CO_2 tax. It is under the control of the modeler, who can tweak it to see what the outcomes will be. For arrows, we use red in the model to indicate negative input or feedback and blue to indicate positive input or feedback. Higher taxes reduce production — there is less money available to the producer — which is why the arrow is red. The second external input is technology. In the basic DICE model the rate of technical improvement is constant but tunable by the researcher. Technology marches steadily upward and improves production (and also adaptation), which is why the arrow is blue.

The cyan box in the middle of the diagram is consumption, the key output. Consumption here is to be interpreted broadly, including all goods and services.

1. An Economic Sketch of Earth

It could and should include such aspects as human pleasure derived from a natural environment with species diversity. (Unfortunately, typical models usually employ only narrower measures.) The ultimate goal of the modeler is to adjust the annual CO_2 taxes to obtain the highest possible time path of consumption.

Digging deeper into the graph offers a few more insights. The production output can be consumed, or it can be used to fund adaptation (such as building dykes or purchasing air conditioning), or it can be invested in future production. But more production also increases emissions. At this point, the earth sciences from Chapters 2–5 come into play (gray boxes) — determining how human emissions change Earth's global temperatures. The damages caused by rising temperatures then feed back and reduce production, and thereby consumption.

There are many nodes and links not included both in this stylized diagram and in the actual full DICE model. For example, the model does not even have a node for population growth, although population growth influences nearly every other aspect of the economic model. Instead, there is some assumed rate of population growth that is simply fed into the model. Furthermore, technology could affect investment, investment could affect adaptation, and so on.

Yet even with the simplified structure, the feedback effects in DICE are so numerous that even Nordhaus says that he often does not understand intuitively how an adjustment of the inputs will change the outputs before he tries it out. For example, consider this feedback effect: if temperature rises, it creates damages which lower production, which lowers emissions, which can then lower temperature in the future, and so on.

The most important and ethically most contentious parameter is not even visible in the diagram. It is the _discount rate_. Discounting is the process of determining how much humanity would be willing to pay today (a concept called <u>present value</u>) in order to avoid $1 of harm in the future (a concept called <u>future value</u>). Different discount rates are also needed depending on how far in the future the harm is expected to occur — in DICE, it could be in a century or later. Discount rate assumptions are necessary to make it possible to add up the welfare of people living in different centuries in order to arrive at one overall "<u>social welfare</u>" measure.

7. MODELING THE WORLD ECONOMIC IMPACT

A simpler verbal version of the discount rate question is this: How should humanity value one extra dollar for a person living today compared to one extra dollar for a person living 100, or even 500, years from now? As simple as the question appears, it is fiendishly difficult to answer — and it is really more of a philosophical than a scientific question. We will explain this in more detail in Section 5. For now, just take note of the fact that the choice of discount rate can have a large effect on the optimal CO_2 tax prescribed by DICE and other integrated assessment models.

2 What Goes In and What Comes Out?

The DICE model considers *only* the next 200 years. Thus, Nordhaus can play with about 200 "knob settings" labelled "Global Tax on CO_2 in Future Year X." They could also be labeled "emission-control incentives." By twisting these control knobs, Nordhaus can determine how various settings will likely change the total social welfare. He fiddles with the knobs until he finds the 200 settings (one for each year) that maximize social welfare.

Nordhaus' latest model considers many scenarios ("cases"), but we focus on just three:

1. A "base" case, which was essentially Earth without any active intervention as predicted in 2010–2015, i.e., before clean technology truly took off. It is now viewed as a more pessimistic scenario.
2. An "aggressive" case, in which the world intervenes to limit the *average* temperature increase to 2°C by 2200. This case is inspired by the Paris Accord, but the limit is softer in that it may be briefly exceeded.
3. A Nordhaus preferred optimal cost-benefit solution, tuned to maximize social welfare, somewhere in between the previous two cases.

The left plot in Figure 7.2 shows the 200 knob settings for the 200 (future annual) global CO_2 taxes. They are stated in terms of dollars per ton of CO_2 emitted. Unfortunately, the optimal CO_2 taxes are also usually called the "social cost of carbon" instead of the "social cost of carbon-dioxide," which is what they really are.[1] (As if the subject were not difficult enough, the experts

---

[1] We mentioned in Chapter 3 that 1 ton of carbon (C) turns into 3.67 tons of CO_2 when burned. When the experts write about a social cost of carbon of $50 per ton, if taken literally, this would imply that it would be $183 per ton of CO_2. However, this is not what they mean. Instead, they mean $50 per ton of CO_2. Grrrr....

2. What Goes In and What Comes Out?

Figure 7.2. Nordhaus CO_2 Tax and Consumption Change

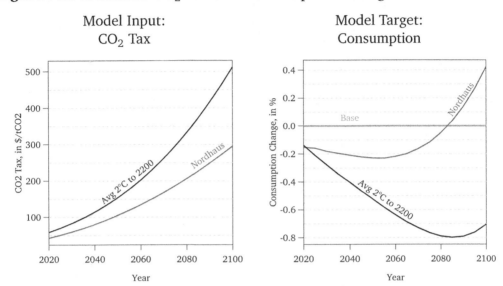

Note: The base scenario is pessimistic and without intervention. The Nordhaus scenario is a welfare-maximizing solution on aggregate consumption (not per-capita). The Limit 2°C scenario is a constrained welfare-maximizing solution, keeping the average temperature by 2200 below 2°C.

have found yet another way to make understanding even simple concepts more confusing!)

Under Nordhaus' parameter settings (especially, but not only, the discount rate), the optimal CO_2 tax starts at about \$40/tCO_2 today, rises to about \$50 by mid-decade, \$100 by mid-century, \$200 by 2080, and \$300 by 2100.[2] By the end of the century, only the most exceptional activities that emit CO_2 would still remain viable. The 2°C-average-limit scenario by 2200 requires imposing taxes more aggressively — reaching \$200/tCO_2 two decades earlier than under Nordhaus' preferred solution. At the end of the 21st century, with

---

[2]Estimation of the social cost of carbon-dioxide remains controversial. Based on a meta analysis, Peter Howard and Thomas Sterner report that when the Nordhaus damage function is replaced by their preferred estimate of the temperature-damage relationship, the result is a three- to four-fold increase in the SCC relative to the 2015 DICE model. Burke et al. estimate significantly higher damages of about 20% of GDP. If they are correct, the SCC should be much higher. This will be an interesting field of research for economists for many years.

7. Modeling The World Economic Impact

its CO_2 tax of \$500/tCO_2, emissions will have already dropped into negative territory. (Not shown, the Paris Accord would require even steeper taxes.)

Both tax functions start reasonably low today but increase ever more steeply. This is because the sudden imposition of too high a CO_2 tax would not allow industries to adapt. A more sudden tax of, say, \$200/tCO_2 next year would bankrupt a lot of businesses that could not switch instantly to cleaner sources of energy. The smooth and predictable rising tax greatly reduces this economic harm.

Pollution Taxes and Welfare

The right plot in Figure 7.2 shows how tax schedules influence the time path of consumption. Any CO_2 abatement imposes some current consumption diminution compared to the base case. However, under the Nordhaus plan, the reduction is fairly mild. It remains under about 0.2% of current consumption. (With global GDP of about \$90 trillion, this is still almost \$200 billion!) Moreover, children born today will still reap some of the benefits of their parents' sacrifices — around 2090, their consumption will exceed what they would have had if no mitigation measures had ever been instituted.

As we mentioned in the previous chapter, if the OECD were to have to pay for the reduction alone (without the help of China, India, etc.), a rough estimate would be that the OECD's GDP-proportional cost would be about 0.5% of consumption in 2050; if the US were to do it alone, its cost would be about 1%.

In contrast, the more aggressive 2°C average limit by 2200 is far more costly, with the largest losses of nearly 1% imposed on our children. This plan is also never optimal in the sense of maximizing consumption. The 2°C sets a target that will make not only us but our children worse off. (This assessment could, however, be wrong if the Nordhaus model is wrong. Then again, all assessments are based on models that could be wrong.)

Figure 7.3. Nordhaus Earth Outputs

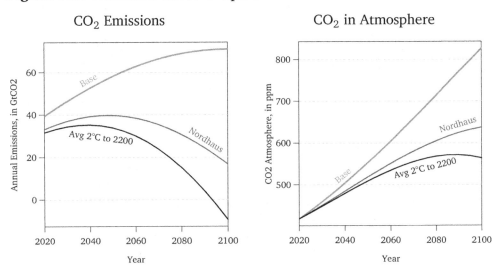

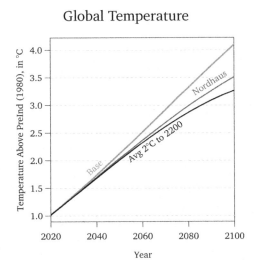

Source: Courtesy of William Nordhaus. Temperature is relative to 1980. (Thus, 2020 is 1°C.)

7. MODELING THE WORLD ECONOMIC IMPACT

Earth Temperature and Others

After Nordhaus has found his best 200 knob settings, his model yields three other interesting outputs. They are plotted in Figure 7.3: human CO_2 emissions, CO_2 concentration in the atmosphere, and the increase in average global temperature.

The plot on the top left shows how both Nordhaus and the 2°C goal curtail emissions in comparison to the base case. In the base case, CO_2 emissions rise throughout the century, before leveling off at about 70 $GtCO_2$/year. Limiting the temperature increase to an average 2°C by 2200 requires pushing human emissions below zero before the end of the century. The Nordhaus plan is more forgiving, pushing emissions to about 20 $GtCO_2$/year.

The plot on the right shows that the Nordhaus plan allows atmospheric CO_2 exceeding 600 ppm for a while, whereas the 2°C plan limits it to about 580 ppm. Both plans are still much better than the pessimistic base case, which has atmospheric CO_2 reaching levels of approximately 850 ppm by century's end and continuing upwards beyond that. Fortunately, this base case is largely technologically obsolete by now.

The bottom plot shows that planetary temperature responds slowly to changes in atmospheric CO_2. Thus, even the aggressive 2°C plan "only" reduces temperature by about 1°C relative to the base case scenario (an increase of 3.2°C instead of 4.2°C). The Nordhaus scenario allows for about a 0.3°C greater temperature increase than the 2°C average-limit scenario by the end of this 21st century. However, both curves continue to rise beyond the year 2100. Visually, they suggest that the average temperature will rise by more than 4°C over a 200-year time frame.[3]

> *sidenote*
> The 2021 IPCC report now includes Shared Socioeconomic Pathways (SSPs). These can be viewed as a step or component on the way to an IAM. They attempt to guess how different government policies, economic growth, and environmental aspects might flow into different Representative Concentration Pathways (RCPs), explained earlier in Chapter 5). For example, SSP 1 paints a low-economic growth picture based on pervasive environmentalist sentiments all over the world, with reduction of inequalities. SSP-5 paints a high-economic growth picture fueled by fossil fuels. In our opinion, the SSPs are poor alternatives to a full IAM. We are not fans of the SSPs.

---

[3]The model's welfare calculation includes only consumption over a century or two. Our understanding is that this is ok, because fossil fuels are likely to be exhausted in about 100-200 years.

3 What is the Scientific Consensus?

The methods of the Nordhaus model have been widely accepted by the scientific community. Even his critics have adopted his approach to modeling the economic impact of climate change. But this does not mean that they agree with his model parameter choices or conclusions. Many of these parameter choices are model inputs that require judgment.

For example, what is the rate of technological change? In its first version "DICE 1.0," Nordhaus' model turned out to have been too pessimistic, especially with regard to clean-energy technology. No one expected it to advance as quickly as it did. Even Nordhaus did not — though economists generally tend to be more optimistic than other scientists about human inventiveness *when money is at stake*.

In addition, it is easy to estimate the cost of reducing emissions *"on the margin,"* i.e., what it would take to eliminate the first Gigatonne of CO_2. But it is much more difficult to estimate how much it will cost to eliminate humanity's total of 30 Gigatonnes.

Furthermore, although economic long-term growth rates have been steady for many decades, what will be the growth rate of the world economy over the next 100 years and what will be the damage caused by global warming? Economists cannot accurately predict economic activity over such long horizons.[4]

The integrated assessment modelers also have to update their models constantly. The world is changing. And the economics and earth sciences components are improving. Nonetheless, there are disagreements. The adoption of different parameters has led different researchers to recommend different global CO_2 taxes.

Table 7.4 illustrates a range of estimates for the first year's recommended tax. (In all models, CO_2 taxes would rise smoothly in future years.)

The most prominent recommendation for the social cost of carbon-dioxide is Nordhaus' updated estimate, which stands at about $50/$tCO_2$ today. The leading alternative take was published by economist Nicholas Stern in 2006

---

[4]This makes it all the more remarkable how confident some natural scientists at the IPCC seem to be about assessing the same economic growth and damages that economists are so reluctant to assess.

7. MODELING THE WORLD ECONOMIC IMPACT

Table 7.4. Various Estimates of An Appropriate CO_2 Tax (circa 2020)

Politics		Model-Based Estimates				Pindyck Survey	
Trump	Biden	IAWG	Nordhaus	Stern	Typ Range	Economists	Climatologists
≤$5	$51	$50	$50	$80	–$15 to $2,500	$80	$120

Note: These taxes are often referred to the "social cost of carbon," though "social cost of carbon-dioxide" would have been a far better name. (No one is thinking of taxing graphite.) The numbers in the table are reasonably representative but not exact. IAWG is the Interagency Working Group, whose estimates are used by the U.S. government for planning purposes.

(in his 700-page "page turner," called the "Stern Review.") As early as 2006, Stern advocated a global carbon tax of $85/tCO_2$, where it still stands today (rising to $100/tCO_2$ as early as 2030). In 2006, this high a recommendation was a shock to policy analysts — a time when Nordhaus was still advocating $30/tCO_2$. Today, the two estimates are no longer far apart.

There are also other estimates developed by international organizations and U.S. government agencies. For instance, the U.S. Interagency Working Group (IAWG), first formed during the Obama administration to help planning for climate change, suggests $51/tCO_2e$ in 2021. There have also been numerous surveys conducted by various experts, including academics. Typically, economists recommend lower taxes than climate scientists ($80 vs $120). However, in line with the theme of our book, the differences among these estimates are unimportant. Both numbers are so high that they mean the same thing for practical purposes: immediate drastic CO_2 tax increases causing CO_2 emission reductions.

In sum, the most reasonable estimates, as of 2022, seem to suggest an appropriate global CO_2 tax in the range of $50–$100/tCO_2$, rising over time. Despite their natural cantankerousness, it's remarkable how many prominent economists agree. An Economists' Statement published in the Wall Street Journal in January 2019 was signed by over 3,000 economists including all living Nobel Prize winners and all past heads of the U.S. Federal Reserve. It advocated a tax similar to that suggested by the basic Nordhaus model. We will return to this statement in Chapter 10.

Where are we now? Ironically, the actual worldwide tax on CO_2 seems to be *negative*. For example, in the United States, the fossil-fuel industry benefits from large direct subsidies — estimates range from about $2 to $60 billion for the $180 billion industry, the equivalent of a subsidy of about $0 to $50/tCO$_2$. Worldwide, the International Monetary Fund (IMF) has estimated that global fossil-fuel subsidies were $500 billion (or $15/tCO$_2$) in 2017.[5]

In light of the world's actual negative tax rate, the differences between Nordhaus and Stern, or economists and climate scientists, no longer seem so large. Arguing about $50 vs. $80 next year is creating discord for no good reason. Where it matters, all agree. The world today has a ridiculously low and harmful tax rate on CO_2 emissions.

4 CO_2 Taxes and Consumption Losses

Most of us do not have good intuition for what CO_2 taxes and consumption sacrifices quoted in percent really mean. Let's translate these abstract figures into more meaningful terms.

Energy Costs

Table 7.5. Price Increases With a CO_2 Tax

	tCO_2e /Unit	Addtl Tax per tCO_2 $0				Price Increase
		$0	**$50**	$100	$200	
Oil, 1 Barrel	0.43	$50	**$75**	$100	$150	50%
Gasoline, 1 Gallon	0.01	$2	**$3**	$4	$5	50%
Coal, Railcar /1000	0.18	$5	**$15**	$25	$35	5×
Natural Gas, 1 MCF	0.055	$3	**$5**	$8	$15	2×
Tree, 1	–0.06	$0	**–$3**	–$6	–$12	(> Planting Cost)

Source: CO_2e/unit emission estimates are from the EPA Calculator Equivalences. The natural gas is emissions at the smokestack and excludes emission on the supply chain. The boldface $50/tCO$_2$ column is the immediate tax recommended by most economists. The $100/tCO$_2$ tax would be reached in short order.

[5] It rises to a misleading headline estimate of $5.2 trillion (a stunning 6.5 percent of global GDP in 2017) if speculative estimates of worldwide pollution and global-warming harmful effects are added.

7. Modeling The World Economic Impact

Table 7.5 shows that at $100/tCO_2$ (which all plans reach in reasonable short order), the price of gasoline would roughly double, i.e., gasoline would cost in the United States what it already costs in Europe. Coal would become uncompetitive given any reasonable CO_2 tax in most parts of the world. In a sense, a $100/tCO_2$ tax on coal might as well be a $1,000/tCO_2$ tax. Natural gas prices would double, thereby doubling winter heating costs for most U.S. households — more if gas leaks are taken into account. (Not shown here, electricity costs would likely also double at $100/tCO_2$.)

Fortunately, sensible adaptation would allow most people to offset much of the tax increases. For instance, people could drive less and in smaller electric cars. New systems could greatly reduce heating and cooling costs. Planting trees would become more profitable, and thus so would probably anything constructed out of wood.

Ideally, governments could and should reduce other taxes commensurately if they can collect necessary revenues through fossil-fuel taxes instead. Unfortunately, real-world politicians rarely end up lowering other taxes, leaving most ordinary consumers with frustrations about ever-increasing taxes. It is difficult to see how the public would support fossil fuel taxes without forced bundling of tax relief.

Consumption Loss

Another way to look at the cost of a CO_2 tax is to add up the expected economic cost and spending reductions (Figure 7.2). We will work with rough characterizations. The Nordhaus plan is the most relaxed approach, suggesting a cost peaking at about 0.2% of GDP in about one generation. A more Stern-like approach would likely peak at about 0.5% of GDP. The even more stringent 2°C path considered by Nordhaus reaches 0.8% loss in consumption in about one decade (when the tax reaches about $100/tCO_2$). Finally, the less economically and more environmentally oriented Paris plan would likely reach about 1.0%. In all plans, we should expect to start out with costs of about 0.2% pf GDP to allow for adaptation.

4. CO$_2$ Taxes and Consumption Losses

➤ Compared To Public Expenditures

The United States has a GDP of about $20 trillion per annum. State and local governments spend about the following:

Roads	Higher Education	Health and Hospitals	Military (Federal)
$200 bn	$300 bn	$300 bn	$800 bn

On the high side of fighting climate change, a 1% consumption loss would be about $200 billion. Thus, the required public expense would be about the same as the current public expense on infrastructure; or two-thirds of the expense of all state and local higher education. It would however be only 25% of our Federal military budget — about the same cost as the U.S. Navy. These are obviously large sums of money.

On the low side, 0.2% would be about $40 billion. Fighting climate change would then cost only about the equivalent of the another *Department of Housing and Urban Development* (i.e., all public housing) or another *Department of Homeland Security* (with its 240,000 employees).

➤ Compared To Private Housing Rent Costs

Instead of public expenses, we can also look at private expenses. The median household in the United States today has two earners bringing home about $65,000/year and pays about $13,000/year in rent ($35/day). Thus, the Nordhaus plan would cost the equivalent of about three days' rent; the Stern plan about ten days' rent; and the Paris-like plan about 20 days' rent.

➤ Global Sharing

Thinking in terms of rents, the above calculations implicitly assumed proportionality. Everyone would pay their 0.2% to 1.0% of consumption to fight climate change. Richer people, like the average resident of Manhattan or Palo Alto, would have to pay more in absolute but not in relative terms. Think an extra $2,000 per year (instead of $500 to 1,000 per year). In China, where rents are about 20% lower, the rent-proportional "global warming tax" would be about $200 per year. In India, where rents are 70% lower, it would mean about $75 per year. And fortunately, unlike rent, households could reduce CO$_2$ taxes by polluting less. That's the whole point of the tax, after all.

However, the United States would realistically have to cover more than its proportional share of the world's belt-tightening. Poorer countries will almost surely not be willing to volunteer a proportional share, especially because the OECD is responsible for most of the CO_2 in the air today. A more realistic and possibly fairer estimate would be twice what we calculated above: think one week for the Nordhaus plan and one month for the Paris plan.

In future chapters, when we discuss potential tax solutions, we will use a "one month's rent" (for an aggressive but incomplete CO_2 abatement plan) as an intuitive benchmark cost. Of course, basing taxation on housing rent is *not* a suggestion for best polity — it is a thought device to put a CO_2 tax in perspective — akin to the Economists' Big Mac Index for measuring price levels across countries.

➤ Summary

Are you an environmentalist worried about global warming? Have you been advocating for action? If so, please answer this one simple question: *how much tax would you be willing to pay?*

5 What are the Best Model Parameters?

How Do We Value Future Generations' Welfare?

Remarkably, the biggest disagreement between advocates of lower and higher CO_2 taxes (from Table 7.4) has nothing to do with uncertain or contentious scientific forecasts. Instead, it is about the previously mentioned parameter that is almost entirely philosophical and subjective: how should humanity value the welfare of future generations — not just in 100 years, but in 200 or 300 years?

In the model, the answer to this question enters as the discount rate. Table 7.6 shows how sensitive the social cost of CO_2 (i.e., the optimal tax) is to different discount rate assumptions. Reasonable variations in the discount rate can swamp the effects of tinkering with almost all other inputs.

Nordhaus assumes that any investments to slow climate change must compete with other investments for profitability. If $1 today can be expected to earn $100 (inflation-adjusted) in 100 years when invested in the real

Table 7.6. Social Cost of CO_2 By Discount Rate in Nordhaus' Model

Assumed Real Discount Rate	Today's $1 in 100 years	Best CO_2 Tax, in 2018-$/$tCO_2$			
		2015	2020	2050	2100
0.1%	$1.11	$970	$966	$917	$665
1%	$2.70	$497	$515	$614	$657
3%	$19	$93	$104	$179	$361
5%	$131	$23	$27	$55	$126

Source: Nordhaus, Nobel Lecture, American Economic Review, 2019, 109 (6): 1991–2014, p.2006. See also IAWG, Feb 2021, Table ES.1.

economy or the stock market, then the same return should be required when investing $1 to reduce climate change — i.e., it should avoid damages of $100. We chose this $100, because it happens to be close to the historical average inflation-adjusted rate of return per year after taxes (about 5%).[6]

Stern rejected this view, arguing that humankind today should treat its future generations more like itself. If $1 today can avoid $1-$3 worth of damages in 100 years, it is good enough to warrant investing this $1 into emission reductions today. His inflation-adjusted discount rate view lies thus somewhere between 0.1% and 1.5% per year.

The Ethical Choice?

The knee-jerk reaction is that Nordhaus offers the egotistical answer and Stern the ethical answer — that is, people today should take as good care of future generations as of their own generation.[7] Humanity should be the stewards of the planet for its children! This ethical high ground is also why the Stern report intuitively appeals to a lot of people — including many scientists. But is it really the *ethically* correct choice?

---

[6] $1 \cdot (1 + 4.75\%)^{100} - 1 \approx \$100$. A fairer number could be $50, because the after-tax return is lower than the pre-tax return. On the other hand, some retirement income and estate transfers are usually tax-exempt.

[7] Both the Nordhaus and the Stern discount rates are sketches of what they actually use. The real deal requires a lot of detailed economics.

7. Modeling The World Economic Impact

Despite two world wars, the average American now has an inflation-adjusted income (GDP) that is about <u>six times</u> higher than it was 100 years ago. The trends in other parts of the world were <u>similar</u>. It is a reasonable guess that in 200 years (about six generations), the average person will earn about 30 times (in real inflation-adjusted terms) as much as the average person today. Would it really be so unethical if these future people would have a standard of living only 20 times higher than our own rather than 30 times? Moreover, with many times our income and wealth, they would be able to afford or remedy damages more easily than we can.

The consensus among economists today is that it was Nordhaus who got the discount rate right. However, many of the same economists also believe that Stern's steeper tax recommendations are better than Nordhaus' for reasons that are not explicit in either original base model — mostly because of uncertainty about the future climate and the possibility of climate catastrophes that the original Nordhaus model did not take into account. We will come back to this issue.

There is also a more practical aspect to consider. Self-interest limits what is politically viable. <u>Miles' Law</u> says that "where you stand depends on where you sit." Recall our warning from the previous chapter that even good taxes can create winners and losers. Even if the sum of humanity today and humanity in the future may collectively be better off, it is also the case that all the losers (those who will sacrifice to a CO_2 tax) are alive today and all the winners (those who will benefit) will be alive in the future. This lopsidedness makes it difficult to build support among the living today to vote to sacrifice. Even the grandchildren of the distant future have not been born yet.

Take a quick self-test. How much would you be willing to sacrifice today to prevent the world's great-great-great-great-great grandchildren from "suffering" in the sense of only having 20-times more than you vs. having 30-times more than you? If you answered "a lot," write a check for one month's rent now made out to great-great-great-great-great grandchildren. (This covers your necessary contribution just this year on behalf of generations 200 years from now. You will have to write it again next year, of course.) Stare at it for a while. If you are still ok with sending it, you are consistent. If you are not ok, you are probably in the majority that would convince themselves that *other* people should pay for lowering CO_2 emissions — just as long as it is not you. After all, is the CO_2 in the atmosphere really your fault or was it the

5. What are the Best Model Parameters?

fault of some anonymous oil barons? The average American is <u>willing to pay about $1/month</u> in higher electricity cost, not $500/month.

Adaptation

Another subjective parameter in the Nordhaus model is the damage from global warming. Simply put, scientists today can only guess what the damage will be. There are no certainties. A big reason is the role of adaptation. Bjorn Lomborg's book <u>False Alarm</u> is highly controversial and provocative, but it makes many good points, too. It points out that some

But you can't blame everything on global warming!

alarmism has been naïve because "the stories assume that while the climate will change, nothing else will." Like most economists (including us) and unlike many other scientists, Lomborg has a lot more faith in human inventiveness when their own hides are at stakes (and a lot less faith when it comes to the capabilities of real-world governments to make good choices).

Historically, Lomborg and the economists have been more right than wrong — although this is no guarantee that this will also be the case in the future. Scientists and environmental activists have consistently underestimated humanity's ingenuity and ability to adapt.

In one of Lomborg's examples, he shows that deaths from climate-related catastrophes have not increased but decreased markedly over recent decades (Figure 7.7). Even though global warming has probably increased the strongest hurricanes, these storms have caused less harm. When you watch news about how climate change has brought about the latest stronger hurricane or drought, just remember that not that long ago, such weather used to kill many millions of people every year. Today, it's "only" a few hundred thousand (too many, of course). Over time, forecasts have become better and people have learned to adapt. If the world becomes wealthier, and thereby more able to adapt, expect less harm, not more.

7. MODELING THE WORLD ECONOMIC IMPACT

Figure 7.7. Disaster Deaths By Decade

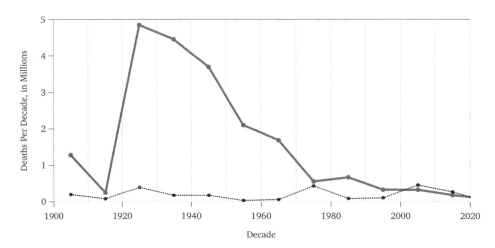

Note: Climate catastrophes (in fat solid blue) are floods, droughts, storms, wildfires, and extreme temperatures. They peaked in the 1920s and have been coming down since. Non-climate catastrophes (in black at the bottom) are earthquakes, tsunamis, and volcanoes, and give an intuitive benchmark. (Their death tolls have remained constant — even though the population has increased greatly, people have been moving out of the way.)

Source: Original Source: Lomborg and International Disaster Database.

Innovation

Activists also often fear that global warming will devastate crops. This factor is often the biggest component of damage estimates incorporated into IAMs. However, there may be much less crop damage than they fear. There is a new potential adaptation. Geneticists can now alter crops to take better advantage of higher CO_2 levels and warmer temperatures. Engineered plants could require less water, fertilizer, and pesticides, and be more nutritious and healthy, to boot. Innovations could even increase agricultural output as the world warms. Obviously, this is also still a guess. We will not know for sure whether this will work until we try.

The relevant subjective parameter in the Nordhaus model is the rate of technological innovation. Innovation can be thought as another form of adaptation. New technology contributes to making our children richer than us. This helps to pay for mitigation of and adaptation to climate change.

5. What are the Best Model Parameters?

Is the relevant technological growth slowing down or not? <u>Some scientists</u> argue that the low-hanging fruits of technology innovation have already been picked, and therefore future growth rates will be slower. <u>Other scientists</u> argue that innovations in fields such as biotechnology and artificial intelligence will keep the economy growing at historical rates. It's not clear. In the two areas most relevant to climate change — energy and biotechnology — the evidence suggests that technological innovation is going strong. These innovations will likely eliminate most CO_2 emissions from the energy sector and beyond. The question seems to be only how fast — sooner or later?

Almost all interested scientists agree that civilization should invest heavily in relevant clean technology and R&D. This could accelerate its progress at a cost that is relatively low. But here is a puzzle. As we explained in the previous chapter, the faster the pace of innovation, the more we should *not* implement new technologies now, but wait just a little longer. In the extreme, if technological progress (especially in energy technology) is fast enough, humanity may even be better off not installing any clean energy today and doing nothing today other than to subsidize and to accelerate research progress. But what if scientists have miscalculated and the technology will not improve as fast as we forecast?

Catastrophic Scenarios

Many first-generation integrated assessment models did not incorporate uncertainty and risk. They can be thought of as having worked only with expected scenarios. Newer models have become better. The potential for worse-case outcomes advises more caution and higher CO_2 taxes. As we pointed out in Chapter 5, our worst fears are not about the likely outcomes and they are not even about a typical "worse-than-expected" outcome. Instead, they are about unknown and unknowable outcomes such as CO_2 sinks that could suddenly be exhausted or feedback loops or tipping points that we do not know about.

And there is another good question: how much should the world invest today to be ready to react tomorrow? There are events that can not be foreseen today. What if humanity triggers some runaway warming process that had not been triggered for millions of years? Or what if a supervolcano were to erupt, an asteroid were to hit Earth, or Sun activity were to diminish or flare up? Do we want scientists to learn how to be ready to reduce billions of casualties?

6 What Else Should Be in the Model?

Models are models — simplified constructs to help scientists understand interesting phenomena. What else could have been put into the models?

Why Is Population Not a Policy Variable?

Our book started with explaining how population is the elephant in the room, having exploded from 1.6 billion people at the beginning of the twentieth century to 8 billion in 2022. But the integrated assessment models largely ignore how the climate or public policies can influence population growth.

There are good reasons for this omission. It is not even conceptually clear how to deal with population growth. For example, should social welfare be measured in per-capita terms or in total population terms? What is the lost welfare attributable to people who were never born? Should the world be considered better or worse off with 15 billion (possibly poor) people than with 1 billion (possibly wealthier) people? If harm were to be apportioned, should people with more children also pay higher taxes to cover their children's future CO_2 emissions?

We do not have answers to these thorny questions. But our lack of answers does not render them less important.

How Does the Model Take Account of Income Inequality?

Climate change and fighting it creates winners and losers. Without emission reductions and a warming planet, some areas of Canada and Russia will become more habitable; other like the Sahel may become uninhabitable. Over the spans of decades, many of those harmed by climate change will migrate, itself a form of adaptation. Over centuries and millennia, even distant migration has been not the exception, but the rule. Of course, migration is also often accompanied by human misery — some short-term, some long-term. Historically, migrating populations often made war on and killed the previous settlers.

Most of the harm from climate change will squarely fall on the poor. Climate change is really only of second-order importance to the rich. They can adapt. They can move or emigrate. The United States can slow the future disappearance of Florida to rising seawater with appropriate water barriers —

6. What Else Should Be in the Model?

just as the Dutch have done for centuries — and, if this fails, Floridians can move. Arizonans can buy more air conditioners.

Such easy adaptation is not available to residents of the Indian subcontinent and Sub-Saharan Africa. The homes of millions of Bangladeshis likely be submerged. Northern India may lose its reliable water supply as the Himalayan glaciers shrink. The African countries in the Sahel could become uninhabitable (unless the rain patterns were to change, as they have in the past). Where could they go? Neither Indian nor African countries will likely be able to afford to move most of their populations to more hospitable environments. And rich countries have raised many barriers to reduce the inflow of poor immigrants. How should they deal not with a few million but with a few billion immigrants from different cultures?

> anecdote
>
> "The disgusting irony of all of it is that the billionaires who have created this global atrocity are going to be the ones to survive it. They are going to be fine while we all cook to death in a planet-sized hot car." — Sarah Silverman, comedian.
> (PS: In fairness, not just the billionaires but all of civilization has benefited.)

Spending Alternatives

Ironically, the strongest ethical arguments *against* fighting global warming also concern the global poor today — and they make us very uncomfortable.[8] Let us explain why we find them as inconvenient a truth as climate change itself.

The United Nations estimates that almost 1 billion people today live in extreme poverty, defined as living on less than $2 per day. It costs only $0.80/day to feed a child in a poor country. For a small fraction of the funds proposed to address climate change, the world could eliminate global child malnutrition. For about $100 billion, we could lift everyone out of extreme poverty — about 5–10% of what the Paris Agreement demands we spend not to stop but just to slow down global warming by a 0.3-0.5°C.

Or take malaria, which kills about 400,000 people and cripples about 200 million people per year. One may quibble about whether eradicating

---

[8]These arguments have again been best articulated by Bjørn Lomborg.

Malaria would cost $50 billion or $100 billion *one time* — but would eradicating malaria not give humanity more bang for the buck than 0.05°C less global warming?[9]

Who Cares?

An even more uncomfortable question to ponder is why the rich people and nations of this world are not already spending a lot more money on poverty today — regardless of how much the world should be spending on climate change. Why does it have to be just a few philanthropists (such as Bill Gates and Warren Buffet) who have stepped in and taken this task onto themselves, while we, the people, have failed? We can only characterize our failure as a great collective shame of the human race. (The ethicist Peter Singer has a lot more to say on this subject.)

7 What Have We Learned from IAMs?

Winston Churchill famously said "that democracy is the worst form of Government except for all those other forms that have been tried." The same holds true for integrated assessment models. What are the alternatives? Intuitive back-of-the-envelope modeling is less likely to offer sensible quantitative CO_2 tax prescriptions. Assuming that "there is no problem" is also an integrated assessment model, though a very bad one, contradicted by the evidence. So is assuming that "the world will come to an end unless humanity fundamentally restructures and makes climate change its top priority." The world will not come to an end. Thus, although IAMs have many deficiencies, they are scientists' best tools for analyzing the collective tradeoffs between economic growth, environmental taxation, and climate change.

Just do not take them too literally please. Despite their complexities, these models remain simplistic sketches of a far more complex world. They give basic advice on what factors governments should consider and at what orders of magnitude. They inform us that a good global social cost of carbon-dioxide is greater than $20/tCO_2$ and less than $200/tCO_2$. The models can confidently tell us that we don't need to debate anything beyond these two

---

[9]Technological progress may allow us to eradicate or alter the specific mosquito that carries malaria for a pittance. The end of Malaria is in sight.

7. What Have We Learned from IAMs?

extreme price points: The world will be better off when it reduces CO_2 when it costs less than $20/tCO_2$ worse off when it costs more than $200/tCO_2$. In between, *if the population of the world could make active choices*, it would have decisions to make. How humanity should be and is making these decisions will be the subject of the next chapters.

A good way to think about integrated assessment models is that they give us pixelated images of the future. As time goes by, the images will become clearer. Humanity should prepare and plan ahead — and do so a lot more intelligently than it has. Here is an analogy. Standing on the ocean shore, you see a dark blue something in the distance. Your great fear should be that it is not just clouds but a tsunami heading for you. It could be a false alarm, or it could be real. If you wait until the tsunami is close enough to see it clearly, it will be too late to flee. The smart thing to do now is to walk back to your car and plan your escape route, even if you may not need to start driving it just yet.

From our perspective, the models tell us that the world is collectively emitting too much CO_2. When any one country reduces its emissions, it has further benefits for the others. Yet the theme of our book is also that IAMs are unrealistic to the point of being irrelevant. They contemplate a global decision-maker optimizing a global choice. However, in the real world, decisions are not made that way — they are made by individual countries. This is why year after year, everyone talks about what should be done, but little is actually been done.

7. Modeling The World Economic Impact

Further Readings

BOOKS

- Gordon, Robert J., 2016, The Rise and Fall of American Growth, Princeton University Press, New York, 2020: A detailed argument that growth will slow because the low-hanging fruit of technology has been picked.
- Lomborg, Bjorn, 2020, False Alarm: How Climate Change Panic Costs Us Trillions, Hurts the Poor, and Fails to Fix the Planet, Hachette Book Group, New York, 2020: A skeptic's view of climate-change mitigation costs, with alternative suggestions for aid. Also, Bjorn Lomborg's website.
- McAfee, Andrew and Eric Brynjolfsson, 2016, The Second Machine Age, W.W. Norton, New York, 2016. A more optimistic take on technological innovation and economic growth.
- Nordhaus, William, 2021, The Spirit of Green, Princeton University Press.
- Sivaram, Varun, 2018, Taming the Sun. An overview of the solar industry and solar technology.

REPORTS AND ACADEMIC ARTICLES

- Auffhammer, Maximilian, 2018, Quantifying Economic Damages from Climate Change, Journal of Economic Perspectives.
- Barnett, Michael, William Brock, and Lars Hansen, 2020, Pricing Uncertainty Induced by Climate Change.
- Cochrane, John H., 2021, Climate Policy Should Pay More Attention to Climate Economics. An advocacy piece on the tradeoff between growth and policies to mitigate climate change.
- Deitz, Simon, and Frank Venmans, 2021, Are Economists Getting Climate Dynamics Right and Does it Matter? Argues that a long delay between emissions and warming leads to optimal carbon prices that are too low and attaches too much importance to the discount rate.
- Levine, David K., 2019, Global Warming: What Sort of Mess Have We Made?: An explanation for economists of the issues with spending too much on climate-change mitigation now.
- Nordhaus, William, 2017, Integrated Assessment Models of Climate Change.
- Pindyck, Robert S., 2019, The Social Cost of Carbon Revisited, Journal of Environmental Economics and Management, 94, 140–160.
- Rose, Ashwin et al, Estimating a social cost of carbon for global energy consumption, Nature. Suggests that global warming will not increase but reduce the net need for global energy.

7. READINGS

- Stern, Nicholas and Joseph E. Stiglitz, 2021, The Social Cost of Carbon, Risk, Distribution, and Market Failures: An Alternative Approach, NBER WP 28472. A critique of standard lower estimates of the social cost of carbon derived from IAMs.
- The Stern Report, with Nordhaus' evaluation and a New York Times explanation by Hal Varian comparing the choices of discount rates in the Stern and Nordhaus models.
- van den Bremer, Ton and Frederick van der Ploeg, 2021, The Risk-Adjusted Cost of Carbon, American Economic Review, uses a DSGE model to derive an optimal SCC under various uncertainty scenarios — with a range from about \$7/tCO_2 to about \$66/tCO_2 and as much as two-thirds of the SCC attributable to uncertainty.

SHORTER NEWSPAPER, MAGAZINE ARTICLES, AND CLIPPINGS

- Tim De Chant, Biden plan eliminates billions in fossil fuel subsidies, 03/31/2021, Ars Technica.
- Economists' Statement, published in the Wall Street Journal on January 16, 2019.
- Garcia-León, David, 2015, Adapting to Climate Change: An analysis under uncertainty provided the original inspiration to our simple IAM schematic.

WEBSITES

- Bralower, Tim, The Economic Costs of Climate Change. Discusses damage components by sector.
- GHG Equivalence Calculator, 2021, Environmental Protection Agency.

Chapter 8

The Wrong Questions

Let's take stock. In previous chapters, we explained most of the contemporary discourse in today's earth sciences and social sciences. A few key observations stand out:

- OECD residents consume about three times as much energy and emit about twice as much CO_2 per person as non-OECD residents. However, the OECD contains only 18% of the world population today.
- The world will soon have more people, use more energy, and emit more CO_2 — though nowadays increasing only in non-OECD countries.
- The planet's temperature is rising. Even though the world will not come to an end, climate change will greatly alter our planet.
- When individuals' actions incur externalities, economics says that a social dictator should curb free-riding by imposing appropriate taxes.
- The integrated assessment models suggest optimal CO_2 tax policies, starting somewhere around $50/tCO_2$ today, rising to about $300/tCO_2$ by the end of the century.

If you think these summarize the world's dilemma completely, we have to disappoint you. You have not even yet confronted the biggest problem! Worse, nearly every scientist, politician, activist, and interested party (including you) is already aware of this problem and will acknowledge it if asked — but then immediately (un-)merrily go back to ignoring it.

The wrong question is "what *should* the world set as its cost of CO_2?" The right question is "what *will* the actual decision makers do," and "how can they be influenced, and what will be the result for the world?"

8. THE WRONG QUESTIONS

1 Problems, Choices, and Outcomes

The first issue to keep in mind is that talking about changing human emissions starts at the wrong point. Emissions are outcomes. They are determined by decisions based on cost-benefit analyses of economic actors and harsh realities. For example, people choose to have children, live their lives, and die. Their emissions are not their primary decision. They are a by-product primarily related to their uses of energy and consumption of food. If we want to influence global emissions, we need to influence whatever goes into their individual decisions. We must analyze and understand choices, not just byproduct outcomes.

The second issue to keep in mind is even more important. So far, we have analyzed the following problem:

The Collective Problem: This requires an economic analysis of what would be the optimal policies from a collective perspective—here what's best for humanity and possibly for the overall biosphere. It is the domain of integrated assessment models.

The collective problem is about quantifying the effects of pollution in the context of global warming. It helps answering questions such as:

- How quickly should humanity ideally get off fossil fuels?
- How bad is it if humanity fails to do so?
- What price should humanity pay for the privilege of emitting CO_2 (the social cost of CO_2)?
- How should humanity best transition to clean(er) energy?

Many climate scientists have focused their careers on answering these collective-problem questions. When Nordhaus and Stern work out the social cost of carbon-dioxide in their integrated assessment models, they are working out answers to collective problems. When you read in the media about the dire effects of climate change, most articles are written from the point of view of the collective problem. The reason why you purchased our book may well have been because you wanted to find out how much the world should reduce its emissions.

Yet, with the exception of the insight that humanity is burning too much fossil fuels, the collective problem is largely just an intellectual exercise. The reason why this is so is comically simple: humanity does not make decisions. We need to analyze a different problem.

1. Problems, Choices, and Outcomes

The Individual Problem: This requires an economic analysis of those who make the decisions — depending on the context, this can be individuals or governments.

The collective benefit exceeds the individual benefit, and currently by a lot (to the tune of $50/tCO_2$). When individual decision-makers choose to reduce their fossil-fuel use, the rest of the world benefits as well. Thus, we know that the world would be much better off with greatly reduced CO_2 emissions; and that cheap mitigation actions would be in the global interest. They would make humanity better off.

Our book considers the primary discussion focus on the collective-choice problem to be a conceptual mistake. It is a common mistake — made by natural scientists, social scientists, politicians, activists, and ordinary people alike. Even economists often discuss the individual problems only as an afterthought to the collective problem. Although Nordhaus, Stern, and almost everyone else have thought much about free-riding, even they have almost always ended up going back primarily to arguing about what the world should do and how much it should tax CO_2.[1]

Many people will now retort to our assessment with "of course we knew this" — and they are correct. The problem has been hiding in plain sight. Yet, the public debates suffer from tunnel vision. They fail to put the horse before the cart. The real problem is not the collective problem but the individual problems. The fact is that the latter (individual) problems largely render discussion of the former (collective) problem moot. Then why are we all spending so much time discussing it? Why are you always instinctively making climate-related arguments that begin with the words *The world should* ...?

Please sequester this mode of collective analysis out of your mind!

It's futile. The world is not the Star Trek Borg.

---

[1] Nordhaus in particular has been trying to offer potential solutions to the global public-goods problem. We will talk about his preferred approach in the next chapter. (Stern has recently focused more on local effects and actions.)

8. The Wrong Questions

> *sidenote*
>
> Environmentalists sometimes accuse others of being callous to the misery that climate change will cause. How may millions are non-activists willing to sacrifice? But are the environmentalists themselves callous to the misery that forced energy and development restrictions would cause? How may millions are the environmentalists willing to sacrifice? And what difference would it actually make if we knew the answers?

2 Can We Go It Alone?

I am a decision-maker. You are a decision-maker. To a more limited degree, so is the United States of America. Its decisions are often the outcome of a political process, which need not produce consistent or rational decisions. But nation states are the largest effective decision-makers on the planet, so we will start with them.

In fact, we will start with just us (the USA) alone. For now, assume that the rest of the world will do whatever it will do (hopefully more). We cannot tell them what to do, nor can we take credit for what they will do.

And please always keep the following inconvenient fact in mind:

> **Earth's climate does not respond to any one's region's emissions. It responds only to the sum-total of global emissions.**

Can the USA Go It Alone?

The crib sheet in the appendix summarizes many important facts in our book. Recall the average energy consumption figures per person per day as of 2022:

OECD	USA	Europe	China	India
141 KWh	232 KWh	109 KWh	98 KWh	23 KWh

Now imagine that the USA could reduce energy consumption to the European (and Chinese) standard within one generation. (The Indian standard — with its large poor rural areas — seems out of the question.) Multiply the largest imaginable relative reduction by the U.S. population of 330 million.

Maybe we could collectively wring out a very, very tough 15 PWh (out of 26 PWh) via forced reductions in our economic activity — and by this we mean

2. Can We Go It Alone?

reductions that could leave us behind the Europeans in terms of standard of living and bring us down about to the Chinese standard of living. (No serious economist believes that achieving this large a CO_2 reduction within one generation via taxes — that is, the good old-fashioned IAMs way — could be achieved without such large economic sacrifices. The idea that speedy drastic reductions in CO_2 are cheap is simply absurd. Only small reductions are.)

Remember, other nations will do what they will do. With a predicted world energy consumption of 260 PWh by 2050, our 15 PWh reduction would leave 94% of primary energy consumption untouched.

It takes approximately 200-250 $GtCO_2$ to change Earth's equilibrium temperature by 0.1°C. If U.S. annual emissions of 5-6 $GtCO_2$e per annum today were curtailed to about 3 $GtCO_2$, it would take about a century to lower the equilibrium temperature by 0.1°C, from 3.0°C coming to 2.9°C coming.

Furthermore, the nature of global warming is such that the beneficial climate effects would take decades to occur and our 6% contribution would be small enough not to cause a noticeable effect in warming. Realistically, our 15 PWh reduction could reduce global warming by about 0.05°C in 2100, out of total global warming of about 3.0°C. It could be measured with satellites but it would not be noticed by the average American. To implement our American belt-tightening program would also require not just one charismatic politician but a series in succession to cover decades of needed commitment. It is difficult to see how this could plausibly happen in the context of the American political system.

Remind us — why are we even discussing the United States fighting solo against climate change in a painful way? This approach seems to border on the absurd.

Can Europe Go It Alone?

The only countries that have achieved reasonably notable reductions in their CO_2 emissions through policy-based belt-tightening interventions are Japan and European Union countries. Although we laud their intent and results, success in these regions is not success in the world (and most of their painful belt-tightening efforts are unlikely to spread to the rest of the world). Japan emitted about 1.0 $GtCO_2$, Germany about 0.7 $GtCO_2$, and France 0.3 $GtCO_2$.

8. The Wrong Questions

The European part of the OECD together emitted <u>4.0 GtCO$_2$</u>. With projected further population declines and more belt-tightening policy changes, the European Union is now expected to wring out a further drop from 4.0 GtCO$_2$ to about 3.7 GtCO$_2$ by 2050e. As welcome as these 0.3 GtCO$_2$ reductions are — especially in the context of reducing not only CO$_2$ but also local particle emissions — they are less than *a drop on hot stone* using the German idiom here (i.e., immaterial) as far as global climate is concerned.

Can the OECD Go It Alone?

What if the OECD were of one mind (and it rarely is)? Could coordinated decisive OECD action based on CO$_2$ taxes directly and effectively shrink world emissions by shrinking their own (OECD) emissions? Still no. The OECD could only slow the growth in emissions. It could not produce a net reduction in world emissions from today's level.

Even if the OECD could shut down all economic activities and return to the stone age, and even if the OECD closed-down industries did not simply relocate to Asia to merrily continue operating there, the world could still not maintain today's emissions, much less reach a zero-emissions future by the end of the century through belt-tightening. Instead, the world would be back to today's energy use by around 2060, i.e., roughly within one generation — and with emissions rising in the following years.

We must not ignore the inconvenient truth that *energy use and emissions are no longer primarily a rich-world luxury problem*. The OECD is already responsible for only about one third of the world's energy consumption and this share is declining. Within one generation, it will be less than a quarter. If we want to make a dent in world problems, we "elitist" Western environmentalists must lose our ethnocentrism.

Regardless of what you consider fair or unfair, the hard fact is that there can no longer be a reduction of global energy use unless non-OECD countries greatly reduce their emissions, too.

However, there is one important argument in favor of the effectiveness of a CO$_2$ tax being enacted in OECD countries even if those countries are responsible for only a small share of world emissions. It relates to the more promising way of fighting emissions via clean-energy innovation. In addition to the direct effect that the tax will have on OECD countrys' own emissions, there is an indirect effect. With a greater incentive to reduce emissions, industries in tax-enacting OECD countries will redirect their research and development towards finding cleaner solutions. Arguably, industries in these OECD countries are also best positioned to invent the kinds of solutions that could greatly reduce the cost of clean energy *everywhere*. Industrialized countries could then export such cheaper cleaner technology to non-OECD countries — and this could make a big difference.

We want to emphasize again that we are *not* arguing against the usefulness of painful carbon taxes. We are only arguing that we believe that they will not be enacted by enough of the most important emitters in order to significantly slow the pace of global warming.

Can't the Non-OECD Join In?

Painful forced constraints on growing energy use in non-OECD countries seem even more unrealistic. The residents of these countries want to escape poverty. Their energy consumption is not a luxury problem. Bangladeshis, Indians, and Pakistanis as well as Africans, South-East Asians, and Latin Americans will put great pressures on their leaders to improve their standards of living — and with it their energy consumption. It seems absurd to imagine that the Indian population would be willing to forego development for the sake of the greater good of the world, much less on behalf of Western climate activists. Climate protests are not likely to meet with great success in India, China, Africa, and beyond — except when they are about *other* countries.

Could the OECD cover the Non-OECD Costs?

The world emits about 40 $GtCO_2$ per year today. By 2050, it will emit about 45 $GtCO_2$ under RCP 4.5 and 60 $GtCO_2$ under RCP 7. Let's assume that the difference of 20 $GtCO_2$ is the removal target. (Recall from Chapter 5 that this would reduce global warming by about 0.1°C by 2050 and 0.3°C by 2100.) What if other countries — like China and India — are unwilling to participate in the reductions? Could the OECD countries take care of the entire 20 $GtCO_2$ alone? Could it pay the others?

Assume for illustration that it costs **$50** to remove one ton of CO_2. Why $50? This $50/$tCO_2$ is the social cost of CO_2 in Nordhaus' integrated assessment model. It is also far below the current industrial sequestration CO_2 cost

8. The Wrong Questions

(of $200–$300/tCO_2$) that some developers are still shooting for — and that the media are often describing with such ringing endorsements as "the world needs sequestration to reach its net-zero goal." However, $50/tCO_2$ is also far above the cost of smarter tree-based sequestration (of $10/tCO_2$). Similarly, simply avoiding CO_2 omissions would also be cheaper than sequestration. The sequestration cost further depends on how quickly and how much we want to reduce CO_2 emissions. Removing the first gigatonne of CO_2 could be dirt-cheap, but removing 20 $GtCO_2$ too soon could drive the cost well above $50/tCO_2$.

For our illustration, we will stick to our $50/tCO_2$. The total removal cost for 20 $GtCO_2$ would then be about $1 trillion per year or 1.2% of world GDP in 2020. We will need to assess how to cover this 1.2% of world GDP. In purchasing power, world GDP is about

	OECD	USA	Eur	Non-OECD	China	India
2022	44%	15%	18%	56%	20%	7%

USA: With 15% of world economic activity, the United States would have to dedicate not just 1.2% of its output, but 8% of its output to reduce global warming from 3°C to 2.6°C in 2100. For perspective, this is about twice as much as the U.S. spends on its military. It is much more than the U.S. spends on its entire education system, primary and secondary schools, colleges and universities combined. The U.S. thus faces a choice: it could either continue educating its residents or redirect its economic activity to reduce global warming from 3°C to 2.6°C in 2100.

We can also calculate per-household equivalents. The CO_2 reduction cost would be about the same as <u>four to six months of rent</u> for a typical two-income household in the United States. Is this a realistic ask?

OECD: If all OECD countries together could agree among themselves to pay for global CO_2 removal, then the proportional cost would only be about 3% of GDP — still a gigantic number but now only about the same as the cost of all higher education. The all-OECD choice would then be between free university education for everybody and reducing global warming from 3°C to 2.6°C in 2100.

We will return to how realistic cost sharing among countries is in the next chapter. Let us just close with the observation that China and India are growing rapidly while the OECD is not. Within one generation, the EIA expects the following:

	OECD	USA	Eur	Non-OECD	China	India
2050	33%	13%	13%	67%	22%	15%
Change	−11%	−2%	−5%	+11%	+2%	+8%

By 2050, China and India together will host more economic activity than the OECD. It is true that all countries will be richer and thus that removing 20 $GtCO_2$ will become relatively more affordable — but will OECD citizens find it more galling to carry the world cost while China and India will have greater GDP, faster growth, and much higher emissions?

If you need further evidence, the Byrd-Hagel Senate resolution of 1997 passed *unanimously* 95-to-0 in the U.S. Senate — itself a near-miracle to see so much agreement across the aisle. It stated that the US should not sign a climate treaty that would "mandate new commitments to limit or reduce greenhouse gas emissions, unless ...[it]... also mandates new specific scheduled commitments to limit or reduce greenhouse gas emissions for Developing Country Parties within the same compliance period."

We are not done yet. Our estimates are in line with the fact that developing nations have recently asked for $1.3 trillion *per year* in climate support at recent climate conferences — *or else* they plan to ramp up their fossil fuel consumption. We note that many of these countries have notoriously high levels of corruption. Who will vote to send money to Congolese warlords? And even if the OECD were to volunteer to pay, how would donors channel the funds in a way that would accomplish their emissions reduction intent? What if they then asked for more?

The key question to us is not whether it would be appropriate or ethical for the OECD to send this much money abroad, but whether it seems plausible to expect it. How realistic would you judge such a transfer? We believe the answer is exceedingly unlikely. There are a lot of things that we think the United States and OECD might do, but "just say no to fossil fuels," pay off poorer countries, and see Earth warm up 2.5°C instead of 3.0°C is not plausibly one of them.

8. The Wrong Questions

Once again, remind us — why are we even discussing this? This approach — of the OECD paying non-OECD countries more than token amounts — seems to border on the absurd.

Money and Mouths

The country with the most concern for climate change and the most public discourse on the subject may well be Germany. In a November 2021 survey, German 14- to 29-year-olds expressed that climate change was their number one concern. How much skin are they willing to put into the game?

Among those surveyed, 60% regularly travel by car and more than 80% cannot imagine life without one. Only 19% are willing to make the sacrifice of life without a car. Only 27% are willing to forego flying — recall from Chapter 2 that air travel is the biggest luxury carbon emission that most of us will rack up within our lifetimes. It's easy to protest against *the other bad guy*. But even when it comes to foregoing what can only be described as superfluous luxuries, their own sacrifices become suddenly much more difficult.

Summary

Although the emissions that matter are not country emissions but world emissions, the decisions that matter are country choices not world choices. Realistically, this means that

1. Climate-related policies that are *much* too expensive for their local benefits will not be adopted. (These local benefits do include reduced local pollution and more clean energy-related jobs.)

2. Climate-related policies that do not induce growing numbers of actual decision-makers to voluntarily adopt them (such as cheaper clean technology) cannot be trusted to be adopted widely enough to meaningfully reduce emissions, CO_2 concentration in the atmosphere, and thus eventually global warming.

3. Climate-related policies that will not be adopted in non-OECD countries will always be very limited in their potential. Real solutions have to be adopted worldwide.

Our suggestions are not as passive as those proposed by ardent free-market proponents. There are a lot of steps that countries could take that would not be greatly against their self interests. They could accelerate the transition to clean energy. (There are many market impediments involved in fundamental energy research, development, and deployment.) There are terrible local health effects of particle pollution that could be reduced with modest fossil-fuel taxation.

Yet our suggestions are also not as active as those proposed by many climate activists. This is not because we are climate-change deniers or fossil-fuel advocates. (On the contrary!) However, this is not the point. Our suggestions are more limited because we believe that "the world" has fewer levers than activists realize. The world is all about the individual problems, not about the collective problem. *Pies in the sky* do not make the air any cleaner.

Further Readings

- Bureau, Dominique, Alain Quinet, and Katheline Schubert, 2021, Benefit-Cost Analysis for Climate Action, Journal of Benefit-Cost Analysis. This recent paper describes some of the challenges in accomplishing emission reductions in a cost-efficient way via policy interventions.
- Nordhaus, William, 2018, Nobel-Prize Lecture. It describes his proposals of climate compacts to reduce the free-riding individual problems. We will return to this in the next chapter.

Chapter 9

Unrealistic Approaches

We share the goal of environmental activists to reduce the world's reliance on fossil fuels. Yet we disagree about the means to achieve this end. In this chapter, we will explain why we believe conventional climate activism — no matter how high-minded — has not worked. This requires brutal honesty. Our task is, after all, to steer an entire planet. If you are an activist, we hope you will hear us  out, even if you won't ultimately agree with us. In this case, we hope we will be wishing one another success.

Global warming is not a problem of this or that country. If it were, it would be much easier to solve. Because climate-change is global, it does not matter whether this or that country reduces its emissions. It only matters whether humanity reduces the sum-total of its emissions. Therefore, within-country battle victories have mostly been Pyrrhic. Worse, we believe the evidence suggests that current activist approaches will not reduce global GHG emissions in the future, either.

Let's briefly count up the approaches that have been central to much climate activism before we discuss them one by one in greater detail.

9. Unrealistic Approaches

- We love the idea of a clean-energy economy. We even sympathize with "just say no to fossil fuels" sentiments. But unfortunately, clean energy cannot entirely replace fossil-fuel energy for at least a few more decades — especially in poorer countries.
- We love the idea of a global CO_2 tax, the center piece of the integrated assessment models of the previous chapter. But unfortunately such a tax is dead not just *on* but *before* arrival. This is not a deep insight but a self-evident one — like "the emperor has no clothes." A global CO_2 tax requires a world government that does not exist.
- We love the idea of international climate treaties. But unfortunately countries and their citizens will not sign onto and enforce treaties that require large painful sacrifices. And the <u>problems of the day</u> (like immigration, jobs, health care, economic competitiveness, terrorism, or wars) make better election topics than carrying through decade-long sacrifices demanded by the United Nations. Climate talk helps politicians get elected. Executing climate sacrifices does not.
- We love the idea of greener cities and electric cars in rich places like California. But unfortunately, the real problems are elsewhere. Many of the biggest potential future emitters have not even come online yet.
- We love the idea of taking responsibility for one's carbon footprint. But unfortunately, personal footprint aspirations are like New Year's resolutions. We consider it counterproductive for activists to place their faiths and our collective future to eight billion such resolutions.
- We love the idea of less poverty and inequality. But these are not primarily climate-change issues. Bundling all social ills into one package is more likely to favor the status quo. It will lead them to fail together and it is a recipe for inefficient spending by the politically connected.

If the goal of climate activism is to feel good about oneself advocating, these approaches may succeed. If the goal is to reduce CO_2 in the atmosphere by meaningful amounts, they will not.

When you are on the Titanic after the iceberg collision, there are a whole lot of things you should not do: You should not be rearranging the deck chairs. You should not be thinking about better ship designs. You should not be proposing more lifeboats. Instead, you should work to make the lifeboats ready. You should pack them as efficiently as possible. And you should worry about the other stuff later — yes, even about the fact that some passengers will go down and die. Sometimes, problems are so difficult that there are no great solutions. Pick the best alternative you have.

1 Why It's So Difficult

We agree that the world should wean itself off fossil fuels as quickly as possible. There is nothing healthy about them. They already kill millions of us with their particulate emissions and devastate the environment.

The task of converting energy and agriculture to carbon-free enterprises will rank among the biggest tasks ever undertaken by humankind. Nothing about it will be small or easy. Unfortunately, many activists seem not to understand what they are asking for. And maybe they do not want to. They seem as much in denial of reality as many fossil-fuel proponents are in denial of climate change. By pretending that the transition is easy or by turning off the spigot too fast, environmentalists run the risk of creating a public backlash that could set back the process for decades.

On the bright side, since the collapse of the modern world due to climate change, we've had no trouble attaining zero carbon emissions.

Even if the world could agree to cut emissions (which it cannot), global warming cannot be stopped. Worse, possible temperature reductions are modest (though not small). International negotiations are about *slowing* global warming by 10–20%, a reduction in expected warming from about 3.0°C to about 2.6°C, over the span of two generations. The world will continue to warm and the Arctic will continue to melt. Today's electorates will notice global warming but they will not notice a difference that their sacrifices will make.

The prize for economic sacrifices today is not the elimination of global warming. It is a modest reduction. The public does not want to hear how much they will have to sacrifice for it for this "little." Which leader wants to explain honestly to the public — especially in poorer countries — what they need to give up to obtain this 0.3–0.4°C? And even this is beside the point. Our book's theme is that the OECD (including the US) could only hope to reduce warming by 0.05–0.2°C with radical decarbonization and 0.02–0.1°C

9. UNREALISTIC APPROACHES

with aggressive decarbonization. These are not our numbers. They are the numbers implied by the IPCC RCPs and the share of world emissions of the OECD. Which leader wants to explain this?

Some environmentalists want to push to overshoot and spend a lot more to go clean even faster. The integrated assessment models teach us that the world is best off with a balance between cutting back too quickly and not quickly enough. The humanitarian consequences of turning off fossil fuels too suddenly would be terrible. There are eight billion people already born, and another 50% in the making. They rely on fossil fuels not primarily out of moral failings but out of necessity. In 2020, fossil fuels accounted for 85% of humanity's primary energy (Figure 2.11). It is not an option for the populations of the world to go back to nature, to the planet that had to support a population of only 1.6 billion at the end of the 19th century. The 8 billion people here today want modern lives, which means more energy — and clean energy cannot give it to them yet.

It's vital now to pursue and promote viable solutions that can maintain public support for a long time. Laying this out clearly is the point of this chapter.

2 Why A Global CO_2 Tax is Unrealistc

To hasten the transition from fossil fuels to clean energy, economists have always strongly advocated a greenhouse gas or fossil-fuel tax. As we explained in our previous chapter, the most common assessments are that this tax should start around \$50/tCO_2 and ramp up to \$200 to \$500/tCO_2 over the decades. Unfortunately, we do not see much chance of this happening — the centerpiece of many environmental debates.

In theory, we agree with our colleagues that if the world had a benevolent dictator — not afraid of voting majorities and lobbies — the global-warming problem could be solved by one global tax appropriately set (roughly) at the social cost of pollution. It would produce the right incentives. Being global, such a CO_2 tax would induce companies and industries to change. They could not lobby or replace the local government to reduce the tax or escape the tax by moving operations to another country.

Activists,[1] environmentalists, climate scientists, and economists have been engaged in seemingly endless arguments about whether the optimal world tax should start at \$30/tCO_2 or \$60/tCO_2, and how quickly it should rise in the decades after. (It is negative in real life.) All of them are living in a fantasy world worthy of the academic ivory tower. There is no world government that could institute a global tax.

Is the threat of global warming so urgent that humanity has no choice other than to establish a world government? Are you ready for one? Are you ready to submit to the consequences? As for us, we are not. Indulge us with an imaginative excursion into why this is so.

Start by asking yourself who should appoint the world government. If it is elected by the people at large, it would surely not reflect the interests of the 1 billion people in the West, mostly ruled by Modernity and Enlightenment. Instead, it would be dominated by coalitions of the 1.5 billion Chinese, 1.5 billion Indians, 1.4 billion Africans, 1.6 billion Muslims, etc. If (Western) democracies are any guide, people would band together in parties that act as

---

[1] A few activists have even been arguing against capitalism and for a new world order to save the planet. Although capitalism has its drawbacks, we shudder thinking about how any alternative would be conducted in the real world. Churchill's statement that "democracy is the worst form of government — except for all the others that have been tried" applies to capitalism as an economic system, too.

9. Unrealistic Approaches

tribes and outvote other cultures and regions. If instead a world government were to be elected by countries as they exist today, it would be dominated by a majority of dictators and demagogues. It could end up like the UN Human Rights Council, which regularly heaps praise on North Korea, China, Iran, Saudi Arabia, Cuba, and other such beacons of human rights and condemns the United States and Israel.

Mister Secretary-General, U.N. Ambassadors, Dictators, Fanatics, Madmen…

Nevertheless, let us assume for a moment that a benevolent dictator, with Western enlightened values, did manage to take world power. On whom should this dictator impose GHG taxes? You may quibble with our numbers, but the rough basics are always going to be as follows.

As we calculated in previous chapters, the cost of aggressive CO_2 abatement today would have to be on the order of about one month of rent. About two-thirds of the world's population — the poor people in most countries, including the West — do not have the resources to afford such added taxes. The average household outside the OECD earns only about $5,000 per year. Most of these households do not have the means to pay one extra month's rent to fight global warming, even given their lower rents. The required contribution could amount to as much as 20% of their disposable income. Poorer nations would almost surely insist that the cost should justly fall more, if not entirely, on the richer countries who have benefited from decades of exploiting fossil fuels.

Yet even the median household in the West earns "only" about $50,000. Its disposable income, i.e., after existing taxes, food, rent, health care, etc., is typically only a small fraction of this income. If the global dictator pushed the burden mostly or only onto households richer than the median, these would have to give up about, say, two months of rent — as much as one-third of their disposable income. You might feel that giving up the annual summer vacation is well worth rescuing the planet, but few households will likely share *your* "warm" feelings for such efforts. The majority would have to agree.

It is good news that polls suggest that very few Americans still believe climate change is fake news. The average American indeed worries about it

2. Why A Global CO$_2$ Tax is Unrealistc

(as do 60% of younger people around the world). It is bad news that the most favored suggested sacrifice by Americans is $1/month! Only 28% would pay $10/month. To share proportionally in fighting climate change would require 10–100 times that much from the average American. Our prediction is that Americans would simply not vote for such change.

Let's open our eyes. From watching news coverage, it seems that climate-change activists tend to be younger, more educated, higher-income citizens of Europe or blue states in the United States. Few of them seem to believe that they should be principally responsible for paying to combat worldwide climate change. Their typical view seems to be that the tax should fall on fossil-fuel companies and individuals who got rich investing in and trading fossil fuels (often from redder states in the United States) and be (somehow automatically) shared by other governments around the world.

I know you all paid your taxes last month, but since then we've had tax *reform*!

They understand neither that the rich countries are not even half the problem nor that there are not enough oil barons to cover any meaningful fraction of the costs of worldwide withdrawal. The burden would have to fall mostly to the middle class in the developed world, plus everyone else in other countries somehow, to shoulder most of the cost. The oil&gas industry will not be rich enough — especially if the plan succeeds.

We would be curious to learn whether even the highest-paid blue-state climate-science professors — who may earn a gross income of $200,000 per year, with post-tax take-home pay of $120,000, mortgage payments of $30,000, food payments of $30,000, health care payments, tuition payments, etc. of another $30,000 — would volunteer to pay an extra $5,000–$10,000 per year in order to reduce global GHG emissions. In the abstract, they will of course agree. But when push comes to shove, would they willingly step up to the plate? Or would they rather find good arguments why it should not be them but others who deserve to pay?

9. Unrealistic Approaches

Consider some further thorny problems. Richer people in the West might support a benevolent world dictator presumably in order to obtain more protection for their great-grand-children through increased climate action. Yet, poorer people would presumably demand more protection for their own children *now*. Should our benevolent dictator, with enlightened and not self-interested values, focus primarily on global warming and future generations or primarily on poverty reductions in today's generation? Why should 1/4 of the world population enjoy wealth in abundance, while 1/2 are poor, and another 1/4 live near subsistence levels? What if the dictator considered it ethical only to tax millions of Westerners more in order to reduce the poverty of billions in Sub-Saharan Africa and India? It's hard to argue against it on moral grounds.

Many of the global poor already scoff at the hypocrisy of the rich, whether the rich live in their own or other countries. One suspects that the poor would consider it "the whining" of rich people (or Western intellectuals), when the latter complain about how their great-grand-children will have to deal with global warming — probably right after taking a flight for this year's summer vacation — if they (the poor people of Nigeria or India) were to double their fossil fuel use. These rich countries had built their economies by aggressively exploiting fossil fuels for centuries. The people of poor countries would look at their own standards of living, at their compatriots in poverty right now, and at the increasing immigration barriers in rich Western countries — and wonder why they should sacrifice *anything*.[2]

But these are arguments akin to how many angels can dance on the head of a pin. The simple fact is that there is no world government on the horizon. A global CO_2 tax, argued for by climate scientists, economists, and activists, exists only in a world that does not exist.

Dreaming of a world with a global CO_2 tax is like dreaming of the world of the United Federation of Planets. It's not our world.

---

[2]In fairness, these are not easy questions. William Easterly has raised good questions about whether international aid has actually been effective. But this is a very subtle consideration that is easily lost.

Big Brother Solutions

Every country in the world has been moving rapidly towards ubiquitous surveillance of its citizens. Governments have never known so much about their people as they do today. Could some of this power be used to encourage more climate-conscious individual behavior?

We are in favor of one form of government intervention — economic incentives — but we are wary about other more intrusive forms. Climate and energy touch on every aspect of our lives. Should governments control, dictate, and punish people that don't do what should be done in the interest of combating climate change?

For better or worse, Big Brother surveillance may work nationally but it is a hopeless approach from a global perspective. There is no global government or treaty that could extend Big Brother from single countries to the entire world. Thus, it's not a viable plan.

3 Why Climate Treaties are Unrealistc

If there is no world government, let's think smaller and more realistically — not the world, but countries (and perhaps institutions). Could climate treaties among sovereign countries do the job?

International treaties can be classified into two categories. The first are treaties that bestow their benefits upon the signers. NATO or the European Union are such treaties. Non-signers do not get the benefits of membership. The second are treaties that bestow benefits upon signer and non-signer alike. These suffer from the public goods problem discussed in Chapter 6. It is only this second kind of treaty that we are discussing here. A CO_2 treaty would be of this second kind. If the world warms less, even non-signers will benefit. (An important warning: we are on the pessimistic side here, though we like to view ourselves simply as being more realistic. We have heard many serious intellectuals express more faith in the viability of international carbon agreements than we can muster. We hope they will be right and we will be wrong. As for you, our reader, hear us out and then make up your own mind!)

9. Unrealistic Approaches

Theory With Heroic Assumptions

Unlike a world government, at least countries and their governments exist. This checks the first box. Countries are real decision-makers that could, in principle, agree on a treaty with a CO_2 tax. Yet to the extent that they act rationally and consistently, countries are also selfish and self-interested, just like individuals. It flies in the face of common sense to imagine that most countries would sign a treaty that left them meaningfully worse off. (Existing climate treaties have never demanded any real sacrifice.) This is not to say that you won't find instances in which countries have acted against their self-interest — just that we cannot expect painful altruism at large scale to win the day.

Allow us to offer a brief overview of the hurdles that a global climate treaty with real sacrifices would face. Let's assume that all countries are eager enough to fight climate change. There are about 200 countries in the world today.

One of the most basic problems is coordination. Have you ever tried to negotiate with more than ten parties at the same time, with each party trying to cut a deal just a little better for itself, hoping to exhaust the patience of others? The proverbial wisdom of Solomon would not be able to cut through the morass.

So it is unanimously agreed...we will all do something about global warming as soon as somebody else does.

But let's assume that you have somehow managed to assemble all important countries in the same room and they have agreed not to play the waiting game (claiming higher-ups will have to ratify any concessions later). There is another problem now. Countries have many concerns beyond climate-change. If country A cares more than country B about climate change, what prevents B from using its cooperation as a bargaining chip with A on something else? For example, China has already begun to

3. Why Climate Treaties are Unrealistc

demand more <u>concessions on other issues</u> whenever the United States has begged for more cooperation on climate change. How much should the United States be willing to give up to China (e.g., on import duties, human rights, intellectual property theft, or social-media interference) for playing ball on climate change? As we are writing this, poorer nations have just introduced their demands for <u>$1-3 trillion</u> from wealthier countries in order to shift away from fossil fuels. India alone is asking for <u>$100 billion</u> per year (for a total of about $1 trillion over a decade).

But let's assume that all 200 countries can somehow manage to coordinate and agree not to link climate-change agreements to other demands. Economics teaches that there is an intrinsic problem that applies to all treaties: The more countries sign on, the greater are the incentive for each single country *not* to sign on. (This is also why cartels and conspiracies don't work with large number of participants.)

This deviating country can pick and choose to attract the best CO_2-emitting industries in the world. These industries will employ its population and pay its taxes.[3] Simply put, each country wants all other countries to sign a climate treaty and tax and reduce their global polluters, but it would not want to sign itself. It is a classic *public goods* problem, where every country wants to <u>free-ride</u> (Chapter 6).

The extreme version of this argument may be taking economists' rationality a little too far from the perspective of the voting public.

First, if countries are all suffering about equal harm and would gain about equal benefit from a climate treaty, they may be more inclined to compromise. Pain and gain can be more easily shared equally. Unfortunately, China and the United States, the two biggest emitters, are among the nations

---

[3]Morals have rarely come first, but it can always comfort itself. Why should it care that its global CO_2 pollution will affect the other "<u>patsies</u>"? They decided their own fates, and you decided your own.

9. Unrealistic Approaches

that will suffer the least consequences from climate change. Thus, why should they agree?

Second, countries are somewhat more inclined to take responsibility for emissions that can be labeled and tagged as "domestic." They may remove CO_2 at their smokestacks, but not out of the air a mile away or 1,000 miles away. Yet, logically, from the perspective of global greenhouse gases, location should make no difference. CO_2 at the top of the smokestack in Indiana is just like CO_2 in India. An extreme version of this argument is to ask who in the United States would vote to pay if there was a new technology that could remove emissions even at the extremely low price of $1/tCO_2$ while China is still pumping more CO_2 into the air every year.

But let's assume that a miracle has made every country sign onto a climate treaty. And most did sign onto the 2015 Paris Agreement. (Fewer signed the 2021 Glasgow agreement.) Under the Paris treaty, countries were required to submit details of their plans to cut greenhouse emissions, called "Nationally Determined Contributions," or NDCs. According to the United Nations 2021 report, the NDCs submitted by countries are explicitly allowing global emissions to keep rising, increasing by 16 percent by 2030, compared to 2010 levels. This is not enough to do better than about IPCC RCP 4.

It's a great treaty, but do we trust us?

How could signing countries make sure that other signing countries are not lax on enforcement? Who is the police? In the end, what determines countries' decisions of whether to emit or not will be the actual penalty for non-compliance. You may or may not be surprised by this, but no international climate accord to date has dared to impose penalties. "Fortunately," the two most recent important international climate treaties, Kyoto 1997 and Paris 2015, did not need penalties. The reason? Even the targets were not mandatory! The result? Almost all countries are behind even their modest voluntary targets. Even Europe — a beacon in the global effort to fight global warming — is already about 21 years behind in reaching its own declared goals.

3. Why Climate Treaties are Unrealistc

Are we too harsh on climate treaties? After all, Kyoto, Paris, Glasgow, etc. were not treaties. They were simply Conferences of the Parties. Maybe we have just not observed a climate treaty yet...

But let's assume not only that a few brave politicians in *some* countries have signed onto a binding treaty, but also that brave politicians in *most* countries have signed on. And now assume that this treaty forces true domestic sacrifices (say, about one month's rent).

How would it likely play out? Will most of these brave politicians survive the next election cycle? Maybe in some countries, but in most? Will the next generation of politicians care more about domestic public opinion related to their handling of current problems (perhaps even climate-related) or about international promises and opinions that won't make any notable difference for 50 years? Even the dumbest opposition politicians will realize that they can promise the electorate to stop abiding and use the one month's rent differently. (If need be, they can point to a few other countries to blame as examples of cheats, whether true or not).

Ultimately, most or all politics is domestic. Democratically elected leaders who have agreed to sign and enforce such a treaty would be less likely to win the next election against skillful populist debaters. It wouldn't be much better in less democratic countries, either. Do you *really* think that China, India, or countries in Sub-Saharan Africa are ready to sacrifice their economic development for a global climate benefit (that will materialize only 30–50 years down the line) if it involves sacrificing their own economic interests now?

But let's assume the signing politicians have survived their next elections. How many countries will be able to sustain such climate-determined voting majorities over future decades — especially during the next recession, when many voters will lose their jobs?

But let's assume that voters are so genuinely determined to fight global climate change that they will not give in to the temptation to elect opposition politicians who promise them less sacrifice — and indeed ignore all other issues on election platforms so that any anti-environmentalist opposition party will never come to power.

9. Unrealistic Approaches

How many of these determined voters would also be in favor of allowing foreign countries to inspect and enforce their country's global CO_2 tax compliance? How would U.S. voters feel about being subject to a worldwide-mandated tax, perhaps enforced by the United Nations or an international climate panel?

We believe it is more likely that the United States would leave the United Nations before it agreed to be disciplined by it. The United States is not unusual in this respect. The majority of voters in most countries remain fiercely nationalistic.[4] They already have instinctive aversions to globalization. They tend to favor politicians who promise independence from global pressures.

> sidenote
>
> Bill Nordhaus, the Nobel-prize winning economist who pioneered the integrated assessment models of Chapter 7, is more hopeful about the idea of global cooperation. He suggests that blocks of countries could force other countries to sign on by imposing extra tariffs on non-signers. He calls this a *climate compact*. But there is no precedent for climate compacts, and we suspect that it would be difficult to set up and sustain one over long periods of time. Sustaining an embargo on global renegades like Iran and North Korea is difficult enough. Sustaining an effective tariff on China, India, and Russia — whose collaboration in addressing the security crises of the day is important — seems doomed from the start. Perhaps the only piece of advice we could offer the designers of such compacts is that any such policy must be formulaic and mechanical, difficult to tinker with, immediate, and difficult to circumvent. We hope a workable climate compact could be established, but we are not optimistic enough to count on it.

Evidence and Practice

Some optimists may consider our arguments to be merely theoretical ivory-tower objections. Should the world place more faith in negotiated arrangements? Take a hard look. How have decades of UN climate conferences changed the world?

---

[4]Most Europeans are in favor of paying money to Brussels to maintain the European Union, but this works because it makes almost all countries individually and immediately better off. One could argue that the richest member countries of the European Union have become somewhat less nationalistic. The rise of populist right-wing leaders in poorer member countries suggests that this sentiment is not universal.

3. Why Climate Treaties are Unrealistc

Montreal Protocol on Ozone?

If you disagree with us, then please tell us: what global environmental treaty that required large national sacrifices has ever worked? Can you please give us just one example?

You may now be tempted to retort that you have one: the Vienna/Montreal 1987 treaty banned CFC chemicals depleting the ozone layer. It was the most successful international climate treaty *ever* — if only because it actually worked! This in itself should be considered a minor miracle—until you look at the cost-benefit estimates. Table 9.1 shows that even going it alone, the United States would have been better off—and not just by a little. Getting other countries to join in was only frosting on the cake.

Table 9.1. In-Time Estimates of Costs and Benefits of Montreal Protocol (in billions)

		Benefit	Costs
US only (–2165)	unilateral	$1,373	$21
	multilateral	$3,575	$21
World (–2060)		€2,220	€200

Source: Barrett (2007), Table 3.2, p79. Original US estimates are from the US EPA (1988). (They are the present value of reduced deaths at $3 million/life, which is why they reach $1.4 trillion.) Original world estimates are from Velders et al (2000).

Montreal worked because it required no sacrifice, the net benefits of the treaty were even larger, and rich countries underwrote some of the costs of the poor countries.[5] It is the opposite for Kyoto and subsequent climate-change conferences. They would require notable reductions in living standards.

We know of no example of a successful environmental treaty that required meaningfully large national sacrifices where parties not signing would have reaped all the benefits of the treaty without bearing the cost.

---

[5] Cass Sunstein's analysis in Of Montreal and Kyoto: A Tale of Two Protocols (Harvard Law Review) lays out more reasons why CO_2 agreements are failing where Montreal succeeded. Sunstein argues that a contributor to failure is that the two worst emitters of CO_2, China and the USA, are not the countries that will suffer most of the damage from climate change.

9. Unrealistic Approaches

➤ Empty Promises

Countries have been making pledges to cut GHG emissions for decades, but on further investigation most of these appear to be primarily public relation exercises. Many are cloaked in complex and relative terms that allow countries to increase emissions. Here are a few examples. If the subject were not so serious, they would make for great chuckles in a comedy club.

China is the world's biggest emitter. It claims it will cut the CO_2 intensity of its GDP by 65% by 2030, compared to 2005 levels. What it does not say is that economies automatically tend to become more energy-efficient as they grow. Of course, CO_2 intensity is not CO_2 emissions. China's targets allow it to emit more over the next few decades than it does today.

The United States EPA proudly declared in 2017 that America was the world leader in emissions reduction. Yes, this is true, but this was also not difficult to achieve. The back story is that the United States was among the least energy-efficient countries in the world, so this reduction was easy and in its own economic interest.

The European Union has pledged a target of lowering emissions by 55% by 2030 compared to America's 52% cut. However, the EU's goal is calculated using 1990 as the base year, whereas America uses 2005. The kicker is that EU emissions had already fallen between 1990 and 2005, so it could do less than America.

Pakistan has pledged a cut of 20% by 2030. These cuts are compared with a pathway where it would take no climate action. That, of course, depends on what is meant by "business as usual." In this case, Pakistan means that emissions can triple this decade.

Brazil has pledged a reduction of 43% below base year 2005 levels by 2030. But the *The Economist* reports that tweaks to its accounting caused its emissions booked for 2005 to increase by one-third, from 2.1 $GtCO_2$ tonnes of CO_2e to 2.8 $GtCO_2$. As a result, Brazil can now emit one-third more in 2030, too. But let's not be too harsh on Brazil. Even if Brazil's emissions rise in line with its pledge, the 2030 carbon footprint of the average American will still be twice that of the average Brazilian.

As of August 2021, half of the world's 20 biggest polluters, accounting for four-fifths of global emissions, are emerging countries. The emerging countries have now pledged climate targets (good), but these targets explicitly provide for their emissions to grow over the next decade (bad).

It is easy to criticize politicians for gaming voluntary agreements with green lip service and public relations. But think about it. Politicians are elected when they can ride public sentiment. While the lights are working, politicians can garner votes with green promises. But make no mistake — if the reliability or provision were ever to fall short, or the cost became serious, public sentiment would shift. Do not expect politicians to follow through with any pledges they or their predecessors have made when the going gets tough. It's easy to "talk the talk." It's difficult to "walk the walk." In late 2021 and early 2022, the price of oil and gas doubled. Surprisingly, even politicians in progressive countries and states (like Scandinavia and California) began to promise subsidies. The jump in gas prices threatened their political support.

➤ Actions

Occasionally, we see climate activists attribute successes to treaties, such as the fact that three-quarters of all planned coal plants were scrapped after the Paris accord. This is like the rooster taking credit for making the sun rise. Most of these coal plants are no longer being built because they have become economically uncompetitive, not because of Paris.

Unfortunately, this may change again. In September 2021, European prices for gas and electricity rose dramatically and supplies for the winter were brought into question just as wind and solar output dipped. This produced a sharp backlash. The energy crisis buffeting the continent placed Boris Johnson and other European leaders in the difficult position of decrying fossil fuels while urgently prioritizing affordable access to them. Even climate-conscious Scandinavians demanded government energy subsidies.

In China, Vice Premier Han Zheng told state-owned energy companies to get hold of supplies at all costs. Government officials stated they were concerned that the squeeze in energy markets, surging prices, and the resurgence of coal would cast a long shadow over the 2021 climate negotiations.

After China was promising emission stabilization in 2020 (leading to a proportional reduction in fossil fuel use) — though always scheduled for as late as 2060 and thus always iffy — China has now largely reversed course. Together with Russia, China has been disengaging from climate negotiations

9. Unrealistic Approaches

starting with the Glasgow COP in 2021. Reuters is reporting that China has announced plans to build 43 new coal-fired power plants and 18 new coal-fired steel blast furnaces. Apparently, China has decided that the benefits in terms of economic development, energy security, jobs in the coal industry, and enhanced international competitiveness outweigh the benefits of reducing GHG emissions. For perspective, remember not only that China uses less energy per capita than the West, but also that China has one of the lowest electricity prices in the world for its industries. These low prices are part of the reasons why so many industries operate there, why demand for more power is so high, and why per-capita emissions are already higher than those in the West.

Nevertheless, one can see a ray of light in the Paris agreement: it set at least some world targets. But the fact is that the world has simply ignored them. The targets required annual reductions. Instead, fossil fuel consumption has marched on as before. Thus, the gap between targeted and actual emissions has been increasing every year.

Our Assessment

We end this section with a mixed assessment. As for us, we have little trust that countries — especially but not only developing poorer countries — will sacrifice their self-interest on behalf of a greater global good that will take decades to show results. We would love to be proven wrong in our skepticism. But there is room for disagreement here. Other experts remain more optimistic.

It is another valid question whether multilateral climate mitigation efforts are worth the time spent on them. On the one hand, negotiations can draw attention to the problem of global warming. They might increase domestic pressures on politicians to do *something*. On the other hand, negotiations pay for a lot of diplomats and consultants and make it appear as if the world is already doing something useful and that change is happening. By distracting everyone, have they prevented more useful steps? It's hard to say. Our view is that it is okay to rearrange the deck chairs on the Titanic, as long as it does not detract from the real rescue operations.

Dreaming of a world in which countries sign onto and enforce a global CO_2 tax is like dreaming of a world without military expenditures. It's not our world.

4 Why Corporate Solutions are Unrealistic

If country treaties don't seem to provide a path we can trust, what about corporate initiatives?

Trust here seems similarly misplaced. Corporations are designed to make money. They are not designed to set policies to combat climate change. It is the government's responsibility to do so. The task of dealing with climate change is a public interest issue, the kind for which governments have been created in the first place. It is the government that has to set the rules that make companies act in a socially responsible way. Abrogating its responsibility and hoping that companies will act against their self-interest is not realistic.

Meet Jason and Pam...they're currently in charge of our climate change research.

Associations

There appear to be some positive developments with industry groups. Many have jumped on the climate bandwagon with seeming gusto.

For example, the Global Financial Markets Association report created by the Boston Consulting Group (BCG) calls for a "globally consistent approach to sustainability reporting" and the "mandatory disclosure" of climate risks and opportunities by financial entities that would be in line with global standards set forth by the Financial Stability Board's Task Force on Climate-related Financial Disclosures. Even as finance and economics professors, we find it difficult to parse much meaning into such a vague and general mission statement.

Perhaps more consequential is that equity giant Morgan Stanley Capital International (MSCI) is about to create global warming ratings for 10,000 firms. Beyond the obvious — some industries have higher emissions than

9. Unrealistic Approaches

others — <u>corporate emissions</u> seem <u>nearly unmeasurable</u>. (What do you think your own warming contribution is?)

Yet there is worse. To our own surprise, these ratings do not mean what we thought they meant. <u>Businessweek</u> explains "there's virtually no connection between MSCI's 'better world' marketing and its methodology. That's because the ratings don't measure a company's impact on the Earth and society. In fact, they gauge the opposite: the potential impact of the world on the company and its shareholders. MSCI doesn't dispute this characterization. It defends its methodology as the most financially relevant for the companies it rates."

Most environmentalists (and many investors) probably will probably not understand this. Did you know what it means to buy a "green mutual fund" based on MSCI rankings? We finance professors did not.

Corporate association green goals will pay for a lot of accounting and consulting firm reports and ratings. They will create a cottage industry for high-priced consultants. And they will also make for great corporate public relations. However, they will not make much of a difference in the global concentration of CO_2 in the atmosphere.

Individual Companies

There are too many examples of corporate advertising campaigns about green commitment that seem disingenuous. Forgive us for not calling out any particular company — there are just too many, and singling out one seems unfair.

Some advocates of corporate responsibility have argued that companies are turning green because doing so will increase corporate value. If they are correct, activism is unnecessary (though it can't harm and may help remind and nudge executives in a greener direction). Why would companies not go green by themselves then? And if they are so socially-minded, why did they contribute to the climate problem in the first place?

Dreaming of a world where industry organizations and companies can tackle climate change is like dreaming of a world in which the grinning cat has not eaten the mouse. It's not our world.

5 Why Divestment Makes No Sense

Many activists, especially university faculty and students, are now advocating for divesting stocks from fossil-fuel companies. However, we are less opposed to these efforts as we are genuinely puzzled about them.

Is divestment intended to speed up the transition from fossil fuels to cleaner technology? In that case, wouldn't it be more useful for universities to invest resources into what they are best at — research and development into relevant clean energy?

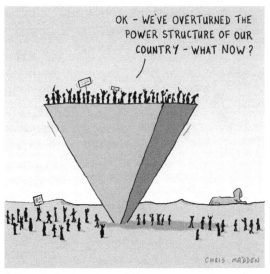

OK — we've overturned the power structure of our country — what now?

9. Unrealistic Approaches

Is divestment intended to get fossil fuel companies to invest more in green technologies? In that case, wouldn't it make more sense to retain the stock and pressure fossil-fuel management at the annual shareholder meetings into change (as has recently happened)?

Is divestment intended to lower the share prices of fossil-fuel companies? In that case, it won't work. Stock markets are so competitive that even a collective simultaneous divestment by all universities together would probably have a value impact of no more than that of a typical day's price change. Any individual university is even less important. And fossil fuel companies also no longer need to raise significant funds in the public market — they can finance their explorations from their revenues.

If this does not puzzle you, how about the following? There are many providers for ESG ratings of companies — but they do not overlap much. Ratings of "goodness" of corporations are more like ratings of best places to retire (that do not overlap greatly from magazine to magazine) than ratings of how good universities are (that do overlap greatly). This is perhaps not surprising. We cannot think of a way to objective measure ESG quality. Tell us — are pharmaceutical chemical manufacturers good or bad? Is Tesla a green company in China or a brown runs because the cars are charged from a coal fired electric grid? And these are the less controversial environmental judgments — now think about more controversial social causes, which are often highly politicized, too.

Again, we are simply puzzled by divestment campaigns. They could not possibly cause a meaningful change in the CO_2 concentration in the atmosphere. Divestment seems like the purest form of climate wellness — a moral stand delivering a warm glow and designed to make the activists feel better rather than an attempt to contribute, however small, to help reduce global warming? If we are wrong, please someone explain to us: how could it possibly work?[6]

Have Our views is what philosophers would call non-deontological. Kant and others were more conferenced with deontological ethics, a framework within which actions are judged to be

An alternative activist goal would be to focus not on divesting but on investing. For example, universities could invest more in clean-energy research, either via venture capitalist funds or their own engineering departments. They could also fund clean-energy projects explicitly at lower required hurdle rates, e.g., in sub-Saharan Africa. Promising cheaper financing to solar installers in Africa could induce larger companies to explore the viability of such businesses. However, we know of no coordinated efforts in this direction.

Dreaming of a world where divestment can tackle climate change is like hoping to change the U.S. government system by casting one invalid ballot.

6 Why Individual Solutions Are Doomed

Climate change is a really, really, really big and really, really, really slow problem. This is why it is so difficult for most people — and this includes researchers — to wrap their heads around it. We need to try to unwrap it. The key points are

1. Your footprint (or lack of it) does practically nothing to change world emissions. Changing your own carbon footprint is no more effective than prayer.
2. Your footprint does not influence enough others, either. Thus, even the indirect second effect of changes in your behavior do not matter.
3. The billions of people that would have to abide by them will not abide by them. (We wish they would, but they won't!)

Carbon footprint solutions are personal wellness, not Earth solutions. For Earth solutions, there are better ways to deploy environmentalist enthusiasm.

---

morally right or wrong, regardless of whether their consequences are productive or not. Ayako Yasudo has argued that ESG investing is deontological (but useless), while impact investing is non-deontological (but useful).

9. Unrealistic Approaches

Carbon Footprints

Small-scale solutions are wonderful *if* they can and will be widely scaled. When you replace one incandescent light bulb with an LED bulb, it is meaningless. But because replacements are in the self-interest of billions of people and LED use is spreading, the effective scale of the LED transition is not the same as the individual scale. Relighting the world produces meaningful change.

In contrast, most carbon footprint writings are best viewed as the wellness version of fighting climate change: They may make their readers feel better, but they have no real effect on climate change.

Two flights, a train, a truck ride, and we camp generating zero carbon footprints!

Even the New York Times is unimportant. When it publishes an article on How to reduce your carbon footprint for its mostly affluent liberal audience, it is meaningless. Even if this article succeeds in getting a few *tens of thousand* more New Yorkers start bicycling to work *forever* — which we doubt — it will make no meaningful difference to the world's global climate problem.

And what about *your* influence? How could *your* decision to bicycle have any influence in inducing billions of humans to bicycle instead of driving? If it can't, it will not make a difference.

It isn't just the New York Times that promotes climate wellness. There have been many bestsellers that have held forth about how to reduce one's carbon footprint. They sell many copies to the faithful. Their advice is rarely followed even by their readers (although we would love to see this happen). These books would be amusing distractions if only the issues were not so serious, if only the beliefs were not so widely held, and if only the diversions would not delay what really needs to be done.

With all the bestsellers, talk shows, and news features devoted to climate change, why haven't most people on the planet voluntarily changed their behavior? Do you believe that they will be convinced to do so by the next few bestsellers? Do you believe that people (not only in the Western world

6. Why Individual Solutions Are Doomed

but beyond) just don't realize how they could reduce their personal carbon footprints or how much they could help the environment?

If this were so, would it be a contribution to print or post 8 billion copies of the next big "personal-carbon footprint" advice book and hand them out freely to people all over the world? We suspect not. In fact, we suspect that it would cost more carbon emissions than it would save.

We think the problem is not ignorance. Most people already know how to change their ways for the global better. They just don't want to change. Economics suggests that not enough people will change long-term behavior if they don't find it in their self-interest. Ergo it is a fallacy to think that voluntary sacrifice could transform the world. (This also implies that they won't go green as long as dirty energy remains a lot cheaper than clean energy.)

Don't shoot the messenger. It's not our fault. We wish it were not so. In Laurence Fishburne's words in The Matrix, all we are offering is the red pill — the unpleasant truth, nothing more.

anecdote

> If it solved global warming, would you give up the TV remote and go back to carting your fat ass over to the television set every time you wanted to change the channel? If that was the case in America, I think Americans would watch one channel forever. — Bill Maher, Comedian. (We disagree. We think the world would warm up twice as fast.)

▶ A List from 2008 by David MacKay

The first question one needs to ask when considering any kinds of solutions is why they have not worked in the past. What has changed that may make them work in the future? This question is easy to answer for clean technology, for example: a number of technological breakthroughs have been and are continuing to reduce their cost. It is easy to answer for local fossil-fuel curtailment in India and China: the public is demanding cleaner local air and willing to pay the requisite cost for government mandates.

It is not easy to answer for voluntary altruistic behavior. Not much has changed with respect to appealing to more climate conscience (again, especially in non-OECD countries.) Even where voters might support forced mandates, they are unwilling to comply voluntarily without mandates.

9. Unrealistic Approaches

In his classic book <u>Sustainable Energy Without the Hot Air</u> published more than a decade ago, David MacKay recommended that individuals adopt the following good practices:

- Put on a woolly sweater in winter and turn down your thermostat (to 15°C or 17°C, say). Put individual thermostats on all radiators. Make sure the heating's off when no one's at home. Do the same at work.
- Read all your meters (gas, electricity, water) every week, and identify easy changes to reduce consumption (e.g., switching things off). Compare competitively with a friend. Read the meters at your place of work, too, thereby creating a perpetual live-energy audit.
- Stop flying. (It is superb irony — though ultimately irrelevant — when environmentalists take even a single airplane flight. Flying is *by far* the best way to quickly and dramatically increase one's carbon footprint.)
- Drive less, drive more slowly, drive more gently, carpool, use an electric car, join a car club, cycle, walk, use trains and buses.
- Keep using old gadgets (e.g., cell phones); don't replace them early.
- Change lights to fluorescent or LED.
- Don't buy clutter. Avoid packaging.
- Eat vegetarian six days out of seven.

None of these recommendations should come as a surprise. They are about as widely known as "eat less sugar and exercise more — it's good for you." They are like New Year's Resolutions or diet plans. Even if we could get every human to read MacKay's advice and pledge to follow it, when the excitement and commitment fades, the old behavior will return. Evanescence won't work for climate change. The necessary behavioral modifications must last multiple lifetimes, because even future generations will have to adopt them; and they must occur not just in the West — whose total carbon footprint is now smaller than China's — but all over the world.

Reading through the list today, the only one that has had widespread and lasting impact is switching lights — and it is not due to environmental aspirations. In the last ten years, technological improvements and mass production have made LEDs both cheaper and better than old incandescent lights. The latter have become obsolete. Of the entire list, only the one due to technological change has worked. Even the modest "sweater" recommendation has largely been a no-go. Comfort and self-interest come first for the world population at large.

6. Why Individual Solutions Are Doomed

Setting an Example

What about setting an example? There are a few select individuals who are so prominent that their actions may influence those of many others. If you are the Pope, the Archbishop of Canterbury, or the Orthodox Ecumenical Patriarch, your appeal may matter — though not in China and India, where their influence would be most needed now.

We can't think of many other individuals with much potential influence. Cristiano Ronaldo has 300 million followers on Instagram. If he put all his effort into it, how many fans could he convince to stop flying or driving? Can you think of other prominent personalities who could sway more than a few hundred-thousand people to change their lives permanently? Even if you can, a few hundred thousand people is only 0.001% of the world population — a drop on a hot stone. You still would need to convince the other 99.999%.

If you are like the rest of us, you are not that important. Frankly, the world does not care what your thermometer reads or whether you eat vegetarian or not or even whether you take another flight or not.

Of course, none of MacKay's recommendations are bad — most of them are outright healthy for you, too, and we can only encourage you to follow them. (And please exercise more, while you are at it.) Just don't think that your actions and examples (and those of your friends) will make a meaningful difference to reducing the CO_2 concentration in the planet's air. The statement "if everyone did it" is another logical fallacy. If everyone does it, it will not be because of you. They will do whatever they do, regardless of you.

What about carbon-shaming others? Fat chance. Social pressure can work in small groups. It does not work for the world. Even if you are the world's greatest carbon-shamer and you can convince everyone you will ever meet and your shaming is so contagious that it also changes everyone that your shamed will ever meet, *it does not matter*. The atmosphere does not care what you eat, where you fly, or what you drive. It only cares about what a billion people eat, fly, and drive.

9. Unrealistic Approaches

sidenote

Carbon-shaming is even more ridiculously unrealistic as a strategy for getting countries to cooperate on global treaties. Do you really think that any Indian prime minister will throttle the development of India because (s)he is being called out for emitting too much CO_2 that harms the rest of the world?

Prominent activists sometimes have to set examples for the sake of public relations. This can help defend their credibility against accusations of hypocrisy. But make no mistake — these are not actions that reduce global warming.

Greta Thunberg is the climate activist with the loudest megaphone in the world. In 2019, she crossed the Atlantic on a sailboat rather than an airplane. How can this choice induce millions to reduce their CO_2 footprints? How many gigatonnes of CO_2 will her trip ultimately reduce, directly or indirectly? A for intent. F for effectiveness.

Bill Gates buys "carbon" offsets for his private jet trips. Why are these purchases even linked to his trips? If it is globally worthwhile to remediate CO_2 in this way, he should be purchasing more credits. If it is not, he should be investing the money into something else that is more efficient — like his Breakthrough Energy Fund.

6. Why Individual Solutions Are Doomed

We need to reiterate what we stated at the outset of this chapter. We are condemning neither activism nor activists. We admire their intent. They are providing a positive externalities to others. They help keep politicians focused on the subject and nudge consumers and companies towards cleaner choices. It is just that we do not trust this sort of activism to make a big enough difference, especially among the 6+ billion people not living in OECD countries.

Local Coordinated Action

In terms of size, localities sit in between countries and individuals. Our own locality is Los Angeles in the state of California. Can local coordinated action reduce global carbon emissions?

Although it is true that changing the entire locality is much more effective than changing yourself, it is still ineffective on a global scale. (Combining it with "if everyone followed our example" does not work for the reasons already explained above.) Put differently, it is true that it is more effective to get everyone in a county of 100,000 to reduce 1% of their emissions than for your household to reduce 100% of its emissions. However, even 100,000 people are a drop in the bucket when it comes to atmospheric greenhouse gases and global warming.

We have a specific example from our own backyard that would merely be amusing if it were not so expensive. The University of California (UC) is a large institution, home to over 300,000 students, staff, and faculty. It is now engaged in a large-scale and expensive electrification effort with the declared goal of fighting global climate change. If this goal is the guideline for decision-making, then it is also a waste of UC resources.[7]

In fact, even if not just the University of California, but all of California spent its next few decades wringing sacrifices from its 40 million inhabitants on behalf of global climate change, it would barely be noticed on the global emissions spreadsheet. (California is responsible for about 1% of global emissions.) Worse, outward migration makes it impossible for a single U.S. state to succeed. Many of California's well-meant policies may already beginning to turn counterproductive in that they may be driving some industries to Texas

---

[7]Careful: our description of calling it waste applies only to the *extra* cost of early electrification over the alternative, not to the *total* cost of electrification.

9. Unrealistic Approaches

and abroad, where they will emit more and possibly invent less. Elon Musk, the most important individual climate changer of our time, has tweeted that he is joining the trek. He is moving to Texas.

Don't be deceived. We have already mentioned that even if the United States and Europe were to reduce fossil-fuel consumption to zero, the world would still be emitting about 30 $GtCO_2$ per year (instead of about 40 $GtCO_2$). The world's problem is now no longer primarily about 1,000 million people who live in the most developed countries, but about the 7,000 million other people. To make a difference, an effective world solution has to be in the self-interest of most of the world's population — or at least not dramatically contrary to their self-interest.

Of course, it would be different if the University of California's electrification were a pilot effort with the effect of speeding up all of California's efforts, which in turn would speed up the United States' efforts, which in turn would speed up the world's. However, this seems implausible in this context of switching its vehicle fleet to electricity. If and when larger domains (like the United States) decide to electrify, it will almost surely not be because UC did it first or made it easier to follow.

The true irony is that the University of California is almost uniquely positioned to make a climate-change impact. It is among the world's premier research and teaching institutions. It could help the world a lot more by investing its limited resources not into its own electrification infrastructure, but into clean-energy research and development, which could then be made available worldwide. This actually could make it cheaper for everyone to use clean energy. What a waste of energy!

> *joke*
> Even the Tottenham Hotspur soccer team is now getting into the game. This surely counts for a goal.

7 Is Climate Change About Social Justice?

Climate change often gets tied to other issues, such as social justice and inequality, but doing so risks derailing all of them and rendering spending decisions more inefficient. Change on any one issue is difficult enough. Changing everything at the same time may be impossible.

Blame

Perhaps you are still grappling with the question of who is to blame for the current problematic state of the world. There is plenty of blame to go around. It is easy to point the finger. Was the fault the accelerating industrial emissions of rich countries? Or was the fault the accelerating population growth in poor countries? (The answer is probably both.)

I think he's saying we should clean up our act...

We already mentioned that half of the world's 20 largest emitters of GHGs (which account for four-fifths of the global total) are now emerging countries. If they will find it in their interests to further grow their fossil fuel consumption, it will effectively spell the end of the global effort to curb human GHG emissions.

Do rich countries "owe" reparations[8] to poor countries? We are not moral authorities, so we won't take a stance. (Personally, we may agree.)

Yet we fear that even the discussion of climate reparations is counterproductive. It would mobilize a big segment of the population of rich countries to end their involvement in cross-country efforts to curb climate change. Few voters anywhere, involved in their own personal day-to-day struggles, will ever vote to pay more because of their parents' past contributions in having

---

[8]The only country that we know to have *ever* voluntarily paid significant sums in reparations for past misdeeds is Germany. We must commend it for this moral stance. If you know of other countries, please let us know.

9. Unrealistic Approaches

driven CO_2 levels up to 410 ppm — be it through past population growth or past industrialization.[9]

Okay! Your turn to take on the baton of climate change!

Who is more to blame is ultimately an irrelevant question. All that truly matters is that the planet now has a CO_2 level of 410 ppm, rising at a rate of about 2.5 ppm per year.

Consequently, there is only one real question that matters and it is the pragmatic one: What can and should be done about moving the needle now, within the limitations imposed by our political, technological, and economic realities?

Worthy Causes

➤ Inequality

Public concern for inequality is also becoming more important. Moreover, the IPCC has shifted its focus away from an exclusively physical analysis to social analysis. New <u>shared socioeconomic pathways</u> (SCPs) have even made global inequality a part of their forecasts. As economists, we know of a lot of good research that has *causally* linked more economic growth to more emissions as a first-order effect (as in the RCPs), but we know of none that have done so for inequality. Inequality is not a principal cause of global warming, although it does affect the distribution of economic harm. To be clear, we think it is plausible that inequality could play a role in increasing emissions — it's just that we don't see the kind of strong empirical evidence that would induce us to adopt the hypothesis.

Wealth redistribution is an important question beyond climate change. Allow us an economic digression. Inequality is philosophically subtle. Assume

---

[9]There is another problem. The poor people in rich countries are poorer than the rich people in poor countries. It is unlikely that the former would want to pay for the latter.

that there was just one government and the rich couldn't simply move away, so it is possible to tax them.

If it is difficult to raise the poor to reduce inequality, is it enough to diminish the rich? Is this in itself a worthy goal? Opinions thereon diverge and emotions (like fairness, despair, envy, or tribal belonging) often come into play.

Then there is the question of efficiency vs. redistribution. How much wealth should government be willing to transfer from the rich to the poor? What if it costs $2 in rich wealth to give $1 to the poor? $10? $100? Or the opposite, $0.50?

The United Nations General Assembly can be described as a large echo chamber, where like-minded parties can find comfort in expressing their misgivings peacefully though largely ineffectively. Thus, it matters little that the United Nation emissions gap report of 2020 states that rich people must cut their emissions by 97%.

It is true that the average household in the United States emits more than three times the amount of CO_2 that the average household in the rest of the world emits. It is also true that most poor people in rich countries emit less than rich people in poor countries. And it is true that bringing Western and Chinese populations back down to Sub-Saharan living standards would help reduce emissions (though not cure the global CO_2 problem by a long shot).

However, if the goal is to be effective and bring both the richer and more emission-intensive countries on board, then the goal must *not* be to have everyone live in equal-sized huts. The goal should instead be to reduce future emissions in the most effective way possible — ideally by taxing larger houses that use more cement and more heating/cooling.

➤ Gender

Similarly, gender issues have become more prominent on climate panels. Gender is of first-order importance in one climate-change-related aspect.

9. Unrealistic Approaches

Reproductive rights and gender equality in third-world countries help slow population growth and emissions. Yet this is controversial and more so in traditional societies. Many religions remain opposed to birth control.

However, some among the United Nations climate change body are now trying to go further and reframe climate change into unequal-harm terms, presumably with the intent of redistribution of resources towards groups it considers more deserving. But harm from climate change is not primarily a male vs. female issue. It's more an issue of poor vs. rich.

Meanwhile, the IPCC vice-chair Ko Barrett has laid out the new gender policy and implementation plan, which demands equal respect. Of course, there is nothing wrong (and everything right) about treating all genders with equal respect, but gender equality on panels is not a principal problem of climate change. A statement is fine. A principal focus is a distraction.

➤ All The Other Ills Of The World

The distracted focus of the ineffective United Nations often leads to more cynicism. Reuters reports that the United Nations now faces a shortfall of $100 trillion fighting poverty, inequality, injustice and climate change — more than world GDP. It is difficult to take the United Nations seriously when it comes to realistic approaches to addressing *any* problem.

But such mixing of activist causes is not limited to the United Nations. It occurs in the United States, too. Democratic lawmakers have suggested that climate change be fought as follows:

> *Supporters of the reimagined [Climate] corps said they intend to ensure diversity among workers and managers, as well as a $15 per hour wage and health care benefits. They envision climate corps workers installing solar panels, weatherizing buildings and providing water and other supplies during heat waves and storms.*
>
> *...[L]egislation introduced ... would require that at least half the members of a climate corps come from "under-resourced communities of need." In addition, at least half the investment would support projects in underserved communities, with at least 10 percent spent in Native American lands.*
>
> *– New York Times, 9/9/2021: A climate corps to build irrigation ditches?*

7. Is Climate Change About Social Justice?

In Chapter 6, we mentioned that economists are less in disagreement with other scientists about existing social ills than they are skeptical that a real-world government can and will solve them. Who in government exactly will decide which communities are most underserved and what tradeoffs between effective climate change reduction and poverty reduction are appropriate?

Fortunately, there are still a few climate-positive changes on which both U.S. parties can agree. The shining beacon here is the Replant Act of 2021 to plant a billion new trees. When James Hansen first testified about climate change, it was a solidly bipartisan issue. Sadly, most of it no longer is. How can this be reversed when foreign trolls fan the partisan fires on the Internet?

Conclusion

"We have to do something" is a common response to the analysis laid out in our chapter. However, this is an emotional appeal, not a rational argument. It is about inputs, not about outputs. It is an aspiration, not a course of action.

Fortunately, we have more to offer than just nihilistic skepticism. There are approaches that are not as uncompromising and high-minded as the proposals that rule the activist echo chambers today, but that do have realistic chances of reducing emissions. They are the subject of our next chapter.

Further Readings

Books

- Barrett, Scott, 2017, Why Cooperate?: The Incentive to Supply Global Public Goods. Oxford University Press, 2007.
- Dressler, Andrew and Edward Parson, 2019, The Science and Politics of Global Climate Change: A Guide to the Debate, 3rd ed. Cambridge University Press.
- Easterly, William, 2014, The Tyranny of Experts: Economists, Dictators, and the Forgotten Rights of the Poor, Basic Books.
- Lomborg, Bjorn, 2002, The Skeptical Environmentalist.[10] Cambridge University Press.
- MacKay, David J.C., 2009, Sustainable Energy — Without the hot air, UIT Publisher, Cambridge, England.

Reports and Academic Articles

- Nordhaus, William, 2018, Nobel-Prize Lecture, mentioning climate compacts.
- Sunstein, Cass, 2007, Of Montreal and Kyoto: A Tale of Two Protocols.
- Teoh, Siew-Hong et al., 1999, Socially Activist Investment Policies on the Financial Markets: Evidence from the South African Boycott.
- United Nations, 2021, Nationally Determined Contributions under the Paris Agreement.
- Vanner, Robin, 2006, Ex-post estimates of costs to business of EU environmental policies: A case study looking at Ozone Depleting Substances, Policy Studies Institute.

---

[10]At the turn of the millennium, Bjorn Lomborg's book The Skeptical Environmentalist was fiercely attacked for predicting that clean technology would make a big difference. In January 2002, Scientific American published a scathing 11-page critique that wreaked more strongly of hot politics than of cool science. Scientific American's tone (but not necessarily their counterarguments) seem inappropriate. Stephen Schneider wrote that "Lomborg asserts that over the next several decades new, improved solar machines and other renewable technologies will crowd fossil fuels off the market. This will be done so efficiently that the IPCC scenarios vastly overestimate the chance for major increases in carbon dioxide. How I wish this would turn out to be true! But wishes aren't analysis." In retrospect, Lomborg turned out to have been correct — and Schneider (who died in 2010) probably would have been happy to eat his words now.

Shorter Newspaper, Magazine Articles, and Clippings

- Albeck-Ripka, Livia, 2021, How To Reduce Your Carbon Footprint, The New York Times.
- Bloomberg: Air Pollution From Fossil Fuels Could Cut Lifespans by 2 to 5 Years, September 1, 2021.
- Carbonbrief, 2015, Tracking Paris Pledges.
- Carbonbrief, 2018, Shared Socioeconomic Pathways explainer.
- Climate Appeals by Medical Journal Editors, following 2015 and 2021 Appeals.
- Coan, K.E.D., 2021, How to reduce our carbon footprint, Ars Technica, 2021/11/09.
- Dyke, James, Robert Watson, and Wolfgang Knorr, 2021, Climate scientists: concept of net zero is a dangerous trap.
- Foley, Jonathan, 2021, Greenbiz, Why the world needs better climate pledges.
- Eos, Transactions, American Geophysical Union, 2021, Half the IPCC Scenarios To Limit Warming Don't Work.
- ISTJ Investor, 2020. Hydrogen vs. Natural Gas for Electric Power Generation.
- Pielke, Roger Jr., 2021, How to Understand the new IPCC Report discusses how politics and science meet.
- Stockholm Environment Institute, 2021 Production Gap Report.
- The Economist, August 7, 2021, How climate targets compare against a common baseline.
- Worland, Justin, 2021, At COP26, It's Domestic Politics, Stupid, Time Magazine.

Chapter 10

Realistic Approaches

In the previous chapter, we were both realistic and cynical about most current activism. We understand that this is unlikely to win us many friends among climate activists. We apologize. We think most activists are well-meaning. Their goals are noble. Their idealism is commendable. We share many of their goals, and we wish that many more of their proposals had better chances of success. Unfortunately, they do not.

In our view, the problem is that most individual and small group activism seems more like a wellness "feel-good-about-yourself" approach to climate change to us than like an effective approach to reducing the CO_2 concentration in the atmosphere. Traditional activism has not made much of a dent. Worse, there is no reason to believe it will be more successful in the future. Right now, most climate activism seems to us like rowing the boat in circles as it slowly drifts toward Niagara Falls.

We may not like it, but we have to live in the world as it is. And in this world, most of the public has little stomach for issues that play out on spans of decades. When climate issues have the public's attention, we think activists should exploit this attention as effectively as possible. This means doing so in ways that build broad and permanent coalitions that can hold up for decades. Otherwise, even victories on the latest divisive domestic issues of the day are only temporary, often only until the next administration and congress come in.

10. Realistic Approaches

1 Basic Requirements for Success

To make progress on reducing global greenhouse gases in the atmosphere, we posit two axioms that realistic solutions must satisfy:

1. The approach must not just work in the West, but throughout the world.
2. The approach must be in both the short-term and long-term self-interest of a good majority of decision-makers and voters in relevant countries.

Because there is a free-rider problem — making individual countries less eager to stem climate change than it would be in the collective interest — each country will only want to enact solutions that are both cheap and have great local benefits. It is a corollary that solutions that fight against economic fundamentals are too expensive to be adopted on a broad basis.

This leaves two promising means to slow climate change. They are the subject of the remainder of this chapter and much of the book:

1. **Economics and Technology:** Fossil fuels are expensive to mine, to ship, and to process. They are also becoming more scarce. This is their Achilles Heel. Meanwhile, clean technology is becoming better and cheaper every year. This progress can be accelerated and coordinated. If clean energy becomes cheap enough, it will be adopted worldwide, independent of ideology. No one had to force people to buy smart phones when they became available.

2. **Local pollution taxes:** Fossil fuels emit not only global pollution, but also harmful local pollution. Unlike CO_2, the resulting smog is locally visible enough to stay on the minds of voters. It also reminds them of numerous serious health problems tied to local pollution. As Chinese, Indians, and Africans become wealthier, they are demanding cleaner air *and* they are becoming willing to pay for fossil-fuel reductions. This desire needs to be organized and channeled.

Both forces operate even in the absence of environmentalism, and we would argue that they are responsible for much of the real progress that the world has made to date. Even the IPCC has now backed away from its earlier most pessimistic forecasts (RCP 8). This is largely due to technological progress that is just starting to revolutionize the energy sector.

1. Basic Requirements for Success

The fact that progress is already happening does not mean that governments and citizens should now sit back, relax, and enjoy the ride. Yes, fossil fuels will eventually fade away even without activism, but the process will be far slower than it should be. Some specific implementations will relatively easily win public approval in the United States *and beyond* (like support for clean air and clean-tech R&D or for a better electric grid with charging stations for electric transportation); others will be more difficult and controversial (like the reduction of fossil-fuel subsidies, the closing of coal plants, and the curbing of population growth).

By focusing laser-sharp on clean technology progress and locally-justified CO_2 taxes, green activism can win over a majority of voters and politicians. We warn against bringing in divisive arguments regardless of whether they are correct. For example, even if the United Nations can back up global warming activism with scientifically valid analysis, their inclusion in a debate will only raise emotions and distract from achieving what is nationally achievable.

Adaptation

An important word of caution: Our book does not discuss climate-change **adaptation**. This is not because we consider adaptation to be unimportant. On the contrary — it is of great importance. The world is almost certain to warm by 2–3°C in any event and adaptation will greatly reduce the associated damages.

Yet adaptation is usually in the self-interest of local decision-makers. As such, the world does not face the same public goods problem as it does with respect of mitigation of fossil-fuel emissions. We do not discuss the issue further, because it goes beyond the declared purpose of our book — moving the needle on global climate change. For more information on adaptation, Robert Pindyck from MIT has recently written a book that focuses more on this important subject.

2 Enact Local Fossil-Fuel Taxes!

In Chapter 9, we dismissed the idea of a *global* CO_2 tax. We also explained why countries will only do what is in their own self-interest. Fortunately, there are good reasons why a *local* CO_2 tax can be in their interest.

The U.S. Economists' Statement

Economists are not (always) as bad as their reputation would have it. We have already mentioned a public statement that over 3,000 prominent economists have signed. It advocates the imposition of a robust CO_2 tax in the United States. The statement reads, "By correcting a well-known market failure,[1] a carbon [dioxide] tax will send a powerful price signal that harnesses the invisible hand of the marketplace to steer economic actors towards a low-carbon [dioxide] future." Does our earlier stance in this chapter not place us at odds with the statement (that we also signed)? Actually, no.

One answer is that we would also sign a petition for peace on earth. We just don't think it will happen or that humanity should count on it. We still believe that there will be no global CO_2 tax that is against the self-interest of countries. This is especially the case for China, India, and many developing countries, representing the more than 6 billion people that already emit the majority of the world's emissions and are on course to emit much more in the future.

Yet the economists' statement is explicitly about a U.S. CO_2 tax and not about a global CO_2 tax. Thus, we need to explain why we signed and why we do not believe that it would harm the US.

---

[1] We explained what a market failure is and how a tax remedies it in Chapter 6, where we tried to turn our readers into economists.

2. Enact Local Fossil-Fuel Taxes!

Competitiveness and Evasion

The Economists' Statement acknowledges that industries can escape to more favorable locales where they are taxed and regulated less. The statement explicitly cautions that tax policy must take this into consideration. We are perhaps a little more concerned about this than our colleagues. But economists agree: the specifics must take into account that CO_2 regulations could place the United States at a competitive disadvantage.

Industry migration is not an overblown ivory-tower concern. Greenhouse gases know no national borders. They are truly global. When they are emitted in China, it is just as bad for Americans as when they are emitted in the United States. Because firms and industries are self-interested, policy-makers better have a healthy fear of unintended consequences.

Firms and industries do not have to move themselves. It is enough if similar businesses spring up in other countries and domestic businesses close up shop. This flexibility of industries to evade regulation is less of a short-term than a long-term problem. It takes a while to leave, which is why declaring local victory early is often misleading. Industrial plants often last for decades. When new regulations are enacted that raise the price of power, they may be unavoidable for already existing plants that can no longer move. But the next round of plants will be built where the plants will be more competitive. And this may well be by different companies in different locations.

The most familiar case of the effect of industrial migration is the erosion of the U.S. manufacturing base over the last half century. It has caused great economic harm in what is now called the rust-belt. Many companies have closed their domestic plants and outsourced their manufacturing to China or Vietnam — where coal-based electricity has often cost half of what it has cost in the United States ($85/MWh in China vs $150/MWh in the US). We may not mourn that Bitcoin-mining — where electricity is the most important input — has largely moved abroad, but we do mourn the loss of the U.S.'s advanced manufacturing supply chains.

We made three billion dollars mining Bitcoin, minus our electricity bill—that comes to $1.61.

10. Realistic Approaches

(And, due to the 2021–2022 logistics crisis, we are now noticing other negative effects of this migration.)

From the perspective of climate change activism, this manufacturing relocation has also not helped global warming. It has reduced U.S. emissions but not global emissions. In fact, to the extent that foreign manufacturing is less efficient, global emissions may have risen.

Local Fossil-Fuel Pollution

Yet, even with our fear that industries may relocate, we still *remain* in favor of local CO_2 taxes. Our primary rationale is *co-pollution*. Co-pollution consists of non-CO_2 forms of pollution that are released when fossil fuels are burned. It is why many peoples (and not just left-leaning intellectuals) are opposed to coal and oil. Voters and politicians, especially in high-density locales, are not revolting against today's emissions of CO_2 or other global GHG gases that disperse over the planet, cause no direct harm to human health, and will contribute to global warming in about a generation or two. Instead, they are revolting against the fossil-fuel byproducts that are *not* diffusing over the entire planet but instead remain local and have immediate harmful effects on them and their children.

These local byproducts have made Delhi, Beijing, or Lagos hazardous without a mask for much of the year. (Not a rank to be proud of.) Large smoke and soot particles are known to kill. Even better, the visibility of these particles helps keep them constantly on voters' minds. And even better yet, sometimes they clear out and remind voters how bad their living environments have become.

The effects are so large that they are worth expounding. As one example, Dr. Arvind Kumar reported to the Economist that when he started working as a chest surgeon in Delhi 30 years ago, nine-tenths of lung cancer patients were smokers and nearly all were men over 50. Now half of his patients do not smoke, 40% are women, and the average age is a decade younger. He regularly sees children with blackened lungs. This has lead Dr. Kumar to conclude "The urgent issue we need to face is not CO2. It is about our own health and the health of the next generation."

2. Enact Local Fossil-Fuel Taxes!

The most prominent harmful co-pollutants are the smallest particulates of fossil-fuel exhaust. PM 10 pollution consists of fine particles that are generally smaller than 10 micrometers, about 10% the diameter of a human hair. They can be inhaled and can accumulate in the body, often in the lungs. In turn, their smallest constituents (PM 2.5) can transfer directly into the bloodstream. It is not an overstatement to call them "murder." Internet sites make it easy even for the poor to look up their local current PM 2.5 levels in real time.

Smog is less harmful but more visible. It is caused by volatile organic compounds and nitrogen oxides, which combine into ground-level ozone.

> *anecdote*
>
> Sure, we can do something about climate change now, but if we find out in 50 years that the researchers made a mistake and that climate change doesn't exist, we would have improved air quality in all major cities, gotten rid of noisy and smelly cars, cleaned up toxic rivers, and destroyed dictatorships funded on money from oil for no reason.
> — Climate-Change Jokes.

Although the health costs of fossil fuels are known to be large, it is difficult to come by precise estimates. Drew Shindell (Duke and IPCC) claims that fossil fuels lead to 250,000 worldwide premature deaths per year. Over 40 years, removing fossil fuels (leading simultaneously to decarbonization) would save around 1.4 million lives. Based on a reasonable value of life, decarbonization could thus save about $0.7 trillion per year, or about $2,000 per person. This is equivalent to approximately $100/tCO$_2$ emitted (although it is *not* the CO$_2$ that is at fault).[2]

---

[2]We are skeptical about higher estimates that add in global climate-change attributed deaths. For example, on the one hand, a UCL study estimates as many as 8.7 million premature deaths in 2018 alone. On the other hand, and showing how difficult attributing deaths to climate-change is, the UK has recently estimated that climate change has saved lives in the last 20 years.

10. Realistic Approaches

With regard to India, the Economist reports that a conservative estimate of lost productivity due to local pollution is $36.8 billion in addition to $11.9 spent on treating illnesses caused by pollution. The sum is equal to 1.8% of Indian GDP — about the same order of magnitude as the cost of converting from dirty fossil-fuel to clean energy.

The relevant cost to American voters is primarily their own personal harm. Scientists have estimated the average U.S. health costs to be around $50/tCO_2$, ranging from about $10/tCO_2$ in Arizona to $100/tCO_2$ in New Jersey. Beyond the health aspects, there are other more aesthetic aspects that many Americans are now rich enough to demand — less noise, smog, and smell; a cleaner environment; etc. It is a valid question whether the harmful local non-GHG externalities from fossil fuels justify a tax of $20/tCO_2$, $50/tCO_2$, or $100/tCO_2$. But it is clear that these health and quality-of-life costs are not small.

How far are these estimates from the effective fossil fuel tax in the United States today? Very far! The current tax is *negative*. Direct producer benefits of fossil-fuel government support alone are estimated to be as high as $62 billion/year (about $10/tCO_2$). A more complete estimate that includes subsidies for exploration, cheap land leases, and non-insistence of cleaning up all spills is between $10/tCO_2$ and $30/tCO_2$. Thus the gap between what *locally optimal* fossil-fuel taxes are vs. what they should be ranges from about $30/tCO_2$ on the low end to $100/tCO_2$ on the high end. Climate change and environmental activism of the collective kind can and should play an important role here, keeping the public's attention on the issues and thereby holding politicians' feet to the fire. It is important to remove fossil-fuel subsidies of all kinds and as quickly as possible. (If need be, cooperating fossil fuel companies could even receive a one-time payment in exchange.)

This is why we strongly support a local tax on fossil fuels, even if no other country adopted one. In the United States (and much of the Western world), the related costs justify an *immediate* tax on fossil fuel on the order of $20-$40/tCO_2$. Imposing such a tax will not reduce but increase the welfare of the United States.

So who would oppose a local tax on fossil fuels? As we explained in Chapter 6, there are many. First, there is the differential effect. Low-density states like Arizona and Nevada may be against it. Wyoming is the top producer of coal in the United States, and it is not even burnt near Cheyenne. Even

2. Enact Local Fossil-Fuel Taxes!

when coal is burned locally, with more area to disperse the pollutants, their population suffers less harm. Thus, from their local perspective, voting against fossil fuel taxes makes sense. Second, there are producer interests. There are still a lot of people working in the coal sector in West Virginia, in the oil sector in Texas, and in the natural gas sector in Pennsylvania. They vote their livelihoods, and it is cold comfort to them if the clean energy transition creates jobs elsewhere in the United States. Third, there are, of course, corporate interests — though the ultimate beneficiaries are not as obvious. Most shares in fossil-fuel companies are held by pension funds — i.e., you — and not by rich villains. And, fourth, there are the politicians. They rely on campaign contributions from lobbies, and energy companies are among the largest contributors there are.

We believe that with preparation, a tax between $20/tCO_2$ and $40/tCO_2$ stands a good chance of social acceptance as long as (1) the public conversation remains centered around residual local harmful health effects, and (2) there are cross-subsidies from tax beneficiaries to those harmed by such a tax — including not only to consumers and employees, but also to fossil-fuel states and companies that sign on.

Importantly, even a net tax as low $20/tCO_2$ tax is probably enough to immediately phase out coal in all but the most extreme cases.[3] A fossil-fuel tax of $20/tCO_2$ may not be enough to solve the emissions problem — not in the United States, not in the world — but it would start moving the needle and it is achievable with proper political finesse.

Like most integrated assessment models discussed in chapter 7, the Economists' Statement recommends an *increasing* CO_2 tax. We are not opposed to it, but we do not think the increases should be on the immediate agenda. First, even with a $20–40/tCO_2$ tax, the transition to clean energy will proceed much more quickly and a lot of fossil-fuel use will disappear by itself. Clean energy is improving; and once fossil fuels lose and clean alternatives gain more economies of scale and network benefits, the decline of fossil fuels will accelerate. Market forces are powerful. Second, once a fossil fuel (like coal) is no longer economically viable, it does not matter whether a higher tax makes it even less viable. Deader than dead is still dead. Third, our concern is about

---

[3] It would also put oil on an even more tenuous basis in its prime use, transportation. However, Americans are very sensitive to higher gas prices. It may be better for political reasons to let the electric-vehicle revolution take hold before imposing the tax on oil.

10. REALISTIC APPROACHES

moving the needle now. Thus, for pragmatic reasons, we really do not want voters to get caught up in complex arguments about future sacrifices and whether 80% is or is not good enough. We would rather focus the discussion on voters' immediate self-interests now and try to placate the opponents.

Of course, we also agree that the world would be better off if the fossil fuel taxes were even higher in order to take account of global warming. However, we fear that too high a tax would erode public support when it is most needed — *now!* Again, a modest net $20/tCO_2$ tax (about $35/tCO_2$ higher than it is today) would be enormous progress and eliminate much of the American contribution to the global warming problem.

The basic problems are much the same in the rest of the world as they are in the United States. A similar dynamic is playing out elsewhere. The IMF estimates that worldwide fossil fuel subsidies amount to about $15/tCO_2$ on average. The first activist step should be to organize locally to put an end to all fossil-fuel subsidies. This would not only deliver significant public savings but would also lower emissions. The tax revenues (or lower subsidy expenses) should be named and visibly bundled with an "energy subsidy" payment to those poor who are most affected by higher fuel costs.

Even China could do this. Coal is usually burned *near* high-density population centers, because electric transmission is lossy and expensive. This "fortunately" makes dirty fossil fuels much more harmful. In China, residents of large cities are on average now not only wealthy enough, but are also so encumbered by daily smog, that $30/tCO_2$ would probably win public support and pay for itself in health cost savings almost immediately. (China still faces two key problem in getting off the fossil fuel train. The first is not the economics of clean vs. dirty energy, but the employment in the coal sector. It may require large-scale subsidies to retrain coal workers. The second is its urgent need for more energy and its lack of time. Coal plants are faster to construct than nuclear power plants.)

Unfortunately, in some poor and non-dense countries, especially in Sub-Saharan Africa, parents have to worry more about providing basic sustenance for their children than about the long-term health effects of emissions. For them, local pollution concerns are probably not a viable reason for reducing fossil fuel use. In the Congo, where health expenditures per person are as low as $19/person/year, CO_2 taxes of $50/tCO_2$ are unrealistic. Realistically, only lower prices will sway them towards cleaner energy.

2. Enact Local Fossil-Fuel Taxes!

▶ Clean Fossil Fuels

Technologies that reduce or filter out more of the harmful copollutants of fossil fuels present a dilemma. Many people oppose fossil fuels not primarily for their global warming effects but for their local environmental effects. Remove the latter and the public cares even less about the former. There is a certain irony that visible local pollution is the best ally in the effort to reduce invisible global pollution. If all fossil-fuel emissions were as invisible as natural gas (which is however not particularly clean, once end-to-end leaks are accounted for), it would be more difficult to rally public support against them.

Pulverized coal plants present the clearest dilemma. They do burn cleaner. But they also reduce the local urgency to eliminate coal plants altogether.[4] And once constructed, the world probably has to live with their emissions for 30–50 years. China is building 250 GW of newer coal plants (about 1/4 of the total U.S. electricity generation) as we write this.

For a similar reason, climate-change adaptation is also a double-edged sword. It provides a way to reduce the local harm and thereby the desire to eliminate the global externality. Shoreline dwellers who can move inland care less about rising oceans. Residents who have installed air-conditioning care less about hotter climates. But their adaptations also reduce their incentives to reduce their fossil fuel emissions on behalf of the rest of the world.

In the end, even more so than local fossil fuel taxes that raise the price of dirty energy, the solution will have to be technologies that lower the price of clean energy.

anecdote

> **Clean coal is a bit like wearing a porous condom — at least the intention was there.** — Robin Williams, comedian.

---

[4]There is another form of clean coal that promises to capture CO_2 at the chimney and inject it under pressure deep underground. However, this technology is still too expensive, and likely will never become economical. Worse, simply "forgetting" to do all maintenance, resulting in less capture, will magically make such plants more profitable producing a conflict of interest.

283

3 Promote Technological Change!

Ultimately, humanity can wean itself off fossil fuels only by advancing clean technologies to the point where they can compete with fossil fuels on an economic basis. Technology is the only truly globally scalable solution. In much of the remainder of the book, therefore, we turn to issues related to the research, development, and deployment of clean-energy technologies.

We are not so optimistic to believe that vested fossil-fuel interests won't be able to delay the clean-energy transition. However, we believe that they will be fighting a rear-guard action once clean technology will become cheap enough. We also believe that watchful environmental activism will serve a useful role in helping politicians resist their interests. Even fossil-fuel interests will eventually prefer to jump on the bandwagon rather than be rolled over by the train. (It's already beginning.)

Pie in the Sky?

Dominant clean technology is not pie in the sky. The technological progress over the last decade has been stunning. The most valuable energy companies today are already no longer Exxon and Chevron, but solar and wind producers like Nextera Energy. The most valuable car company, by far, is no longer Toyota, but Tesla.

The cheapest source of useful energy *in the history of humanity* is today's wind and solar power. When their power is available, no fossil fuel can match its cost. In a cosmic sense, this is not a surprise. Solar and wind plants can generate electricity without the need to mine and transport fossil fuels. And when (not if) the utility-scale electricity storage problem will be solved (Chapter 12), clean electricity will also become cheaper than the lowest-cost fossil fuel electricity on a 24/7 basis.

Realistically, we expect it will take two decades to invent and refine clean technologies, and another two decades to replace the existing fossil fuel infrastructure. This is because already-built fossil-fuel plants can still produce cheap electricity on the margin (Chapter 6). Thus, it often makes economic sense to keep running them. But as pieces of the fossil fuel infrastructure age out, they will be replaced by cleaner, better alternatives. It is at the top of our wish list to stop the building of new coal plants asap — unfortunately, it is not on China's or India's wish lists.

3. Promote Technological Change!

We are less worried that humanity has so far not installed a lot of clean energy and especially energy storage (at least compared to new fossil fuel plants). Batteries that used to cost $1,000 ten years ago now cost $100. Solar power that used to cost $250/MWh now costs under $30/MWh. Wind power that used to cost $150/MWh now costs $30/MWh. When building green power plants required large subsidies, each one was as painful to pull off as pulling teeth. Today, for the same price that 1 unit would have cost ten years ago, entrepreneurs can profitably install 10, 20, or even 30 units. It could even be that humanity should optimally *not* deploy more batteries and other green technologies for another few years. They could be so much cheaper that entrepreneurs could install 100, 2,000, or 30,000 units instead at the same cost. But these are decisions we can leave to the expert entrepreneurs.

Setting an Example

In Chapter 9, we dismissed the idea that individuals or countries can systematically induce large numbers of others to follow by "setting examples" or "shaming" them. When you bicycle to work, it won't make 1.4 billion Indians more likely to bicycle. They will bicycle more only if they find it in their own interests to do so, regardless of your good example.

However, example-setting can work with technological solutions — though not by moral suasion. A country can lead the way by trail-blazing the adoption of clean technologies, which drives down the early adoption costs. (We called them FOAK – first-of-a-kind – in Chapter 6.) Once the technology cost has fallen enough, other inventors and adopters will more likely follow, because technological imitation is typically much cheaper than invention — if only because imitators can see what has failed and what has worked.

The best historical example of this dynamic were German subsidies for the deployment of wind power. They were expensive. In 2020, the average German consumer paid $370/MWh for electricity, while the average American paid $150/MWh, and the average Chinese and Indian paid $90/MWh. Germans paid dearly for the privilege of being the first mover in the large-scale adoption of wind turbines — the equivalent of a carbon tax of about $200/tCO$_2$.

The main beneficiary of German policies was not Germany but the world. And the main benefit for the world was not the reduction in German emissions. The benefit was that Germany shouldered much of the initial development

10. Realistic Approaches

cost that drove wind power to where it is today: the lowest-cost source of electric power in the world! Germany's initial costly widespread adoption of wind has done and will continue to do a lot more for humanity than just a puny 0.1 $GtCO_2$ reduction of its own emissions. Unfortunately for Germany, the largest producer of wind turbines today is *not* Germany. Germany is a player, but the main manufacturers now sit in China, the United States, Denmark, and Spain.

We have stated before that the only viable interventions are those that are cheap enough to be economically viable. Government support for R&D is among them. Bjorn Lomborg estimates that every $1 spent subsidizing investment in clean technology produces $11 of benefits. Although his estimate may be on the high side, it is likely in the ballpark. With the OECD responsible for about half of world GDP, it's worth it for us.

Deployment

In addition to subsidizing research, it is also in the interest of governments to help the deployment of new technologies. For instance, there are coordination problems. The incentive to build new solar farms depends on development of the grid. Electric cars require charging infrastructure, and so on. In addition, outdated regulations have to be updated to ease the transition to clean energy. Although they will have to overcome some entrenched lobbies, it is overall in the interest of politicians and individual country governments to help solve these problems.

4 Recommendable Activism

To summarize, here are our two recommendations for climate activists that want to make the world a better place. They should focus on messaging the following:

1. *Clean energy technology is in your own interest. It will create well-paying jobs and prosperity, for you and for your children.*

   In detail, we need our government to improve and subsidize the relevant research, development, and deployment of clean energy technology. A reasonable tax on fossil fuels is another effective step in the right direction. We need not only more but also better government support and coordination. Do you want foreign countries to eat America's lunch?

2. *Reducing fossil fuel use is in your own interest. They are harming your health today.*

   In detail, the negative health effects of fossil fuels can be similar to those of cigarettes, except you personally do not have the option to quit. Fossil fuels could even be killing your children and elderly parents right now. What price would you put on their health?

The two messages work together. A fossil-fuel tax can provide more incentives for the development of clean energy technologies. Clean energy makes it more affordable to live with a fossil-fuel tax.

We would further suggest that other issues — perhaps even global warming itself — should be mentioned only in passing. The less talk there is about international institutions, the better. Initiatives focusing on approaches that cannot work on global scales to reduce the CO_2 concentration in the atmosphere — such as corporate responsibility, divestment, individual efforts, carbon footprints, shaming, or social justice — should be supported only if they do not needlessly antagonize and distract attention from initiatives that will work.

Now that you have heard us out, feel free to disagree.

10. Readings

Further Readings

Books

- Barrett, Scott, 2017, Why Cooperate?: The Incentive to Supply Global Public Goods. Oxford University Press, 2007.
- Dressler, Andrew and Edward Parson, 2019, The Science and Politics of Global Climate Change: A Guide to the Debate, 3rd ed. Cambridge University Press.
- Easterly, William, 2014, The Tyranny of Experts: Economists, Dictators, and the Forgotten Rights of the Poor, Basic Books.
- Lomborg, Bjorn, 2002, The Skeptical Environmentalist.[5] Cambridge University Press.
- MacKay, David J.C., 2009, Sustainable Energy — Without the hot air, UIT Publisher, Cambridge, England.
- Pindyck, Robert S., Climate Future: Averting and Adapting to Climate Change, 2022, Oxford University Press.

Reports and Academic Articles

- Dominici, Francesca, Michael Greenstone, and Cass Sunstein, 2014, Particulate Matter Matters, Science, This article explains how different methodologies can be used to assess the health costs of particle pollution.
- Nordhaus, William, 2018, Nobel-Prize Lecture, mentioning climate compacts.
- Sunstein, Cass, 2007, Of Montreal and Kyoto: A Tale of Two Protocols.
- Teoh, Siew-Hong et al., 1999, Socially Activist Investment Policies on the Financial Markets: Evidence from the South African Boycott.
- United Nations, 2021, Nationally Determined Contributions under the Paris Agreement.
- Vanner, Robin, 2006, Ex-post estimates of costs to business of EU environmental policies: A case study looking at Ozone Depleting Substances, Policy Studies Institute.

---

[5]At the turn of the millennium, Bjorn Lomborg's book The Skeptical Environmentalist was fiercely attacked for predicting that clean technology would make a big difference. In January 2002, Scientific American published a scathing 11-page critique that wreaked more strongly of hot politics than of cool science. Scientific American's tone (but not necessarily their counterarguments) seem inappropriate. Stephen Schneider wrote that "Lomborg asserts that over the next several decades new, improved solar machines and other renewable technologies will crowd fossil fuels off the market. This will be done so efficiently that the IPCC scenarios vastly overestimate the chance for major increases in carbon dioxide. How I wish this would turn out to be true! But wishes aren't analysis." In retrospect, Lomborg turned out to have been correct — and Schneider (who died in 2010) probably would have been happy to eat his words now.

Shorter Newspaper, Magazine Articles, and Clippings

- Albeck-Ripka, Livia, 2021, How To Reduce Your Carbon Footprint, The New York Times.
- Bloomberg: Air Pollution From Fossil Fuels Could Cut Lifespans by 2 to 5 Years, September 1, 2021.
- Carbonbrief, 2015, Tracking Paris Pledges.
- Carbonbrief, 2018, Shared Socioeconomic Pathways explainer.
- Climate Appeals by Medical Journal Editors, following 2015 and 2021 Appeals.
- Coan, K.E.D., 2021, Lower your footprint — How to reduce our carbon footprint, Ars Technica.
- de Chant, Tim, Fossil fuel combustion kills more than 1 million people every year, study says, Ars Technica, 2021/12/16.
- Dyke, James, Robert Watson, and Wolfgang Knorr, 2021, Climate scientists: concept of net zero is a dangerous trap.
- Foley, Jonathan, 2021, Greenbiz, Why the world needs better climate pledges.
- Eos, Transactions, American Geophysical Union, 2021, Half the IPCC Scenarios To Limit Warming Don't Work.
- ISTJ Investor, 2020. Hydrogen vs. Natural Gas for Electric Power Generation.
- Pielke, Roger Jr., 2021, How to Understand the new IPCC Report discusses how politics and science meet.
- Stockholm Environment Institute, 2021 Production Gap Report.
- The Economist, August 7, 2021, How climate targets compare against a common baseline.
- Timmer, John, US government sees renewables passing natural gas in 20 years, Ars Technica, 2020/01/30.
- Worland, Justin, 2021, At COP26, It's Domestic Politics, Stupid, Time Magazine.

Part

The Technology Problem

Chapter 11

Leaving Fossil Fuels

Even after record years for green energy, the world today still runs approximately 85% on fossil fuels. We shall thus start the technology part of our book with an introduction to the advantages and disadvantages of coal, oil, and natural gas vis-a-vis three important clean alternatives: hydrogen, nuclear power, and batteries (most likely charged by wind and solar farms). We will also caution to keep a cool head when it comes to taking in all the propaganda. This caveat applies both to clean-energy objections from the fossil-fuel side and overly exuberant and unrealistic technology forecasts from the clean-technology side.

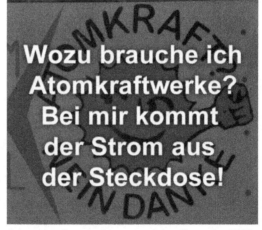

A popular German bumper sticker from the 1980s: "For what do I need nuclear power plants? At my place, electricity comes out of the socket!" Beneath: "Nuclear Power, No Thanks."

11. Leaving Fossil Fuels

1 Ongoing Growth

The scale of the energy transition poses huge challenges. If humanity simply wanted not to increase fossil-fuel consumption further — too modest a goal — then clean energy will "only" have to cover the future increases in energy needs. Nevertheless, as we explained in Chapter 2, where we also mentioned that nameplate power and fossil-fuel inefficiency in the conversion to electricity are roughly similar, this still means that clean energy will need to grow by a factor of 15 *within the next 30 years*. In numerical terms, that translates into an increase from a measly 8 PWh to 117 PWh, with about 110 PWh just to prevent increases in fossil fuel use. If humanity also wants to retire coal, it will require growth of 150 PWh, a factor of 20. If the goal is to retire all oil, gas, and coal, it will require 240 PWh, a factor of 30. The latter two numbers represent real growth rates between 9% and 12% per year. Think about what it would take to increase anything by a factor of 30 — say, your income or bank account.

Right now, clean energy is growing by about 15% per year — but it's easy to grow from a low base. Growing at 12% per year consistently over thirty years is not impossible, but it's a tall order. Doing it much faster seems both economically and physically impossible.

2 Fossil Fuel Advantages

Fossil fuels have taken over the world because they have some important advantages. You should understand them first. Besides the sheer challenge of replacing such large amounts of energy, it is also the case that fossil fuels are different from clean replacements — their energy is better in some respects, worse in others.

The most important advantages of fossil fuels are:

1. An existing infrastructure to collect fossil fuels, send them to the desired destination, and use them efficiently (especially in highly-developed countries). This includes wells, storage facilities, refineries, distribution networks, and devices that run on fossil fuels, such as cars, ships, trucks, planes, heaters, factories, and furnaces.

2. Abundant availability at low cost, even after figuring in logistics costs.

2. Fossil Fuel Advantages

3. Low shipping costs for gas (pipelines) and oil (tankers), but not for coal.
4. A high energy density for oil and gas, both by weight and volume. This is important not only for shipping, but also when used to power engines. Airplanes, cars, and trucks simply perform better when light and small.
5. A good safety record, with limited potential for disasters.
6. Near-perfect efficiency when burned for heat. However, fossil fuels have poor efficiency when generating electricity or producing kinetic energy.

These characteristics need to be judged relative to those of the clean alternatives available — principally hydrogen, batteries (charged from wind and solar), and nuclear energy.

We will discuss technologies in more detail in the next chapters, but for now, let's outline how clean energy differs from fossil fuels.

Figure 11.1. Flows of Energy Through U.S. Economy (ca 2020).

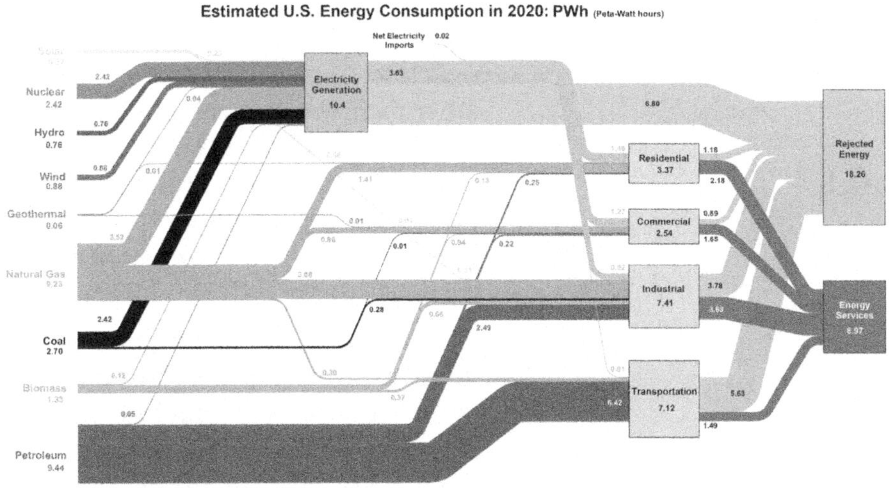

Note: Two-thirds of all energy ends up as waste heat.
Source: LLNL Flowcharts 2020.

Figure 11.1 is a beautiful visual representation of U.S. input and output energy flows. The far left side lines up primary input energy. The far right side shows the outputs.

11. Leaving Fossil Fuels

First, the "good news" from the perspective of clean electricity as a substitute. Two-thirds of all the primary energy inputs today, mostly from fossil fuels, end up as waste heat without ever providing any services. This is the case even for electricity generation This means that it should not be too difficult and expensive to replace fossil fuels with cleaner sources for electricity generation. The same holds true for transportation, where almost 90% of the fossil fuel energy only produces waste heat instead of kinetic energy.

Now the bad news. Fossil fuels are intrinsically superior in generating heat. This is what our one-trick ponies do well. And heat production accounts for at least half of residential and commercial use, and much industrial use. Yes, electricity can also generate heat, but it will have to be incredibly cheap to compete with near-ideal chemical heat sources.

3 Hydrogen

We've reduced emissions by 78%!

Hydrogen is the energy carrier that is the closest potentially-clean substitute for other fossil fuels. It contains energy in the form of chemical bonds. It is so similar that it can even use most of the existing natural gas pipeline infrastructure (with some alterations to reduce corrosion). But hydrogen is not exactly the same.

First, there are differences in energy density. Hydrogen is lighter but requires more space (even in liquid form). It can store almost 40 KWh per kg, which is about three times higher energy density *by weight* than oil and gas (and eight times higher than coal). Great! However, hydrogen can store only 2.8 KWh per liter. This is only one quarter the energy density *by volume* of oil and one half the energy density of gasoline. Not so great.

Despite drawbacks, hydrogen has an almost assured future: It will likely become the preferred clean fuel for airplanes. Airplanes have no grid connection and weight matters. However, the necessary increase in fuel tank volume will require airplane and power train redesigns.

If you feel queasy about flying in a hydrogen airplane, this is probably because you have watched the 1937 Hindenburg disaster. A spark ignited leaking hydrogen and caused a massive fire. It has given hydrogen a bad rap that it has never overcome. Yet with modern technology, hydrogen can be just as safe as fossil fuels. The real problem was not even the disaster (airplanes have had many worse disasters) but the spectacular film record. Even this footage was misleading. Most casualties on the Hindenburg were from people jumping out of the gondola. Hydrogen burns quickly and upward from its envelope. The gondola, where the passengers were located, was below the envelope. Passengers who simply waited until the gondola descended walked away scot-free.

As of 2021, hydrogen has "only" one major practical drawback, but it is deadly: for the same amount of energy, when created from clean sources via electrolysis (rather than from natural gas), hydrogen energy costs about ten times more than natural gas. Over the next 30 years, the cost differential is likely to decline from this factor of ten to a factor of two.

Even if hydrogen becomes economically viable compared with natural gas, it will still likely not be used in all the same ways. Fossil fuels are a cheap way to store massive amounts of energy that will eventually be turned into electricity. That is, fossil fuels today are mined, then stored and finally burned when needed. Without a breakthrough technology,[1] the round trip (make hydrogen from electricity, store hydrogen, make electricity from hydrogen) will remain more expensive than the alternatives of either using batteries, or storing heat.

Therefore, we can predict that in this century hydrogen will be important in transport applications that have no close access to the electric grid (specifically, airplanes and ships), but not in utility-scale electric energy storage less than one day or in automotive transportation. If electricity generation continues to drop in cost into the next century, hydrogen could eventually become worth catalyzing. Unfortunately, we won't be around to find out if our prediction

---

[1] H2Pro is promising a generational leap — great progress but probably not enough. Other processes could improve on catalysts. We hope our skepticism will turn out to be incorrect. However, there are also issues with containing hydrogen, which can ironically then contribute to global warming.

will come true. Perhaps a good approach is first to focus on reducing the price of clean electricity, then to focus on the cost of electrolysis, and only thereafter to invest in uses of hydrogen.

We will return to hydrogen production and uses in Chapter 13.

4 Nuclear Power

The next clean alternative is nuclear power, the ultimate Promethean fire. Uranium and thorium are powerful and dirt-cheap energy sources, but they also have serious drawbacks.

▶ Safety

Nuclear plants may be among the safest plants ever designed by engineers, but they have such exceptionally catastrophic potential that safest may not be good enough. Despite extensive regulations, there have been major nuclear accidents about every 30 years. (Chernobyl was the worst.) The estimated rate has been 1 core-melt down per 3,704 reactor years. This actual rate is far higher than what engineers had designed the plants for.

To prevent human error, the power station only employs trained seals.

You can see that this is a problem when you start counting: there are about 500 nuclear power plants operating in the world today. This means one meltdown every 8 years or so somewhere on the planet. The odds are better-than-even of having one such accident in the United States about every 25-30 years.

We can even assess the extra expected accident cost that we should attribute to existing nuclear reactors. Over-the-envelope estimates for the cost of the Chernobyl meltdown range from $200 billion to $800 billion, for Fukushima meltdown about $500-$800 billion. With about 400 nuclear plants in the world, economists should budget about $5 billion for expected damages — roughly doubling the construction cost per plant.

4. Nuclear Power

> What's the most terrifying word in nuclear physics? Oops! — joke

It is also likely that the wider public (and some experts) will always doubt whether nuclear plants can be trusted. (And this adds further political risk, too!)

Think about the construction incentives when not everything can be triple-checked and inspected. In practice, Contractor shortcut and profit motives will trump extremely-low-probability accident fears, so every screw needs to be triple-checked. But who checks the checker? We cannot forget John Glenn's famous quote: "I felt exactly how you would feel if...you were sitting on top of 2 million parts — all built by the lowest bidder."

Despite excellent engineering arguments about how safe nuclear plants are or can be, it remains plausible that future unknown unknowns will cause new types of unforeseen nuclear accidents. Each accident will be a little different and then engineers will fix the problem — but potentially catastrophic once. Our failure to advance nuclear technology — instead having been stuck in a time warp in which iterative improvements were impossible due to regulatory-cost reasons — may have had good intent but has not made nuclear plants safer over the long run.

Somewhat ironically, the public has been more forgiving of coal plants, which have been much more harmful. They kill thousands of people with their relentless pollution every year — but they do so with more consistency, less individual-death attributability, and most importantly, with less bang on the evening news.

▶ Regulation

Good regulation of nuclear plants is difficult. Regulators are on the hook if something goes wrong (as they should be), but they get no reward when the plant is running. In the United States, no new nuclear plant has been *both* designed *and* built since the inception of the Nuclear Regulatory Commission (NRC) in 1975.[2,3]

---

[2] Unlike the FAA, which has an official mission to help airplanes operate, the only mission of the NRC is to protect the public and environment. Zero plants is clearly safest. A highly biased perspective appears in the American Action Forum. It still is interesting reading.

[3] By 2022, 66 nuclear reactors with pre-1975 designs had come on line in the United States, the most recent in 2016. No new designs were both approved and built since 1975. This may

11. Leaving Fossil Fuels

The Nuclear Regulatory Commission has imposed not only stringent and smart restrictions (good), but also many stringent and <u>stupid</u> restrictions (bad). It is an excellent <u>question</u> whether the current regulatory approach and thicket of regulations have been making plants safer by adding more safeguards, or less safe by making design iterations so expensive that improvements have not been worth inventing and installing. Our view is that if a new design itself is passively cooled and intrinsically not subject to the risk of a meltdown (and ideally also to chemical explosions), then a different regulatory regime should apply. In this different regime, small alterations should no longer require year-long reviews. Is bureaucratic reform possible under the scrutiny of a hostile press and public? We don't know.

Although the unfriendly approach of the NRC has contributed to the high costs in the United States, the slow demise of nuclear power cannot primarily have been the NRC's fault. We know this because companies have also not been racing to install nuclear power plants outside the NRC's jurisdiction, either.

➤ Nuclear Waste

The public is also concerned about the <u>nuclear-waste disposal</u> problem. However, this problem is solvable. It was created in large part by stupid government commitments. For some reason, governments had promised to take care of the waste, thus giving nuclear power operators little incentive to reduce it. It is cheaper for nuclear plant operators just to mine new dirt-cheap uranium and hand the spent fuel to the government. Yet much of this nuclear waste could be reused thousands of times more with reprocessing in breeder reactors. (This is not magical, because the waste remains highly energetic). Of course,

---

be changing. In November 2021, <u>Terrapower</u> announced its intention to build a completely new design in Wyoming — a first in decades. Estimates are that it could come together for under $1 billion for an 0.5 GW power plant — about half the power of a typical nuclear power plant but only about 10-20% the cost and with more intrinsic safety.

4. Nuclear Power

new breeder reactors are no longer being built, either, making reprocessing not only expensive but also currently impossible.

➤ Economics

But ironically, the biggest threat to nuclear power is no longer actual safety, perceived safety, poor public relations, excessive regulations, or radioactive waste disposal — although all of these have contributed to nuclear energy's malaise. Instead, the biggest threat now is economics.

It takes about 10 years and $10 billion to build a traditional nuclear power plant — and construction costs have always seemed to run many times over projections. Thus, the only way nuclear power plants would likely be built nowadays would be with guarantees by electricity regulators that they will buy power from new plants for decades at a committed price that covers the construction costs — itself recouped from taxes on captive consumers.

By the time we'd lobbied the government, got planning permission, raised capital, put the job out to tender and built it, we didn't need it any more!

The more basic problems are (1) the low price of natural gas in many countries; and (2) the looming potential of cheap clean energy. Who wants to invest their own money into a $5-10 billion nuclear plant that will take 5-10 years to build and then needs to earn money over a 30-year lifespan? The invention of better, clean energy storage could obsolete the plant before construction is even finished. It would turn it into a "stranded asset." And this is ignoring the political uncertainties if the electorate demands the shutdown of all nuclear plants after a nuclear accident somewhere in the world (as happened in Germany).

But what if clean-energy technological progress stalls? The world's 500 nuclear plants are on average over 30 years old. Most will shut down within our lifetimes. Extending their lifetimes, with stricter safety inspections and guidelines, should be under active consideration. The choice of how long to operate plants has no clear safety cutoff. Where is the line? If mid-age nuclear reactors are operating at a rate of one core meltdown every 3,704 years, should we shut down old-age plants at one meltdown every 3,600 years? Every 3,500 years? (Or maybe we should shut them all down!)

▶ Newer Designs?

The existing nuclear power plants are at the end of their lives. Could we design and build a newer generation of better, safer, and cheaper nuclear plants?

All new plant designs face one big hurdle: First Of A Kind (FOAK) plants and nuclear plants specifically are much more expensive than Next Of A Kind (NOAK) plants. We won't fully understand new designs until we have built a FOAK. But safety considerations need to be paramount, which drives the cost of such FOAK nuclear plants to uneconomic levels.

Intrinsically Safe: New designs would have to improve safety by two orders of magnitude. Even nuclear fusion plants, despite their completely different physics, are *economically* really just like nuclear fission plants. (Uranium fuel costs have always been trivial.) Fusion differs primarily in that it immediately turns off when disturbed even slightly, which thus renders fusion intrinsically perfectly safe.

To be two orders safer, better fission reactor designs should probably no longer be based on active pressurized water cooling that can lead to a chemical explosions and radioactive contamination when cooling fails.[4] The newest operating plants such as the Westinghouse AP1000 have passive emergency cooling, which is already much better than earlier designs. Fourth generation pebble bed reactors do not melt down (or release radioactivity) even if all active cooling breaks down. Gravity will disperse the pebbles which will end the nuclear fire. China just put the first such reactor into service.

Smaller Reactors: An important problem is that countries often need energy "yesterday" (or at least asap). China is building coal plants rather than nuclear plants not just for employment and cost reasons. Instead, coal plants can be built within 4 years, while nuclear plants take about 10 years. China needs energy *now*.

---

[4]Contrary to popular perception, nuclear reactors cannot explode like a nuclear bomb. Instead, they can explode chemically when they are not appropriately cooled, because their heat then generates flammable hydrogen as a byproduct. Once the hydrogen explodes, the core can melt down and becomes like a "dirty bomb" — which is not really a bomb in a conventional explosive sense, but a source of highly toxic pollutants that are difficult to contain.

4. Nuclear Power

Yet, perhaps there could be a nuclear alternative. Many countries are now working on smaller reactors that could be mass-produced and shipped on trucks. Small nuclear reactors have already been used on military ships for decades, but they do not suffer from the problem that hazardous material could fall into the wrong hands. Nevertheless, just in the last two years, small reactors have made good progress:

- The DOE is funding companies to develop 1–10 MW reactors that could always stay on their trucks and drive to where they are needed. Think about this: 1,000 trucks could power a small U.S. state!
- The NRC has approved a more traditional but (first) smaller, modular, and potentially factory-mass-producible NuScale reactor for civilian use.
- Britain started the approval process for a new kind of nuclear mini-reactor built by Rolls-Royce — one could say the "Rolls Royce of nuclear reactors."
- France has announced its own prototype small modular reactor, again with the goal of being able to mass-produce them more cheaply in the future.

This is just a selection. There are also other promising tiny and safer nuclear power designs on the horizon.

➤ Our View

Worldwide, the construction of traditional nuclear power plants has slowed to a crawl. In the United States, since 2000, only one new reactor has come on line. Construction is only a little more active elsewhere around the world.

Yet nuclear plants could play an important role in reducing global emissions — if they were only a lot safer and a lot cheaper. Without top-to-bottom changes in *everything* — from regulatory processes to mass production to safety to operations to nuclear waste handling — it's not likely to succeed at large enough a scale. This is neither a lament nor a rejoice. It is simply our factual assessment.

Reasonable experts can and do disagree about whether new plants will live up to the promise. Bill Gates believes they will, while the Union of Concerned Scientists thinks otherwise.

Let's keep our fingers crossed that some new designs will prove to be safer, cheaper, better, and produce less waste. The world could surely benefit from better clean base-power technology. But let's also not be too eager and blue-eyed about what will prove to be a tough road ahead.

5 Batteries

We will discuss wind and solar power generation (and energy storage) extensively in the next chapter. They are already the cheapest sources of power today. Their Achilles heel is that they generate power not on demand but only when nature cooperates. Fortunately, batteries can be charged by wind and solar power when electricity is cheap and abundant; and discharged on demand when electricity is expensive and scarce.

anecdote
> **Abraham Lincoln, 1860:** "Of all the forces of nature, I should think the wind contains the largest amount of motive power ... Take any given space of the Earth's surface, for instance, Illinois, and all the power exerted by all the men, beasts, running water and steam over and upon it shall not equal the 100th part of what is exerted by the blowing of the wind over and upon the same place. And yet it has not, so far in the world's history, become properly valued as motive power. It is applied extensively and advantageously to sail vessels in navigation. Add to this a few windmills and pumps and you have about all. As yet the wind is an untamed, unharnessed force, and quite possibly one of the greatest discoveries hereafter to be made will be the taming and harnessing of it."

Batteries are intrinsically completely different from chemical and nuclear energy storage. They cannot store energy in every chemical or atomic bond. Thus, they have low energy density. Even the best lithium batteries provide only about 0.25 KWh/Kg and 0.5 KWh/L (compared to 12–15 KWh/Kg and 5–11 KWh/L for oil and gas). For transportation, this limitation is partly compensated for by the fact that electric engines have 90% (or more) efficiency compared to 25% for combustion engines. Similarly, for grid-based and near-grid-based electricity storage, the low energy density of batteries is not very important. Their high input-output efficiency makes up for it.

However, for some applications, batteries are as wrong an economic solution as hydrogen is for electricity storage. The least suitable application is heat. Burning fossil fuel for heat is too efficient and cheap.

However, even if fossil fuels were banned, batteries would still not be the right solution. The alternative to storing electricity in batteries and making heat later is making heat first and storing the heat in thermally isolated containers. The latter is far cheaper and more easily scaled. This is not only the case for home heating, but also for industrial furnaces.

The second and greater problem with batteries is a central subject of our next chapter: their fixed cost structure. More energy storage for batteries means manufacturing more batteries. This is expensive. In contrast, more energy storage of fossil fuels simply means a larger tank. This is why there is only about 100 GWh of battery storage on the U.S. grid. That's enough to power the U.S. electric grid for about 10 minutes. To cover just ordinary days (and without growth of demand) will require at least 50 times as much capacity. If the world were to electrify transport and heat, too, it would probably require 100 to 200 times as much capacity. Currently, batteries are too expensive to take over electricity storage at this scale.

6 How To Read Technology Forecasts

At this point, you are probably as enthusiastic about wind, solar, and battery technology as we were when we started writing this book. (We still are, but a little more cautiously so.) A lot of pundits are painting an exciting energy future ahead. Not a week passes without more great news on some invention. And the progress of clean-energy technology has consistently outperformed even its most optimistic predictions. But before you buy into all the clean-energy propaganda, let's take a step back and explain why you should remain excited in general but not in the specifics.

For example, Agora is one of our favorite battery technology candidates. It already has a prototype for a CO2 consuming "redox flow" battery, whose emissions are primarily bicarbonates. These are costly chemicals used widely in industry. Agora could revolutionize the world. What could possibly go wrong? Plenty! The devils are in the details, and there are many details before the technology can be mass-deployed — if ever. Foreseen or unforeseen problems could throw a wrench into the gears (though batteries have no gears). Agora could fail to solve the toxic bromine byproduct problem. The owning partners could fall out among themselves and litigate rather than develop. Or the founders or CEO could be incompetent. Or their sales department could be incompetent. Or the money could run out in the height of a financial crisis. Or another battery technology could obsolete them before they can even start. Or government regulation and red tape could kill them. Or an accident, possibly with great publicity, could set them back. Or electricity demand could stagnate. Or lithium car batteries could last virtually forever and simultaneously back up the grid. Or some other countries could wait for the first Agora product, disassemble it, reverse-engineer it, and produce it more cheaply in mass. (Litigation over property could well take decades to resolve, by which time Agora could be bankrupt.) More concerning, Agora is a technology firm, and they will need global manufacturing partners and chemical commodity partners. And so on.

Maybe you should replace the super platinum-iridium-cadmium batteries in your laser pointer with ordinary alkaline batters?

The right way to think about Agora and other battery technologies is that even the most promising truly new technology (i.e., that is not just a small improvement on existing lithium batteries) has perhaps a 1-in-10 chance. (A 1-in-10 chance of revolutionizing the world is no small feat!)

However, the future for humanity is far more promising than just Agora. There is not just Agora, but maybe two dozen battery developers with innovations of various kinds. Any one of them has only a small chance of success. But one or two of them will almost surely hatch.

Put differently, we would not put all our eggs into Agora's baskets. But we would take a bet that within 10–20 years, today's conventional Lithium batteries will either last 10,000 cycles and be an order of magnitude cheaper or they will no longer be the dominant form of utility-scale electricity storage.

7 The Politics of Defending Fossil Fuels

We just advised caution about clean-energy propaganda. We would advise twice the caution about better-funded fossil-fuel propaganda. The fossil-fuel industry and its employees are not taking the clean-energy transition in magnanimous resignation.

Their most prominent approach has been to support surrogates who sow FUD ("fear, uncertainty, doubt"). The goal of their campaigns is to discourage customers from buying into newer and better alternatives. Historically, the fossil-fuel industry has not just been prolific in providing energy, but also in spreading misinformation — the subject of Michael Mann's book *The New Climate War*.

Today's fossil-fuel proponents are delighted when they can raise environmental objections to wind, solar, and nuclear power. Although their objections are typically correct and indeed require consideration, they are mostly red herrings. There are no intrinsic show-stoppers preventing eventual large-scale deployment of clean energy. Let's go over a few of them.

Energy Density

Their most important objection to wind and solar is their low intrinsic energy density. It is true that physics limits the area density of wind turbines to about 3 Watts per square-meter and the equivalent area density for solar cells to about 10 Watts per square-meter. (There is room to improve this solar number. Moreover, there is plenty of space. Offshore wind alone could probably provide enough power for the entire electric grid of the United States, though at a higher price.) Because of the low energy density of wind and solar power, critics note that to provide 4 PWh of energy (the current annual electric energy demand of the United States) would require an area twice the size of Massachusetts — about 20,000 square miles. This observation is true.[5]

However, keep the size of the problem in mind. This area would supply the entire electric energy demand for the entire country. Figure 11.2 shows the required area in a more appropriate perspective, courtesy of Bill Nussey. The Mojave desert alone could meet the entire electricity generation demand

---

[5]Actually, fossil fuels only provide about 2.5 PWh of these 4PWh, and even the area for 4 PWh is overstated, given newer and more efficient solar cells.

11. Leaving Fossil Fuels

Source: Bill Nussey, 2018, at freeingenergy.com. Note how small Massachusetts is in comparison to Western states. Requiring even this large an area is not a problematic constraint. There is more than enough space all over the United States. (And either sun or wind are in abundance almost everywhere on the globe.) Note that because of transmission costs, placing so many solar cells in Nevada would not be anyone's first choice.

of the United States.[6] Of course, Nevada is not the best spot for all of the U.S. electricity generation given the costs of transmission.

Comparing the estimated required 20,000 needed square miles, Nussey also points out that the oil & gas industry leases about 40,000 square miles from the Federal Government (though they do not even use it all!), that about 13,000 square miles are impacted by surface mining, and that about 30,000 square miles are used to grow corn for ethanol. Agricultural land in the United States covers about 900,000 square miles, about 45 times the 20,000 square mile area. Furthermore, wind farms can be built on land that can still be used for agriculture and are even more efficient in mountainous terrain; and solar farms are optimally located in areas where there is little agriculture or forests. Suitable locations exist almost everywhere even near major population centers in most countries.

Nevertheless, although doubling or tripling the 20,000 square mile area to cover 4 PWh of electricity would still not be a problem, the United States

---

[6]Of course, environmentalists will object that lizards' and turtles' habitats will be adversely affected, but until the environmentalists can present a better constructive electricity alternative, these concerns should not be enough to stop construction, merely enough to modify specific plans. It is also not clear whether the extra shade won't be of natural benefit to many species.

consumes a whopping 30 PWh in primary energy (i.e., all energy, not just in electricity). Electricity is a higher-quality source than fossil fuel if it has to be converted to kinetic energy, but it would still require an area more like 100,000 square miles (five times the square in the area), *plus* a requisite area for energy storage, to transition all power to clean electricity. This is a taller order than just transitioning electricity generation — but it is not impossible.[7]

Many similar surrogate objections are variations on the theme that the required scale is just too large — for example, a recent variation on this theme claims that the United States would have to build one new solar and wind installation every other day. This assertion seems frightening — until you realize that 7 new power plants per U.S. state per year sum up to 350 power plants. It's one new plant per year for every 1 million people. The United States is a big country.

Fortunately, all of this is a tall order that we do not even need to contemplate for at least another decade or two — area density and growth will not be limiting constraints for decades. Instead of focusing on the debate how much the world should, could, or need to ultimately cover, smart governments should instead focus on how to best improve the grid and develop wind and solar generation in order to move the needle now.

Scarce Ingredients

Another objection from the fossil-fuel lobby is that wind and solar farms require resources and energy to build and install and this is bad. Ars Technica (2021) has a wonderful rebuttal of a typical set of fossil-fuel shill claims trying to knock clean technology.

One version of this argument is that many clean-energy technologies need more rare raw materials (such as lithium, nickel, graphite, cobalt, and rare earths) than the world is producing now. This objection is, in fact, correct.

There will be battery price fluctuations related to shortages of ingredients for today's battery chemistries. It likely won't be lithium, which is actually the cheapest part of the battery, but cobalt and nickel used for anode and cathode. But these shortages should be temporary.

---

[7] Deserts can similarly provide most of China and Africa (and to a lesser degree India) with power, although with high transmission costs. China has the potential to meet 13 times its electricity demand with solar power.

11. Leaving Fossil Fuels

From an economic perspective, cobalt (and nickel) just happen to be the best materials at the moment. For almost every needed ingredient for batteries, there are already many alternative materials on the horizon — manufacturers are simply using the chemistries that are cheapest at the moment. This is especially the case for stationary utility-scale batteries that exist in labs but still have to be developed and deployed.

The world has also just not needed these materials in large quantities for a long time, and it will take a while to find and open new mines. Here, the free market and profit motive will work wonders. In the long run, ample natural availability of ingredient elements, mass production, and competition will almost surely continue to drive down battery prices. Mining companies are already exploring actively for new sources.

The skeptics have one good point, though — a mine in the US can take 7–10 years to approve, more if faced with well-funded NIMBY (not in my backyard) lawsuits. They can effectively delay and sometimes outright stop the transition. But this is a self-inflicted wound that could be treated. We will come back to how to handle this concern appropriately in our final chapter.

Recycling

Still another version is that turbines and panels will have to be retired at the end of their lives. These claims are again true. The green industry has not yet worked out how to recycle its devices — the industry has been busy and has not yet built enough wind, solar, and battery devices even to worry about large-scale recycling.

Although it is true that mining materials for wind, solar, and batteries will have adverse environmental consequences, for comparison, a lot more mining is required to keep fossil-fuel plants going. The desolation and pollution spawned by many coal, oil, and natural gas fields are comparably devastating. In comparison, if worse comes to worst, at the end of their lifespans in 30 years, we can just bury turbines and solar panels in shallow graves or landfills. Unlike coal, oil, and gas infrastructure, obsolete turbines and panels are not hazardous waste. Even lithium batteries are comparably harmless.

But it probably won't come to this. When there are enough end-of-life installations, someone will probably find a new use for them. In 1990, there were three billion car tires in disposal sites (over 1 billion in the United States).

There are no more tire mountains today, not because the environmentalists raised the alarm, but because used tires are now a valuable raw material for the construction of cement, flexible surfaces, etc. In fact, used tires are expensive now.

More likely, industry will be able to reuse some and discard other parts. And in any case, there are no externalities that could not be priced into the construction and disposal of clean energy — especially if the United States were to institute a fossil-fuel tax and sensible environmental regulations.

In a cosmic view, recycling objections to clean energy should even be welcome. Thinking about these issues early on is a good idea. For example, there will indeed be environmental impacts associated with the transition. How else could the world provide the energy for 8 billion people? Companies could build solar cells, windmills, and batteries designed for easier recycling, especially if the government forced them to take back the residuals at the end of their lifespans. (The race to recycle is already on.) The government could also plan better in terms of where and how to foster specific clean-energy solutions.

Unfair Competition

An even sillier claim is the complaint about unfair competition and subsidies to clean energy. Although it is true that clean energy is now subsidized in many locales, the sum-total does not remotely come close to the subsidies that the fossil-fuel industry has enjoyed for over a century and is continuing to enjoy. As already mentioned, the IMF (itself no anti-capitalist green institution) has assessed the worldwide externalities and subsidies to the fossil fuel industry at more than $5 trillion per year.

In any case, ultimately, the transition will be unavoidable. At current extraction and usage rates, fossil fuels other than coal could be depleted in about a century.

8 The War on Climate-Change

Given news coverage of public concern about climate change, is it the case that the world is now at war with climate change? Allow us to be cynical for a moment. Who exactly views climate change as a coming apocalypse? Climate change seems to be primarily a niche concern of middle- and upper-income people living in richer countries. By and large, most people go on with their lives instead of thinking about future generations. For most, the national soccer team or personal relationships seem more important. The press is mostly an echo chamber. It writes what its audiences want to hear and audiences self-select. Most of the audiences of climate-change websites are the people who are already concerned. And even most of them have other more immediate problems to worry about.

Table 11.3. Annual GDP, Spending, and CO_2 Removal Costs, ≈ 2020

		In Trillion-$	Per Person
Global	GDP	$ 84.71	$10,870
	Health Spending (9.9%)	$ 8.40*	$1,078
	Defense Spending (1.5%)	$ 1.98	$254
US	GDP	$ 20.94	$63,480
	Health Spending (17%)	$ 4.00*	$12,200
	Defense Spending (5%)	$ 0.78	$2,380
	... minus RU ($0.062) and CH ($0.252)	$ 0.46	$1.400
5 $GtCO_2$ (US) Removal Cost @ $50/$tCO_2$		$ 0.25	
15 $GtCO_2$ Removal Cost @ $100/$tCO_2$		$ 1.50	
30 $GtCO_2$ (World) Removal Cost @ $200/$tCO_2$		$ 6.00	

Explanations: The figures are approximate. Per-person numbers are based on a global population of 7.8 billion and a U.S. population of 328 million (2020). For perspective, the per-person per-year emissions of India are 1.5 tCO_2, of China 7.5 tCO_2, and of the US 15 tCO_2. The global average is about 5 tCO_2. Key Sources: OECD, SIPRI, and OECD.

8. The War on Climate-Change

Table 11.3 shows where government spending actually goes. As a whole, Western countries are not putting their money where the media's mouths are with regard to climate change.

Take the United States, for example. We emit about 5 $GtCO_2$ per year. There are plenty of opportunities to remediate or switch technologies to remove at least the first tonne of CO_2 at $50/$tCO_2$ or less. Multiply 5 $GtCO_2$ by $50/$tCO_2$ and you get a total cost of about $250 billion per year, or about $750 per U.S. citizen.

Yes, $750 per person per year (or $3,000 for a family of four) is a lot of money, especially considering that the median income is only about $35,000/year per capita and $63,000/year per household. But $750/year is also only $2 a day. And it is "only" about one-quarter of our military budget. The fact is that countries are not at war with emissions. They are at (a low-flame) war with one another.

We share the obvious wish to redirect the world's military spending to better causes. But, for better or worse, as we explained in the previous chapter, the world is not a decision-maker. Thinking in terms of global welfare is a conceptual error. Countries are the decision makers. And realistically, they will not redirect their military spending towards environmental spending.

As of 2021, American energy-related spending remains small and almost incidental. Not surprisingly, the U.S. Armed Forces spend more on nuclear weapons than on clean-energy technology. But the same is true even for the so-called Department of Energy! The National Science Foundation did offer modest support, but much of that spending funds university overhead rather than specific energy projects. Clean-energy R&D could sure benefit from more funding administered in a better fashion.

But the fact is that our voters and politicians have spoken, and they do not view climate change as the apocalypse. They prefer to support their militaries rather than the war on climate-change. And even if the current U.S. or any foreign administration is willing to direct more funding to environmental issues, the next administration may not be. And, at the rate that our voters' views can be changed, we may get around to committed large-scale pollution fighting in, say, 100 years. By that time, clean energy will likely be so cheap that voter intervention will no longer matter.

The United States is by no means alone, either. Take Germany, among the most environmentally conscious countries on Earth. (The European Union in general is the world exception.) With the Green Party in government and without any enemies on its borders, Germany's typical green spending is approximately $12 billion, about 25% of its spending of $50 billion on its military.

Go beyond Europe and the rich West, and there is almost no spending on green initiatives and lots of spending on militaries. And as we explained in Chapter 3, the world's most urgent task now is to convince China, India, and Sub-Saharan Africa to curtail their emissions growth. The United States and Europe are no longer enough.

Thus, our recommendation for realistic environmentalists is this: Advocate moving on actions that we can take *in our own countries here and now*, that have a good chance of winning and maintaining political support *in our own countries here and now*, and that are likely to have staying power *in our own countries and beyond here and now*. Let's leave the Utopian proposals to later.

Realistic solutions should be such that, once implemented, it will be cheaper and more convenient to continue with them and not to return to the old fossil-fuel way of life in the next electoral cycle. Again, this means focusing on low-hanging fruit. And the sweetest fruits are those that require only getting a process started, without having to pay forever to keep it going — in other words, let's convince our governments to view themselves as catalysts rather than cops.

Conclusion

Don't believe everything you read. Clean technology is exciting. However, it is not yet ready to fully take over the world. If activists want to change the world for the better, they should push the most intelligent and effective policies for and within each country. The most obvious one is accelerating clean technology research and development for countries' own sake (and surreptitiously for the world's sake).

The immediate next steps are really all that activists should care about right now — *moving the needle now*. Let's worry about the grander proposals for full decarbonization only after we have made good progress on the first steps.

Further Readings

Books

- Epstein, Alex, 2021, The Moral Case for Fossil Fuels: Revised edition, Penguin Books, New York.
- Macavoy, Paul, and Jean Rosenthal, 2004, Corporate Profit and Nuclear Safety: Strategy at Northeast Utilities in the 1990s, Princeton University Press.
- Mann, Michael E., 2021, The New Climate War, Hachette Book Group, New York, NY.
- Smil, Vaclav, 2016, Power Density, MIT Press. Explores the consequences of the power density in different energy sources.

Reports and Academic Articles

- Department of Energy, 2019, The Ultimate Fast Facts Guide to Nuclear Energy.
- Eash-Gates,Philip, et al., 2020, Sources of Cost Overrun in Nuclear Power Plant Construction Call for a New Approach to Engineering Design, Joule. Summary in Timmer, John, 2020, Why are nuclear plants so expensive? Safety's only part of the story.

11. Readings

Shorter Newspaper, Magazine Articles, and Clippings

- Barnard, Michael, Fukushima's Final Costs Will Approach A Trillion Dollars Just For Nuclear Disaster, CleanTechnica, 2019/04/16.
- Barnard, Michael, Agora CO2 Redox Battery Wins Global Deeptech Competitions & Has 1 Year ROI. CleanTechnica, 2021/08/14.
- Bousso, Ron, 2021/09/20, Special Report: BP gambles big on fast transition from oil to renewables, Reuters.
- Brady, Aaron, Global crude oil cost curve shows 90% of projects through 2040 breaking even below $50/bbl.
- Tim De Chant, France to cut carbon emissions, Russian energy influence with 14 nuclear reactors, Ars Technica, 2022/02/14.
- Tim De Chant, Sensor-driven turbine platforms could unlock 4,000 TWh of offshore wind
- Ghosal, Aniruddha, 2021, 'Nothing Else Here': Why it is so hard for the world to quit coal. Associated Press News.
- Johnson, Doug, The decreasing cost of renewables unlikely to plateau any time soon, Ars Technca, 2021/10/03.
- Laville, Sandra, 2021, G7 nations committing billions more to fossil fuel than green energy.
- McGreal, Chris, 2021, Big oil and gas kept a dirty secret for decades. Now they may pay the price, The Guardian.
- Mills, Mark P., 2019/08/05, If you want renewable energy, get ready to dig, Wall Street Journal. (The substance is correct, but Mr. Mills does not compare how toxic it is to dig up and process fossil fuels into electricity.)
- Myers-Briggs ISTJ, 2020, Hydrogen Vs. Natural Gas For Electric Power Generation.
- Porceng, Nathan, 2022. features Sola Talabi's defense of small nuclear reactors (powering Navy vessels for decades).
- Terrapower's natrium (molten salt) reactor.
- Timmer, John, 2021, Pure nonsense: Debunking the latest attack on renewable energy, Ars Technica, 2021/03/01.
- Timmer, John, China's solar power has reached price parity with coal, Ars Technica, 2021/10/12.
- Timmer, John, US regulators will certify first small nuclear reactor design, Ars Technica, 2022/07/30.
- Your Batteries Are Due for Disruption, New York Times, Sep 12, 2021. (Sila for silicon-based anodes.)

Chapter 12

Electricity

The future of energy is electricity. It intermediates the two key sources of clean energy, solar and wind power. This chapter explains how electricity works today and how its scale can be increased.

1 Why Electricity?

Electricity is the most versatile form of power. It is the jack-of-all-trades. It can heat and light homes, power cars, drive industrial processes, and power all our gadgets. It can be easily, cheaply, and efficiently converted into other forms of power. Table 12.1 shows conversion efficiencies. For example, the electric motor in a Tesla has an efficiency of about 90% (compared to only about 25% for a gasoline motor). Even refrigeration, a difficult thermal conversion, can be accomplished at greater than 50% efficiency.

In contrast, fossil fuels are one-trick ponies. They can only generate heat efficiently. For anything else (such as kinetic or electric power), fossil fuels require further, rather inefficient, conversions, typically reaching no better than 30–40%. (If power comes in electric rather than fossil-fuel form so that conversions can be avoided, humanity may need only half as much primary power as it does today.) In addition, electricity can be routed with a transmission grid to the place where energy is needed at the speed of light, although there are losses in the transmission process.

Best of all, the basic technology required to transition from a fossil-fuel to a clean electrical economy is already available today. Transportation was the

12. Electricity

Table 12.1. Energy Conversion Efficiency

	\multicolumn{5}{c}{Energy Type}
	Cool	Heat	Kinetic	Light	Chemical
Conversion From Electrical	Refrig 50%	Heater 100%	Motor 90%	LED 50%	Electrolysis 70%
Conversion To Electrical	... Turbine ... 50-90%		Generator 90%	Solar Cell 20%	Fuel Cell 60%

Note: Chemical means either fossil or hydrogen. Basic Source: Wikipedia

largest remaining undisputed domain of fossil fuels, but Tesla jump-started the electric car industry in 2012 with its Model S. All major car makers have announced that they will stop making combustion-based cars and light trucks within a decade. Governments are following, too. California will only allow clean cars and trucks to be sold by 2035.

Although technological breakthroughs are always welcome (and indeed likely), the green electric transition will require no moonshots with uncertain probabilities of success. There are just engineering, economic, and business challenges, and we already know that they are solvable. It remains only a matter of research, development, deployment, implementation, coordination, and scale.

Unfortunately, the world is also not yet fully ready for 100% clean energy or even just 100% clean electricity. Our chapter will explain why. The world will be ready soon, though, and there is already a lot that can be done today.

2 Not All Electricity Is the Same

Electricity is always just electrons, but from an economic standpoint, not all electricity is the same. For instance, electricity in the Sahara, where it is inexpensive to generate from solar energy, is not as valuable as electricity in Germany, where it is needed for industry. (It is too expensive and lossy to string power cables from the Sahara to Germany.) Almost everywhere, electricity at 6–8 pm (when demand typically peaks) is more valuable than electricity at 4–6 am (when demand typically troughs). Furthermore, the cost of electricity at the generation plant is only about half of the cost of delivered retail electricity. Someone needs to be paid to build and manage the plants, store power, maintain the grid and transmission infrastructure, handle billing and collections, and so on.

Different technology mixes will also dominate in different locations. Geothermal power can work in California or Iceland. Wind power can work in Chicago and Great Britain. Solar power can work in Phoenix and Mexico. Hydroelectric dam power works in Oregon and Norway. But these technologies may not work elsewhere. In contrast, other technologies, like nuclear power or batteries, can work everywhere.

Suppliers and customers also need to consider that both electricity supply and demand are constantly changing. The allocation problems are so complex that not even the smartest and most benevolent government could plan them perfectly. It's a patchwork of educated guesses.

Shortest-term, there is predictable daily demand variation. Electricity demand usually peaks around 8pm. But weather patterns (and with it both supply of and demand for power) can change, some predictably, some unpredictably. A heat-wave can increase the demand for air-conditioning services. A cold-wave may reduce the available wind capacity. Medium-term, there are seasonal differences in supply and demand — summer and/or winter usually require more power than spring and fall. Long-term, plants have to be built today with lifespans of thirty years or more. Better technology may arrive and obsolete the plant. People may move to different locales. Bitcoin mining demand may increase or decrease. Investing in large power plants involves large, risky decisions and is not for the faint of heart.

3 Basic Electricity Provision

Let's explain how power works today. The United States has approximately 1.2 TW of generation capacity. The largest power sources are natural gas (45%), coal (20%), wind (10%), nuclear (10%), hydro (10%), and solar power (5%). (We will explore these numbers in greater detail in Table 12.11, where we also provide numbers for China, the world, and predictions for 2050.) Plants don't run all the time, so the power mix is not representative of the energy mix. Instead, of the 4 PWh we consume per year, gas covers 40%, coal 20%, nuclear 20%, wind 9%, hydro 7%, and solar 2%.

U.S.	NatGas	Coal	Wind	Nuclear	Hydro	Solar
Power	45%	20%	10%	10%	10%	5%
Energy	40%	20%	9%	20%	7%	2%

In the last decades, no new coal and nuclear plants have been built in noteworthy amounts; only gas, wind, and solar plants.

It is beyond fascinating how electricity actually manages to arrive at your house. There are thousands of electricity suppliers — some public, some private — with all sorts of different technologies, each with its own generation costs and location relative to the electric grid and local regulations; and of course, there are hundreds of millions of customers. The electricity generators synchronize their power into an irregular interconnected grid, and the consumers tap it whenever they want it. The grid operators are the intermediaries. They route electricity over transmission lines, some over half the distance of the United States. However, transmission lines have limited capacity and are very expensive to build and maintain (and they lose some power in the process of transmission, too), so there are never enough transmission lines. Keeping electricity supply close to electricity demand saves a lot of money.

The most important aspect of electricity relates to daily use patterns, followed by seasonal patterns. Let us explain the system in more detail by describing the daily patterns first in California and then in the United States.

3. Basic Electricity Provision

California

Figure 12.2. Demand on March 21, 2021 in California

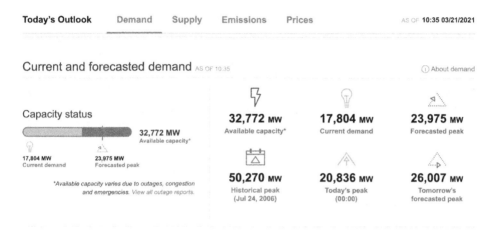

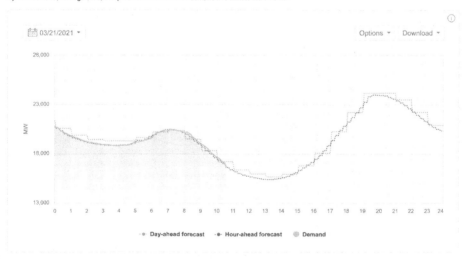

Source: Today's Outlook from CAISO.

The California Independent Systems Operator (CAISO) is the independent non-profit company that operates the grid. Every day, CAISO publishes its

12. Electricity

anticipated demand on its public computer system, OASIS. CAISO then buys power via both one-day-ahead and real-time auctions from a large number of providers, who all compete to provide power for the lowest price.

Figure 12.2 shows the demand in California on a random day (March 21, 2021). California typically has nice mid-day weather in Spring, so demand around noon in Spring is modest, between 15 GW and 25 GW. (California also needs to be prepared for days when it needs more than 50 GW of power, such as on hot summer days when a lot of air conditioning and cooling are required.) Demand is highest in the early evening, around 8pm, when people are at home and doing chores. There is also healthy demand all night long, including but not limited to lighting, refrigeration, electric cars charging, and some industrial plants. California's use pattern is similar to that observed in many places around the world.

On this particular Sunday, California expected peak power needs of about 24 GW. This demand was covered by available capacity of 33 GW, of which about 10 GW would never be switched on. The graph also shows discrete time slots when plants were scheduled to start up or shut down (according to forecast demand). Not shown in the figure, on weekdays of the same week, demand was typically about 3 GW higher, with similar day/night use patterns.

Figure 12.3. California Generation Mix, March 21, 2021

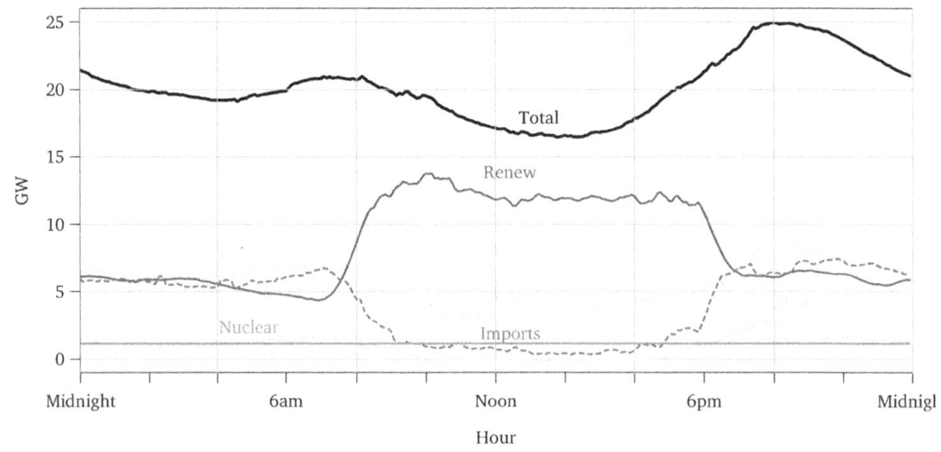

Source: CAISO Oasis.

3. Basic Electricity Provision

Figure 12.3 was recorded the following day. It shows when and how the electricity was ultimately generated. The 10:35am forecast for the rest of the day proved pretty accurate (as CAISO forecasts usually are). Energy provision was lowest at about 17 GW in the early afternoon around 2 pm, coming down from a smaller high of 22 GW at about 8am and peaking at 25 GW around 8pm. Wind and solar covered most of the demand during daylight hours, with wind covering about 5 GW during the night and solar covering about 10 GW during the day. Gas and electricity imports from other states (many from coal plants elsewhere) covered about 5 GW during the night, but not during the day. Nuclear was small but steady throughout the day. Hydro-electric dams began releasing water in the late afternoon.

United States

Each state and country has not only its own energy mix but also its own daily and seasonal demand peculiarities. For example, Florida has a lot more demand in the summer, Alaska in the winter. So let's expand our perspective to broader regions.

Year	Baseload			Base/Dispatch	Intermittent			
	Nuclear	GeoT	Coal	Gas	Hydro	Wind	Solar	Other
California 2019	9%	5%	3%	34%	18%	10%	12%	9%
USA 2018	20%	1%	19%	40%	7%	9%	2%	2%
World 2020	10%	0%	37%	24%	16%	5%	3%	5%

California is naturally blessed with a lot of clean energy — not just nuclear and geothermal energy ("GeoT"), but also hydro, wind, and solar. However, the table is a bit misleading, because California sources between 20 and 40% of its electric power at night from out-of-state imports, presumably generated by fossil fuel sources elsewhere. (Another 35% of power at night is natural gas.) In contrast to California, where coal is expensive, coal is cheaper in China. Hydropower is more plentiful in Northern California and China, but not in Australia. And so on. Yet not everything is determined by locales and economics. Plant construction costs tend to be more or less similar everywhere. Nuclear power costs are particularly similar worldwide, because few localized resources are required to build and run one. Worldwide, countries and states also create their own specific issues when they go their own ways on subsidies, regulation, politics, etc..

12. Electricity

Figure 12.4. United States Electric Power Generation, Sep 2021

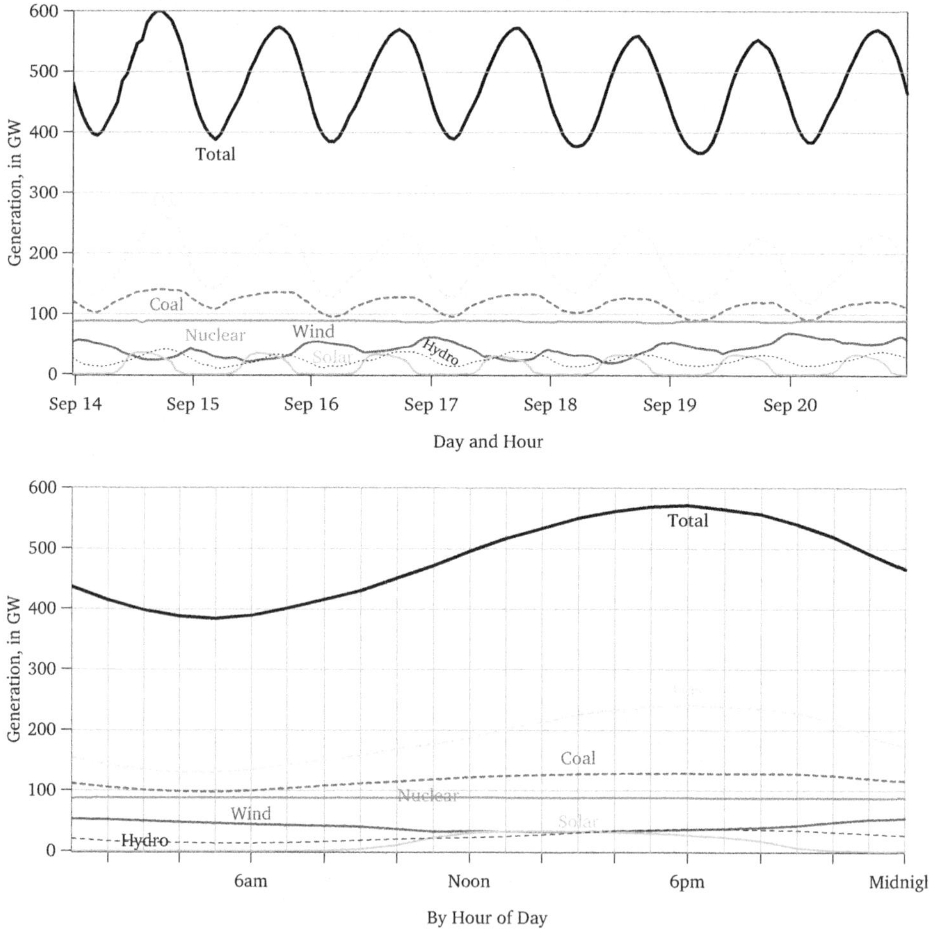

Source: EIA Dashboard. Fossil fuels are dashed. Hydro consists of about 80 GW of conventional hydropower and 20 GW of pumped storage.

Figure 12.4 shows the daily electricity generation for the entire United States on a fairly ordinary week in September 2021. U.S. power generation peaks around 6pm (EDT) and troughs around 5am. (This is also roughly the case in local time, too, because relatively more people live on the East Coast — a fact that is then reflected in the national patterns.)

Covering a larger area than California, the U.S. supply and demand seems even more (boringly) predictable. However, this is a little misleading, because it is not economical or possible to transmit large amounts of power across long distances. Thus, solar power in Los Angeles is not useful in Seattle. This means that grid operators must forecast a lot more than just this overall U.S. demand pattern. In particular, wind is nowhere near as steady as the graph suggests. It is true that wind blows relatively steadily during the day *somewhere*, but this somewhere changes around, often unpredictably so.

As in California, nuclear is always steady; solar is always daytime only. Coal, wind and hydro are more steady. Natural gas is the most important single source at any time of the day.

4 Base, Intermittent, Dispatch Power

The two energy mix figures illustrate that not all power is used in the same way: nuclear power is steady, gas power goes up and down, and solar power is day-time only. For this reason, power is sometimes classified into one of three different kinds (admittedly with some overlap):

1. **Baseload power** is from sources that are basically always running. Nuclear power plants are the ideal example. Once built (and staffed), they can supply power at very low marginal cost. They are also expensive to shut down and restart. It makes no sense to turn them off, other than for very rare maintenance.

    Coal plants were also primarily developed as a source of baseload power. However, because the fuel has become relatively more expensive, operators no longer run all the plants all the time but ramp them up or down with on demand. Coal was used to supply <u>almost</u> all electricity in the early 20th century.

2. **Intermittent power** is primarily wind and solar. In large scale, it is the newest and still least important kind of electric power on the grid, but

12. Electricity

it will eventually become the most important one. Intermittent power sources operate only when nature cooperates. Not only are there times of the day when they (predictably) cannot generate power, but they may also be (unpredictably) off for entire days or weeks. This variability makes intermittent power the least valuable form.

Solar power is the ultimate intermittent source. The sun does not shine at night, and it's moody even during the day (except in California). In the Northern Hemisphere, there is also less sun in the winter than in summer, when days are longer.

We already mentioned that wind looks more steady in the aggregated U.S. graph in Figure 12.4 than it should. In real life, from the perspective of where it is needed, wind is quite intermittent, too. However, unlike solar power, wind often blows at night. It is thus more regularly available, but often also less predictable.

Both solar and wind have modest fixed installation costs. However, their marginal post-installation costs are unbeatable — they do not even require fuel. The United States now can obtain up to about 100 GW of peak power each from wind and solar farms, about 1/10 of its installed total generation power. This is about the same magnitude as nuclear power and not far off from coal power. However, given its intermittent nature, wind and solar cannot be replacements for those two. Their power is less available and far less valuable. By 2050, wind is forecast to offer peak capacity of about <u>400 GW</u>, solar a whopping <u>1,000 GW</u>.

3. **Dispatchable power** could also be called "stored power." It is electricity that can be delivered on (short) notice at the operator's discretion. Dispatchable power was always needed for handling above-average demand, as in the evening. However, with the arrival of large amounts of intermittent power, dispatchable power is now becoming far more important.

    Its flexibility makes dispatchable power the most valuable and most expensive form of electricity. The big economic problem with all dispatchable power plants is that they sit idle much of the time. Thus, the ideal dispatchable power would have low fixed costs and high capacity, but it could tolerate higher marginal (fuel) costs. We will cover storage in greater detail in Section 6.

If electricity demand were constant over time, we would only need to compare baseload power on the one hand against combined intermittent-plus-dispatchable power on the other hand. That is, it would make no sense to team up baseload power with intermittent power. But with varying demand, the two typically have to work in combination. Moreover, the distinction need not be as clear over the long run. Even base power plants could be converted into dispatchable plants using a heat reservoir. For example, instead of converting the thermal heat from a nuclear plant immediately into steam and hence into electricity, the plant could instead heat up a molten-salt reservoir that could then be tapped on demand — of course, with an efficiency loss.

5 Technologies For Generation

The fixed construction costs of plants are often the biggest cost component of electricity generation. Once built, the fixed cost is sunk and becomes largely irrelevant. Thus, the single-biggest cost difference among electricity plants is not what type they are, but whether they have already been built or not. With the exception of coal, the cost of fuel ranges from modest to trivial.

Ergo, in the United States, it still makes economic (not environmental) sense to run an already-built coal plant, even if no sane investor today would build a new coal plant. A coal plant can cost $1 billion and has to be profitable for 30 to 50 years. Who wants to build a new coal plant when it is clear that wind and solar plants plus battery storage will be cheaper within about a decade or two, even in the absence of a fossil-fuel tax? Existing coal plants are now just finishing off their 50-year design life.

When we think about electrification of the economy, we have to think in terms of decades. Thus, it is more important to take the perspective of building the next generation of plants rather than worrying about what plants are running and aging out at the moment. Consequently, it is the economics of new plants that matters for determining whether the world will move to clean energy or not.

12. Electricity

The Levelized Cost of Electricity (LCOE)

The standard measure for the cost of a new plant is the "levelized cost of electricity" (LCOE). This calculation seeks to include everything — from capital construction costs, to the time-value of money, to operating costs,[1] to fuel costs, and so on. In economic terms, the LCOE calculation is based on the present value of all known and projected cash flows (i.e., appropriately discounted by interest rates and summed up).

Beyond the limitation that the LCOE does not matter after plants have been built, it has a second problem: The LCOE relies heavily on projections of the future. For example, if someone were to invent a newer and better technology, one's own plant may suddenly become obsolete. In this case, the construction cost can no longer be amortized over many years, which means that the true LCOE will turn out to be much higher. Unplanned obsolescence is riskier when both the investment cost and budgeted life-spans are high. Many older coal plants have become functionally obsolete much earlier than anticipated – a fact that has made their realized LCOE skyrocket relative to their planned LCOE.

Table 12.5 is our best attempt to piece together reasonable cost estimates from many sources (especially the National Renewable Energy Laboratory (NREL), the U.S. Energy Information Administration (EIA), the OECD/IEA, Lazard) and others. The table provides both current (very near-term) and projected future LCOE estimates. The cost figures should be broadly representative for many places around the globe. Of course, this statement is not to be taken too literally. Solar power is cheaper in the Sahara, wind power in England, and geothermal power in Iceland.[2]

---

[1](Operating costs could further be divided between quasi-fixed costs [what it costs to staff the system] and variable costs [what it costs in fuel, wear, and tear to generate another MWh].)

[2]For example, regional US Variation (Table 3) suggests a range of $30-$40 per MWh for nuclear power, $70-$100 per MWh for battery power, etc. The summary comparisons in the OECD/IEA report across countries are interesting, too.

5. Technologies For Generation

Table 12.5. Cost Estimates For Electricity Generation in 2026, 2020$/MWh

	Plant type	EIA Estimates for 2026 (in 2020$s)				2050 Frcst
		Runs	Capital	Operation	LCOE	
Intermittent		Average Residential Solar Panels:			**$100**	$30
	Solar Panels	30%	$25	$10	**$35**	$15
	Wind, onshore	40%	$25	$10	**$35**	$20
Baseload						
(Scarce)	Geothermal	90%	$20	$18	**$35**	
	Nuclear	90%	$50	$20	**$70**	$60
(0.5 tCO$_2$/MWh)	Natural Gas CC	90%	$10	$30	**$40**	$45
(1 tCO$_2$/MWh)	Coal	85%	$45	$30	**$75**	$65
	Biomass	85%	$35	$55	**$90**	$85
Dispatchable, Limited Capacity						
(Scarce)	Hydropower	60%	$40	$15	**$55**	
Dispatchable, Unlimited Capacity						
	Natural Gas (more below)	10(–50)%	$50	$150	**$200**	

Note: These estimates are our heuristic summaries of information published by the U.S. Energy Information Administration (EIA), the National Renewable Energy Laboratory (NREL), and the OECD/IEA, all quoted in real dollars. Solar panel costs are photovoltaic (i.e., based on the cells that you also see mounted on rooftops and that convert light directly to electricity) in utility-scale farms. CC are combined-cycle plants. Hydropower is partly base, partly intermittent, depending on water availability. The levelized cost of energy (LCOE) summarizes all capital and operation costs, as well as transmission costs. (Transmission costs today [usually in near proximity] typically run about $1/MWh for base 24/7 electricity, $3/MWh for intermittent electricity, and $10/MWh for batteries.) In favorable locations, solar can already be installed at an LCOE $20/MWh as of 2022.

Wind and Solar

Table 12.5 shows that utility-scale solar cells and on-shore wind power in utility-scale installations are already the cheapest sources of electricity, at about $35/MWh. (This price already reflects the fact that solar equipment is idle more than half the day when no power can be generated.) Better yet, their costs are still falling. And it seems almost unreal, but the price for utility solar power is expected to fall to $20/MWh as early as 2030 (and less in Western states). When the sun is shining, electricity will cost only a third of what most generated electricity costs today and a sixth of what retail customers are paying today and still make good money for the builders! By mid-century, daytime power could become almost free, perhaps to be sold at a flat service fee to retail customers, similar to how landline telephone service is sold today. This scenario has been called "energy too cheap to meter." And this future is almost here, too. In Western U.S. states, the LCOE has already been quoted as low as $20/MWh as of 2022! It is often cheaper to build and run solar than to buy NatGas fuel for an existing power plant.

Residential solar panels are much more expensive (at $100/MWh) than utility-scale solar ($35/MWh), partly because installation is more expensive, partly because each house needs some additional equipment (such as an inverter to feed unused power back into the grid). However, rooftop solar avoids many other non-generation costs (long-distance transmission, administration, etc.). This is partly why retail electricity costs about $120/MWh at your house today — but it also works at night! In a fairer apples-to-apples comparison, residential solar is probably about 30% more expensive than utility-scale solar, not more than twice as expensive. The future will also be a race — will roof-top solar (with local batteries) or industrial-scale solar become cheaper faster? The answer may well depend on the location.

Just as Germany jump-started the wind-turbine sector, with strong subsidies and at great cost to its consumers, California is now trying to jump-start roof-solar generation. Besides forcing the grid to accept electricity from roof solar at high prices, California will require rooftop solar on all new construction. Economists remain skeptical whether economies of scale could bring down the rooftop price of solar so significantly that it will become a predominant technology. It is not impossible. The future will tell. If the experiment succeeds, the most important beneficiaries of Californian competition, scale, and learning about rooftop solar will almost surely not be California alone but the wider world as well.

Natural Gas

The cost of natural gas is low enough that it can be used almost all the time in many places around the world. It is often the cheapest source of *baseload* electric power today. The best 24/7 generators are "**combined-cycle (CC)**" natural gas turbines, which use exhaust heat for a second pass at power generation. They can run 24 hours a day and are designed to take advantage of this 24/7 mode of operation. The cost is $40/MWh. A fossil-fuel tax could change some of the economics of gas plants. A good rule of thumb would be to add about half the proposed tax on CO_2/tonne — perhaps more if the tax is smart enough to penalize methane leaks at the well rather than just at the generator. With a $50/tCO_2$ tax, the gas price would thus be between $65/MWh and $90/MWh.[3] Still, gas would remain competitive for quite a while. Gas supplies about 40% today and will supply 50% of the electric power in the United States around 2050.

There is also a second form of gas-powered generation that is more like dispatchable power. To cover electricity demand during off-times, intermittent sources can be combined with gas turbines. This is often done in regions where natural gas is more expensive. Moreover, plants that do not operate 24/7 are more expensive per MWh generated. The last line in Table 12.5 shows that dispatched power from such "**peaker plants**" is a lot higher. It is somewhere between $150/MWh to $200/MWh. We will come back to dispatched power in the next section.

Geothermal Power

The cheapest *clean* baseload power can be geothermal ($35/MWh) — heat that comes from the sub-surface of the earth. It is produced from mile-deep wells that tap into the radioactive heat coming from the center of the planet. Unfortunately, drilling these wells and creating the infrastructure to extract energy from them (primarily sending water down the hole and retrieving steam from it) is cheap only in a few limited locations. Otherwise, geothermal energy is rare and expensive. California is blessed because it sits on the Pacific Ring of Fire, where geothermal power is viable and cheap. This is also the case in some other countries, like Iceland, the Philippines,

---

[3]The $25/MWh add-on estimate is also in line with the cost of Carbon Capture Sequestration.

and Indonesia. However, geothermal power is currently too expensive for utility-scale electricity generation in most locations.[4]

Technology may or may not cure this sometime soon. Quaise Energy, an MIT spin-off, is investigating a new method of drilling super-deep holes which could make reaching 500°C temperatures viable in almost any location on earth. If you have ever seen a volcano, you understand what awesome power is available for the tapping!

Tidal Power

Earth has yet another untapped energy source: the flow of water due to tides. The easiest way to tap this *very large* power source will probably be in tidal inlets. The New York Times reports that Nova Scotia is very close to installing the first clean multi-GW power plant in the Minas Passage — the "Everest of Tidal Energy in the World." Although tidal power is not base-power or dispatchable power, it is also not synchronous with other clean power and thus more valuable. It is too early to say how much tidal power will be able to contribute to humanity's energy needs.

Nuclear Power

The next-cheapest clean baseload source is nuclear power. It is more expensive at "maybe" $70/MWh. Yet, this is guesswork because few, if any, new nuclear power plants have been designed and built in the United States and Europe for many decades.

One advantage of nuclear plants is that they can be constructed almost everywhere in the world. They are also often the *only* viable clean alternative where there is not enough wind or solar power. After a nuclear reactor has been built, it provides cheap base electricity at $20-$30/MWh. (The fuel itself is dirt-cheap; most of the operational costs stem from staffing and other regular expenses regardless of operation.)

We have already discussed nuclear power in the previous chapter. There are many problems: safety concerns, disposal of used radioactive fuel, potential for nuclear weapons proliferation, political and popular opposition, regulatory

---

[4]Home builders can also often install geothermal heat pumps in houses that extract heating in winter and cooling in summer from coils that are laid just a few meters below ground.

5. Technologies For Generation

costs, slow and expensive construction, and so on. But the killer problem is economics. It makes no sense to construct a new nuclear power plant in many places (such as the United States) when natural gas can provide base power for \$40/MWh. Even at a \$50/tCO$_2$ fossil fuel tax, nuclear power would have a tough time competing. The niche of nuclear power has thus been mostly in locales where natural gas is not available in abundance, such as in France.

Even if natural gas were taxed severely, the economic problem of nuclear energy would still not be solved. There is still the fact that a new type of energy storage could also quickly make wind and solar power dominate nuclear power in terms of cost.

Coal

Coal plants used to produce most electricity just a few decades ago. Today, they produce only about as much power as nuclear plants, about 10% of the U.S. power supply. In China, it is close to 65%!

Table 12.5 shows that new coal plants are not only unpopular in the United States but also already obsolete at "maybe" \$75/MWh. No one has built coal plants in the United States for at least a decade. If someone did, they would almost surely not operate for the 30 years that coal plants have operated in the past. Even the fuel is too expensive, as it needs to be mined and transported in many locations. Already-built coal plants still remain running at \$30/MWh (although even they are already idling much of the time). Coal plants will disappear from the U.S. grid within a few years — but unfortunately, not from grids worldwide. China and developing countries are still building them in large scale.

Of course, the true social cost of coal electricity is much higher than the generation cost in Table 9.2. The left-most column in the table shows that each MWh of coal produces about 1 tCO$_2$. With a \$50/tCO2 fossil-fuel tax, the economics would kill American coal even for most plants already built.

There are many ironies here. The free market helped coal dominate, at first because it was cheaper than alternatives, later because coal pollution was not taxed appropriately. Nowadays, the free market has abandoned it. Even ignoring environmental concerns, there are simply better and cheaper alternatives in most locations today — if not natural gas, then nuclear power. Coal plants were essential to humanity's past. They are now the enemy of the future.

12. Electricity

Coal's survival in much of the world now depends on the opposite of a free market, wherein governments maintain obsolete regulations and/or are catering to coal mining lobbies and employees. (Worse yet, coal has become an irrational rallying point for some nationalist parties.) It is no longer enough for clean technology to be cheaper than coal. Clean tech also has to overcome the vested and legitimate interests of people whose livelihoods depend on coal.

In the United States, Donald Trump won the swing state of Pennsylvania with 48.2% over Clinton's 47.5% in 2016, partly because he supported coal miners — even though there were only 20,000 left (among 6 million voters). Politicians ignore fossil-fuel lobbies (and farmers) at their own risk.

Unfortunately, as for other countries, many are still building coal plants. As already mentioned, China is by far the worst problem. Its coal plants produce about 30% of the global CO_2 emissions, because about 60% of its electricity comes from coal. In total, China is currently planning or constructing about 250 GW of new coal plants (equal to about one quarter of the *total* U.S. generating capacity). These plants will lock in decades of emissions — a globally devastating plan. Why? After all, by the time the plants will be ready to open, solar power with storage should be cheaper in China than coal. The best explanation is that China now has about 2.5 million coal workers. This is down from about 5 million just a decade ago, but still enough to scare the party.

In terms of coal electricity generation, India is about to overtake the United States and become the world's second problem. Other strong builders include Turkey, Indonesia, Vietnam, and Bangladesh, constructing 20-30 GW, each.

If anyone has a good idea about how to stop or throttle coal plant construction in China (and India), this is the time to speak up. The impotence of global institutions and climate negotiations to meaningfully reduce coal-plant construction activities of these countries only reaffirms our views. Much cheaper and better green technology is still our only hope. We even believe that it could be in the self-interest of the United States and Europe now (though not necessarily for the inventors) to share their best nuclear-plant designs with

Figure 12.6. Coal Power Plant Status, in GW

	Operating	Construction	Permitted	Announced
OECD	501.0	16.0	5.0	3.9
USA	232.8	-	-	-
EU27	117.8	12.2	-	-
China	1,046.9	96.7	43.0	72.1
India	233.1	34.4	11.7	11.7
All others	≈280	≈37	≈20	≈24
World	2,067.7	184.5	78.9	111.8

Source: Global Energy Monitor Global Coal Plant Tracker, February 2022. The tracker excludes Costa Rica, Estonia, Iceland, Lithuania, Luxembourg, Norway, and Switzerland for OECD; and Cyprus, Estonia, Lithuania, Luxembourg, and Malta for EU27.

all countries for free — taking proper nuclear proliferation precautions, of course.

Finally, if you think the world climate meetings in Scotland in November 2021, marked the beginning of the end for coal, think again. In December 2021, the Wall Street Journal reported that despite efforts to slash carbon emissions, global coal-fired generation is expected to rise 9% and hit a record by the end of 2021. The main drivers of the growth are China and India which together account for roughly two-thirds. As we have said throughout this book, when energy provision is at stake, countries will do whatever is in their own economic interests. That is why continued work to make clean energy cheaper and more reliable is critical.

12. Electricity

6 Tech for Storing Electric Energy

Dispatchable power is stored energy in a form that is ready for quick release as electricity. The first shoe to a clean future has already dropped. Wind and solar are the cheapest source of electricity today—and they are still getting cheaper. Storage is the second shoe that has to drop.

We have already discussed the most important dispatchable power source in the United States — natural gas. In addition to providing base power, some gas plants function primarily as peaker plants — plants that run only when there is a high demand. Today, natural gas is still marketed to the public as the cleanest fossil-fuel form of electricity (ignoring methane leakage in the transmission). In the future, we expect gas to be marketed as the fossil fuel that makes wind and solar power possible.

If you are like us, when you first hear "electric storage," your brain probably starts to blink "batteries." This turns out to be wrong when it comes to utility-scale storage on the grid. Instead, dispatchable power comes in different forms and fulfills different purposes. Most of it is pumped water (hydroelectric dams). It covers about 95% of the world's storage, which is roughly 170 GW of power and 9,500 GWh of storage. Batteries are less than 5% of this. Roughly speaking, currently all of humanity's electric storage could power the world for only about 30 seconds, and batteries for a measly 3 seconds. But batteries are the most exciting new technology, so we start with them.

Lithium Batteries

All of today's best batteries are based on the element lithium. Unfortunately, upon contact with the humidity in the air, lithium catches fire. Lithium batteries also tend to heat up a lot in operation. These two problems make lithium batteries hazardous and finicky, and prevents manufacturing large lithium batteries in giant pools. Instead, lithium batteries need to be manufactured into many small packages, which are expensive to make and need good care, feeding, and cooling. On the plus side, they are highly efficient in the sense that almost no energy is lost in the charge-discharge round trip and remarkably lightweight. Not too long ago, it was a sensation when Tesla bet the farm on a Gigafactory capable of producing more than 1 GWh of batteries per year. (They are now at more than 35 GWh/year.) In 2021, nobody bats

6. Tech for Storing Electric Energy

an eye when Koch industry, a fossil-fuel giant, announces plans to build a 50 GWh/year factory.

The cost for a utility-sized battery storage farm can be summarized by the following rough dollar figures:

Acquisition Costs of Battery Packs	$120,000/MWh
Costs of Battery Packs, incl. Wear&Tear	$250,000/MWh
Installation and Integration	$250,000/MWh
Rough Farm Cost	$600,000/MWh

After they are installed, the batteries are charged and discharged many times, which is why these figures are three orders of magnitude larger than the LCOE figures in Table 12.5.

The first figure of $120,000 is based on the cost of the physical chemical Lithium battery cells. On average a battery pack cost about $120/KWh in 2020, with some quotes already down to $100/KWh, others still at $150/KWh. The price of Li-Ion battery packs has been falling by about 10% per annum. It was about $1,000/KWh ten years ago. It will be solidly under $100/KWh before 2023, and potentially will reach $50/KWh by 2030 — and this is without any technological quantum leap discoveries.

The second figure takes into account that batteries don't last for very long. They wear out. We have seen lifetime estimates of 500 to 2,000 cycles. The most common operating pattern of battery farms is to charge every day and discharge once fully at peak time (8pm). Thus the battery packs need to be replaced every 2-4 years. Do basic math and it follows that if a battery farm can work for three years (about 1,000 days) and can charge/discharge fully once a day, then the LCOE from pure battery decay is indeed roughly on the order of $100/MWh. Over the full lifetime of a battery farm, it is not the $120,000 but the $250,000 that is more meaningful, because it includes not only batteries bought today but also in the future, over many generations of batteries.

The third figure is based on the cost of a battery farm. Integrating cells into the electric grid requires planning, housing, integration, inverters (devices that synchronize electricity to allow it to be connected to the grid), maintenance, operators, safety equipment, insurance, taxes, land, capital costs, the strategic know-how to buy electricity when it is cheap and sell it when it is expensive, etc.[5]

Roughly speaking, the true cost of a battery farm with 1 MWh capacity is about $600,000. It will run for about 10–30 years. This is a guesstimate. Even government estimates of LCOEs for lithium-battery provided energy can vary wildly, ranging from $150/MWh (at the EIA) to $350/MWh (at the PNNL) today. Recall that battery wear alone can account for about $100/MWh. The reason for the differences in the EIA and PNNL estimates is that they assume different expected lifetimes of battery farms as a whole (and thus different durations to amortize the non-battery costs). The EIA assumes 30 years, the PNNL only 10 years.

If the PNNL is right, the farm price is high today because storage technology will improve even faster, thus rendering today's batteries obsolete sooner. This means that technological progress in energy storage (including batteries) lowers today's expected price for tomorrow. In this case, the best response is to delay the aggressive installation of batteries. The low installed base in the United States (only about 1.7 GW of batteries at the end of 2019, with perhaps an average capacity of 10 GWh, for a grid of 450 GW and 4,000,000 GWh) and the slow installation pace may seem depressing, but this is because the true situation may actually be quite the opposite.

◇

There are already important lithium-ion battery breakthrough technologies on the near horizon. The biggest cost problem today is the wear and tear. As already noted, current lithium batteries can only charge about 1,000 cycles. More cycles are not important for your $1,000 cell phone, for which extra cycles would be nice but not crucial. Many cell phones break before their batteries do (around 3-4 years in typical use), and the battery replacement cost is only one tenth that of the cell-phone itself. But the economics of battery farms is all about expected battery lifespan. The wear-and-tear cost component looms large.

---

[5]Storing more energy requires more batteries (think $1 million per 10 MWh); providing more power just requires a bigger inverter (think $40,000 per MW).

6. Tech for Storing Electric Energy

The most important progress for battery farms will be batteries that can last for many more cycles. Graphene electrodes allow batteries to charge faster and last 2,500 cycles, but they are expensive. Tesla has already announced that its next generation lithium chemistry will charge 5,000 to 10,000 times. There is no law of nature that limits this number, either. Future engineering could push it to 20,000 or even 50,000 cycles. At this point, the fixed battery costs would become less important, because most people could use their car batteries not only for driving, but also for home grid storage.

However, it seems unlikely that the battery-farm based price will ever go much below an LCOE of $50/MWh. This is because of the integration cost of battery farms. They, too, will come down when storage farms are mass-produced (estimates suggest reductions from $250,000 to $200,000), but this is not as fast as the battery pack prices themselves. This situation mirrors the one for solar farms, where the prices of the solar cells are becoming less and less important, leaving most of the cost to installation and operations.

Bringing down utility battery costs will require a lot of mundane fine-tuning on each cost aspect. If batteries and wind/solar are co-located, the fixed costs can be shared. It makes a lot of economic sense to combine solar, wind, and batteries on the same farm to reduce overall cost. A DC rather than AC based transmission system could further reduce cost. And so on.

Other Batteries

If lithium batteries sound exciting, wait until you hear this. There are altogether different battery technologies that could obsolete lithium-ion batteries for utility-scale storage. Some of these batteries weigh a lot more (and are thus unsuitable for a car), but weight matters little for utility-scale storage. For example, a flow battery is akin to a giant pool of electrolyte with anode and cathode sticks inside. Increasing the battery capacity means increasing the size of the pool and sticks. The potential energy capacity of such batteries kept in large ponds could go far beyond those of racks of finicky lithium-ion batteries. VFlowTech is already scaling up manufacturing of a Vanadium redox flow battery with claimed LCOE of $100/MWh. Honeywell is building a 400 KWh flow battery pilot plant for 12-hour usage (with secret chemical composition) and plans to scale to a 60 MWh plant in 2023.

There are probably another dozen different battery architectures in advanced research stages. Corporations are investing $12 billion into battery storage in 2021 alone. Some technologies seem like magic — such as Aluminum-Ion batteries, which can charge in seconds and store multiple times what a Lithium-Ion battery can provide. One publicly-traded startup claims to have a battery that charges by converting rust into iron and discharges by converting it back. It claims that this could bring down the cost per KWh by a factor of three relative to lithium-ion batteries before the decade is out.

Our previous chapter advised caution. It is wise to remain skeptical about any one particular technology. Technology that works in the lab is a far cry from technology that works in the real world. However, there is no scientific reason why any one of these new technologies could not make a giant leap over current lithium battery technology. As we wrote, a good way to think about the new battery technologies that claim to have solved the problem is that each has a probability of success of less than 10%. It is only because there are dozens of potential breakthroughs that we are optimistic. Yes, there is a chance that none of them will work out, but the smart money bets on odds. In our minds, the odds are that within 10-20 years, either lithium-ion batteries will cycle more than 10,000 times or another battery technology will replace lithium as a utility-scale storage solution.

We close with another irony. Battery farms are relatively small and scalable investments, but battery R&D is not. The biggest risk today to spending billions on developing better batteries are batteries themselves. Any one new promising battery technology could be made instantly obsolete if another battery technology turned out even better and thereby stranded one's own R&D investment! We wouldn't put our money betting on any one technology, but we would put our money betting on at least one of them getting us there.

Hydro

Batteries are far from the only dispatchable storage. In fact, they currently work well only in niche applications.[6] Batteries are simply not economical yet compared to most storage alternatives. They are a long way from being able to supply a full night's worth of electric power almost everywhere.

---

[6]Batteries can take over many niche tasks. In particular, they can come online within 10 seconds to smooth out quick spikes of energy and thereby stabilize a wobbly grid. This allows operators not to have to over-provision as much electricity. They can also supplement other dispatchable power at peak times or when there is not enough transmission capability.

6. Tech for Storing Electric Energy

Their most important shortcoming is capacity scaling — and this is the primary consideration for energy storage. Think of the upper reservoir of a hydro-electric dam. Its energy is determined by the amount of water in the reservoir; its power is determined by the number of turbines. Increase the reservoir basin and you have more water and thus more energy capacity. For natural gas, the power is also limited by the turbine size, but its energy is practically unlimited as long as it is connected into the U.S. gas pipeline system.

In contrast, batteries have hard energy capacity limits. If all that is needed is 1 hour of 1 MW of backup power, batteries are already cheaper than natural gas. The farm needs just a few batteries. If 10 or 100 hours of 1 MW of backup power is needed, the farm needs 10 or 100 times the number of batteries. The price of gas dispatch power per MWh at a rate of 1 MW is the same for 1 MWh, 10 MWh, or 100 MWh of energy. Right now, it appears that batteries and gas are about equally expensive for 4 hours. This is not enough to cover a night's worth of electricity. Batteries have similar installation costs as dams per MW of power, but again fall short in terms of energy when the upper basin can hold a lot of water.

Thus, it is hydropower and not batteries that is the most important clean energy store today—by far. There are different kinds of hydropower. To be dispatchable, there has to be a dam. Conventional dams hold back water from a river in an upper reservoir and release it when needed. <u>Pumped-storage</u> hydropower (commonly called pumped hydro) means that a pump can push the water back up above the dam when electricity is cheap. This is not very efficient for each round trip, but it is easy to do at large scale. The United States has about twice as much conventional storage as it has pumped storage.

<u>Hydropower</u> is the largest source of storage today, with about <u>20–30 GW in the United States</u> and 130 GW in the world. It accounts for <u>about 2.5%</u> of U.S. generation capacity. In terms of energy capacity, think roughly half an hour's worth of U.S. needed energy (compared to all grid-scale batteries, which could muster a few seconds). Thus, hydroelectric dams sit somewhere between batteries and gas. Reservoirs can hold enormous amounts of water. However, the water is not infinite. Once the water has been released, the upper reservoir needs to recharge, either by pumping water back up or by waiting until the rivers naturally refill it.

Unfortunately, hydro-electric power is not only expensive but also in very limited supply. Sites are limited by terrain, geology, and water availability.

12. ELECTRICITY

Figure 12.7. Hydro Power and Energy

	Pumped	Dammed
USA	30 GW / 250 GWh	100 GW
World	180 GW / 1,600 GWh	900 GW

Source: Wikipedia and Wikipedia. Beside the fact that it provides on-demand energy, the world generated about 17% of electricity from hydro, which is about 4,000 TWh per year. (The third category, flowing hydro-power is not even dispatchable.) Note: Other estimates suggest as much as 550 GWh of U.S. pumped storage.
For perspective, the world is expected to reach battery storage of 135 GW / 450 GWh by 2030 — about one quarter of the world's pumped energy storage. However, unlike pumped water storage, battery installations are suitable for installation almost everywhere and are growing rapidly.

Nevertheless, the world could install more than four times the existing capacity. Getting more power by building out hydro can cost as little as $10/MWh and as much as $250/MWh — and this includes fixed costs. On average, hydro power newly built these days has an LCOE of $50/MWh. And once built, the marginal cost of hydro has been estimated to be as low as one-tenth that of batteries: not $100/MWh in battery wear-and-tear, but $10/MWh. Figure 12.3 shows that CAISO brought hydro power online around 5pm, throttled back around 9pm, and turned it off around 10am.

There are also other drawbacks. When there is a drought, hydro loses power. In California, this is already a serious threat. And dams can have negative effects on the environment, as well. Many environmentalists are fiercely opposed to them. They have a point because dams often have a significant impact on the natural environment. Nothing comes for free. As we economists would argue, one needs to weigh the costs against the benefits in each case.

6. Tech for Storing Electric Energy

Other Solutions

There are also other ways to store electricity that are not (yet?) in wide use. There are mechanical solutions, such as one that involves driving a train up a slope, and another that uses a crane to lift and lower blocks. The two most interesting large-scale options are, however, geothermal storage and compressed-air storage. Both take advantage of underground caverns, many created by century-long oil and gas extraction.

Geothermal storage could warm a substance like molten salt and extract the heat energy (as steam) on demand. (Earth further donates some extra energy in the form of radioactive heat coming from the planet's interior. It could even make sense to augment this heat further with a human reactor deep underground.)

Compressed air is similar to pumped hydro but more experimental. Air is compressed into underground caverns when electricity is cheap, and let out (like a balloon) when it is expensive. As with hydro dams, air-storage caverns and plants are expensive to construct and require suitable underground rocks and caverns.

The round-trip energy leakage is much higher for hydro, geothermal, and compressed air than it is for batteries, but their energy capacity potential is also much larger. And again, ironically, the biggest problems with all large-scale storage schemes are their high fixed costs and the risk of batteries. Who wants to build out energy storage at a cost of many billion of dollars when it could become obsolete if a better battery were to be invented?

A smaller-scale solution is chemical storage. Hydrogen could be electrolyzed when electricity is cheap and stored. The previous chapter explained why this is particularly useful for off-grid needs, such as in airplanes or ships. In stationary applications, hydrolysis is not (yet) economically viable. The round-trip efficiency of about 35% remains too low. It may become viable when solar and wind power cost $20/MWh and chemical engineers find better catalysts. We will return to hydrogen in the next chapter.

Table 12.8 shows rough estimates for the cost of storing large amounts of electric energy today. All technologies can store energy for somewhere between $100/MWh and $200/MWh. It is much cheaper to capture 1 MWh from the sun at noon at $25/MWh than it is to store and retrieve the same MWh.

12. Electricity

Table 12.8. Electricity Storage Choices, LCOE per MWh, ca 2020

		2020 100 MWh	2020 1 GWh	2030 100 MWh	2030 1 GWh	Efficiency
Battery	Li-Ion (EIA)	....	$120	....		
	(PNNL)	$350	$340	$250	<$240	85%
	Redox Flow (PNNL)	$220	$180	$210	<$210	
Hydro	(Pumped)		$130		$130	80%
	Natural Gas	Peaker Plant: $80–$200, including fuel.				
Other	Compressed Air (CAES)		$105		$100	50%
	Hydrogen		$200		$150	

Explanations: These are energy storage devices where both input and output are electricity. If the end use is heat, heat storage is likely to be cheaper. The energy cost to charge is not included. Figures assume one full charge-discharge cycle of once per day. Efficiency is the fraction of energy that is regained from input to output. The PNNL $350/GWh cost reflects Lithium battery prices of about $140/KWh in 2020, expected to fall to about $50/KWh by 2030; and a 10-year lifetime.

Source: PNNL 2020 Report and EIA 2021 Outlook.

Beyond Daily Storage

Most of the previous discussion centered around the provision of regular night-time electricity. However, wind and solar power may fail not only at night, but for days at a time.

For example, in January/February 2019, a polar vortex over the East Coast for about one week took out about 10 GW wind-power from the grid's 30 GW load. The electricity price shot up from its usual $50/MWh to $200/MWh and more. On some days, conventional generation had to step up from its typical 60% to 99% coverage (with record profitability for generators in the process).

Energy storage for polar-vortex-like events would need neither the immediate response of batteries nor their near-perfect input/output efficiency. The power provision could come on line more slowly and have higher variable cost (as long as fixed costs are really low). The grid would hopefully not have to resort to vortex-emergency power very often.

6. Tech for Storing Electric Energy

The economics of cell-based battery capacity for rare but long-time power provision is brutal. Recall that capacity expansion is not just a matter of enlarging a reservoir or pool, but a matter of purchasing more expensive cells. It is unlikely that batteries will become economically sensible for this purpose within a few generations. It also makes little sense to build other high-fixed cost installations (underground caverns, dams, etc.) for such unusual events.

Because of such rare cases, a litmus test requiring 100% green energy makes no economic sense. It would be so expensive that it could wipe out public support for the transition. We therefore believe that it is enough when 95–99% green energy provision can be sensibly achieved. The only viable economically sensible "last-resort" alternative for long-term storage (and for decades to come) is natural gas or hydrogen with their near infinite capacity and run-time. Even otherwise crazily expensive and dirty Diesel generators (at $100-$200/MWh) could have a very rare role to play.

But the message of our book is to stop arguing about whether decarbonization should be 80% or 100% — the world is so far from 80% that the arguments are currently irrelevant. The world should instead focus on *moving the needle* to 80% asap and worry about the final 20% later.

Many other interesting developments are coming out of left field, often seemingly mundane improvements over existing designs. Standard radial flux alternators (generating electricity from turbine engines) have efficiencies of 90%. Axial alternators can push energy losses down to 2–3%. This can improve the economics of round-trip converting mechanical power to energy and make all sorts of alternative energy storage viable. A new way of drilling non-mechanically may just have opened up access to very deep holes to tap the power of magma.

Speculating for fun as in science fiction, what will ultimately be feasible in terms of storage cost? The world will probably need many types of energy storage. It could be that the best storage method has not even been invented yet. If some storage technology could solve the problem of large-scale provision when there is a run of low intermittent power days (i.e., offering long-term storage at high capacity), this technology could also solve another related problem: long-term seasonal storage. Electricity demand is highest in winter, then summer, then spring and autumn. If there was such a viable technology to store energy in spring and autumn, the economy would need less generation capacity in the first place. Builders could tolerate even more energy losses as long as fixed installation costs were cheap enough and if the new technology could hold sufficient amounts of energy. This scenario may eventually become feasible with some aforementioned underground solutions, such as molten salt. A cost of $50/MWh in a few decades seems achievable. At this cost, wind and solar power could compete economically with all but natural gas. Push it to $30/MWh, and wind and solar would become dominant.

We admit much of this remains a dream. Molten salt storage as well as many other potential energy storage solutions all call for more research and development. Put the brightest minds on the biggest problem of the world today and give them enough money to experiment and come up with better potential solutions. Let's create the conditions that could allow us to get lucky!

No Electricity Out

The above solutions discussed the storage from the perspective of electricity in, electricity out. For many applications, this is not necessary.

The cheapest and best solution to the problem of energy storage is to avoid having to store electricity in the first place. This is not a crazy idea, and it does not require either more base power or a return to the stone age. Instead, it mostly requires passing the right incentives on to electricity consumers. We will come back to this theme a few times in the rest of the book.

Much electricity ends up being converted into heat or cold. In such cases, it is usually cheaper to transport electricity to the destination first, where it is used to heat or cool a substance (often water or oil) in an insulated container. This heat/cold can be released later when it is needed. It even has a fancy technical term, "sensible heat storage."

There are sensible heat storage solutions both on industrial and small scales. Industrially, underground caverns and insulated furnaces can hold large amounts of heat. In homes, most of today's water heaters already work this way. Sensible thermal storage (together with heat pumps) could take over much of both residential heating and cooling; and do so both for water and air heating and cooling. This will work well *if* electricity is extremely cheap when wind and solar generation is at a maximum mid-day. With the right incentives, electricity consumers will no longer want to buy as much energy at night.

7 Transmitting Electricity

Remember the polar vortex? Why did the rest of the United States not send power from elsewhere to its East Coast? It's because it is expensive to build a grid that can shuttle electricity over large distances. Even now, without such expensive long-term transmission, about $1-$3/MWh of the $120/MWh that retail customers tend to pay for electricity is due to transmission costs. If the grid had to be capable of carrying a lot more intermittent power over longer distances and be smarter, the transmission cost could increase to as much as $5/MWh. The grid itself is a fascinating topic to study. Let's carry you away for a moment.

A cable that can transmit about 2 GW of power costs about $2-$6 million per mile to build and install. (It also loses about 1% of power every 500 miles.) A coast-to-coast 2 GW line would cost about $10 billion and lose about 10% of its power. To transmit 80 GW (about 8% of today's power generation but a smaller fraction of the future's much higher electricity generation) would cost $400 billion — an almost unimaginably large figure. It is the equivalent of about one year of the entire revenues of the electric power sector of the United States today. Even if the cable were to last a century, the building cost is still *very* expensive. Clearly, whenever local generation and use are feasible, it is a lot cheaper. At $400 billion, one may as well build another 20 nuclear power plants on the East Coast instead.

Design and Capacity

Today's U.S. electric transmission grid remains both primitive and chaotic, but sophisticated in its many patching mechanisms that make it work. This is because our grid was never truly designed. Grids began as private efforts in the 19th century and grew organically during the 20th century to handle primarily connections that ubiquitous and relatively local coal plants needed to provide power to customers.

Thus, the U.S. grid has never been operated by one centrally coordinated agency, but by many private operators within different states and regions, with strong links to their local providers, customers, and politicians. This arrangement worked well when local supply and local demand were tightly paired. However, the problems today are changing.

With intermittent wind and solar power, a lot of extra energy may need to be shuttled around. This alone could double the necessary wires. Moreover, wind and solar power also require greater coordination, because the grid will have to be ready to transmit when these generators want to send power, not when the operators and customers want power. The grid will also have to be ready to allow connections to newly built wind and solar plants. Too much power and the wires could fry.

In some situations, transmission could be a substitute for storage. Instead of storing electricity in Los Angeles when afternoon demand is low and supply is high, it could be transmitted to New York City where the opposite (with its 3-hour time difference) is the case. Wind power is almost always available somewhere in the United States, but not always where it is needed. It needs to be shifted around. For another wrinkle, AC transmission lines are great when power comes from rotating engines (like windmills) and when there are many on-ramps and off-ramps, but it is not efficient for DC power-based generation (like solar) and not over very long distances. The United States may thus want to build a new DC network. The DC power line from Washington state to California proves that this can be economically viable at least in some cases.

Smarts

Operators have only modest real-time intelligence. They know how much electricity was needed in the past and they can learn a little from how stable the frequency and voltage are at a few sensor points. However, by the time they learn of problems, e.g., a plant that goes offline, customers may already have suffered consequences. Because the operators' guesses are inaccurate, their best option is to provide too much electric power, so that the grid will not brown out.

A better solution would be a "smart grid" that could entail all sorts of real-time measuring meters and switches that allow grid operators to improve their routing of electricity to meet demand. Even more importantly, smart meters could allow customers to signal how much power they will need in the future. With better intelligence, operators could waste less electricity.

With price signals, customers could balance their demand throughout the day. The grid could let consumers know when the price is lowest so they could charge their cars or do their laundries — or, more likely, let their cars and laundries know. The ability of devices to adjust their demand and signal it back to the utilities would reduce the volatility of electricity demand, the volatility of price, and the total price itself. A smart grid would reduce the need for storage and backup power.

Linked Grid and Technology Coordination Problems

A large expansion of the U.S. grid is a necessary precursor to a clean energy future. If the economy is to electrify other work that is not electricity-based today, then the grid needs to have a capacity of at least twice of what its capacity today is. More distant regions need to be interconnected with fatter wires. Some estimates suggest that a region-by-region clean solution without dramatic changes in long-distance transmission capabilities would cost \$135/MWh (about three times today's price), while a grid-supported national solution could reduce the cost to only \$90/MWh (about two times today's price).

For many other issues we discuss in the book (e.g., generation and storage), we can worry about this decade — about "moving the needle now" — rather than about future decades. However, this is not the case for the grid. The grid has to be planned and upgraded asap in its transmission capability, its coordination, and its smarts in order to be ready for the subsequent construction of more intermittent generation and storage.

The upgrading process will be difficult, because it will involve hundreds of interest groups — generators, customers, storage, politicians, regulators, lawyers, environmentalists, and so on. Today's grid operators are already pretty good and quick when it comes to sending power on to neighboring regions' operators on a daily basis. They are not so good and quick when it comes to planning and approving large new transmission infrastructure over years and decades, especially across larger distances. It is not clear whether private and regional operators will be capable of engineering the large and rapid changes that are in the public interest and not necessarily just their own.

And if all this was not difficult enough, the changes must also not jeopardize what the fossil-fuel grid largely already delivers today — though with terrible consequences for the environment and public health. Most of the time, despite its mechanical nature prone to breakdowns, fossil-fuel generators have delivered reliable electrical energy that is critical for the operation of a modern economy. Interruptions, such as those recently in Texas, are costly.

But upgrading the grid while maintaining its reliability is a delicate and tall order – it's like upgrading an airplane in flight.

8 Earth's Economic Energy Problem

In the 20th century, almost all electric power came from baseload power in the form of coal, and almost all transportation power came in the form of oil. The future belongs to wind and solar power, supported by a better grid with dispatchable energy storage. Wind and solar power are ready. They are already the cheapest sources of energy *ever*. They are so cheap that further improvements are no longer of first-order importance. The grid and storage are not ready. The grid just needs some grit to improve it. It can be made ready, though at a non-negligible cost. The key remaining real problem is storage. It remains an order of magnitude too expensive. It is not an overstatement to characterize humanity's energy problem as little more than an energy storage problem. Solving it will mean solving the world's dirty energy problem. With cheap storage, wind and solar technologies will rapidly eliminate the need for fossil fuels worldwide without the need for much further intervention.

Perhaps the storage problem has not already been solved because it was not so important, so urgent, and so potentially profitable in the past. We hope that some people will get very rich solving the problem — capitalism at its best.

Base or Intermittent/Dispatch Power?

To understand at least the outlines of the tradeoff today vs. where it has to be, we want to compare how much it would cost to supply all of the United States with natural gas electricity on the one hand vs. with solar/wind/stored power on the other hand.

Our goal is not to be exact. We are not grid operators. The actual operation of the grid is beyond human comprehension. In real life, the grid operators model, simulate, and predict the grid on large computers. When it looks as if power will fall short soon, they offer to pay more, thereby inducing higher-cost providers to come on-line. Over longer time horizons, with technological and construction uncertainty, the operators' allocation tasks become even more difficult. If they get it wrong, a few years later, woe is us. (China is seriously afraid of running out of power, which is why they are building so many new coal plants.)

Instead, our goal is to present back-of-the-envelope calculations and only for daily provision needs. We will round aggressively, because our calculations are far from exact and make many simplifying assumptions — such as assuming that electricity on the West Coast is the same as on the East Coast (even though the United States cannot transmit a lot of power across the continent) or assuming that storage incurs no losses. We will not allow wind to blow at night or allow the sun not to shine during the day. We will assume that intermittent wind and solar power will always be working like clockwork from 10am to 6pm, i.e., for 8 hours a day, and not a minute more or less. You have been warned.

The United States needs about 520 GW of power during those 8 working hours, i.e., $520\ GW \times 8\ h \approx 4\ TWh$ of energy. It also needs a further 450 GW during the remaining 16 hours, i.e., $450\ GWh \times 16\ h \approx 7\ TWh$. Ergo, these 7 TWh need to be generated and pushed into storage during working hours and pulled from storage when needed. To charge storage of 7 TWh in 8 hours plus deliver 4 TWh immediately requires about 11 TWh of generation during the time when wind/solar are available. Dividing 11 TWh by 8 hours suggests that 1.4 TW of wind/solar power generation can satisfy immediate power needs and charge the storage reservoir.

The example's first rule of thumb is thus that the United States would require a nameplate (peak) capacity of wind and solar output of about 1.5 TW. which is about two to three times the immediate retail customer daytime power requirement of about 0.5 TW. A second rule of thumb is that it would require energy storage for about 20 hours.

Figure 12.9 shows a more realistic assessment. The law of diminishing returns (from Chapter 6) is at work: About 80% of the grid could be covered with solar and electricity storage of about 10 to 20 hours of electricity. To reach 90% would require about 100 hours. To reach 100% could require 1,000 hours or more — the equivalent of "murder" from an economic perspective.

Let's move on to cost. Ignoring storage cost, generating 11 TWh for a price of, say, $33/MWh from wind and solar would cost about $350 million. The same 11 TWh from natural gas at a price of, say, $50/MWh, would cost about $11\ TWh \times \$50/MWh \approx \$550$ million in generation costs. The difference of $200 million is our clean energy-storage budget.

12. Electricity

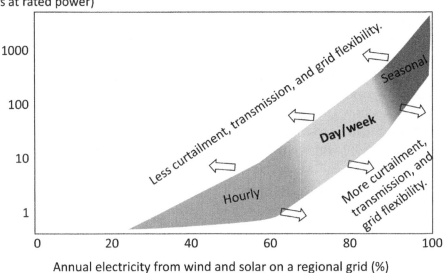

Figure 12.9. Required Energy Provision

Source: Albertus, et al., 2020, Joule. Recall from the introduction to Section 6 that the world today has only about 30 seconds worth of storage, 95% of which is pumped hydro.

If scientists and engineers can find a way to store 7 TWh for less than $200 million, it's curtains for natural gas on ordinary days. This suggests that technology would need to bring down storage cost to $200 million/7 TWh ≈ $30/MWh. Ziegler at al. estimate that storage costs would have to drop to $20/MWh to allow a 100% clean system — but $150/MWh could be enough to make it to 95%. *We are already there!*

Now look back at Table 12.8 for today's storage technologies. Unfortunately, current technologies can't make it down to $30/MWh. It costs more like $100–$200/MWh today for storage at scale. Existing hydroelectric storage could do it at $30/MWh, but only if construction costs are ignored and only if magic created an unlimited supply of hydro-electric storage dams.

Under realistic current storage costs, say $150/MWh, the cost of a clean-energy U.S. system would balloon the electricity cost to $350 million, which comes to a cost of about $1,400 million in total. This is roughly three times the cost of a natural-gas electricity system — think 5–10% instead of 3% of

8. Earth's Economic Energy Problem

GDP. (Double this if you want to electrify the transport and heating sector.) Nuclear power would be much cheaper — but, like wind/solar/storage, its economics cannot compete with natural gas.

Simply put, completely clean energy is still too expensive, despite the low price of wind/solar power. Even if generation were free, it would not be enough. The storage-cost problem still kills the universal clean-energy solution on an economic basis. A fossil-fuel tax could help somewhat, but even $50/tCO$_2$ wouldn't be enough to completely wipe out natural gas.

Today, there is really only one way to transition to a clean energy economy — shifting energy demand towards day-time hours. When no storage is needed, no one needs to pay for it. As mentioned, we will come back to plans to shift energy use below. In the very long-term, storage technology breakthroughs will hopefully change the arithmetic.

Now look back at Figure 12.4. The United States is still ramping up natural gas generation even when wind and solar are still working. That is, expensive peak gas generators are still being dispatched due to demand increases, not due to supply decreases. Not all but a lot of this extra mid-day power could be replaced by wind and solar energy almost immediately. Instead of 100 GWh generated by wind and solar today, it could probably be 200-300 GWh. *Moving the needle* could get us to 200 GW in short order. Market forces are already working. In 2021, new installations added 30 GW of wind and solar power.

Now or Later?

Economically, it is too soon to transition *all* of the grid by installing 7 TWh of battery storage. The public is unlikely to stomach electricity prices 2–3 times what they are paying now. Moreover, the clean energy cost is getting cheaper by the year, so waiting just a few more years makes sense. And many other storage technologies also look promising at utility-scale needs, perhaps reducing the cost to $60/MWh within a decade or two. At this point, countries can contemplate whether the full clean-energy sacrifice is worth it. More importantly, for the world' sake, it has to be not just the United States and Western Europe that make sacrifices. A clean-energy solution will also have to become feasible all over the world. Remember if only the West were to move away from fossil fuels that would not be enough to reduce CO$_2$ in the atmosphere .

If there ever was a role for governments to subsidize R&D in the social interest, clean-energy storage research is it! This is where collective clean-energy sacrifices should be directed.

9 The Business Perspective

Let's look at the decision of an entrepreneur today. She is not concerned about how to migrate all of California, the United States, or the world to clean energy. She is more concerned about the economics today — how she can make money by selling electricity or building new plants. She need not install many hours' worth of batteries to cover all 12 hours of nighttime demand. All she needs to do is to work out whether she can make a profit from buying electricity at the lowest day price and selling it at the highest day price — say, buying 1 hour's worth at 1pm and selling it at 8pm. In an ideal free market, competition between entrepreneurs is such that the rest of us are getting them to sell electricity to us almost at what it costs them. This is the socially positive aspect of capitalism at work.

The decision entrepreneurs face regarding what type of new plant to build depends not only on the demand pattern but also on all the other plants on the grid. If there is a lot of intermittent power, it makes more sense for her to build dispatchable power, and vice-versa. If there already is a lot of base power, then building solar power only makes sense if there is excess demand during daylight hours. How can the grid operator direct what plants entrepreneurs should build? Or how can the entrepreneur decide?

Fortunately, for the most part, this is not a decision that regulators need to make. It's a decision that they can leave mostly to market forces. And all the entrepreneur needs to do is to look at the price of electricity. If it is usually high during the day, she can build a new solar plant. If the price varies too much, she can build electricity storage (and thereby make the prices more similar).

Let's look at the business case from the perspective of a California entrepreneur. (Of course, entrepreneurs elsewhere have different problems. For example, gas is more expensive in many other parts of the world.) Figure 12.10 shows the electricity price on the same day for which we graphed the provision in Figure 12.3. Our entrepreneur could have sold her electricity at those prices, which came from multiple auctions conducted by CAISO.

9. The Business Perspective

Figure 12.10. California Price, March 21, 2021

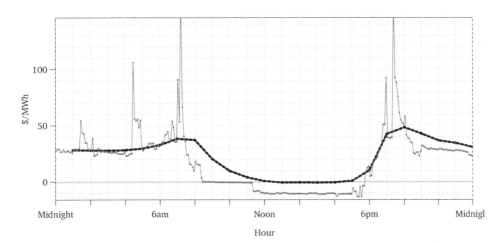

Source: CAISO Oasis. The black line is the 1-day ahead prices, averaged for Southern and Northern California. The blue line is the 5-minute auctions. Note that the day-ahead price of electricity was $0 from noon to 5pm in Southern California.

The smooth black curve shows prices auctioned one day ahead. The electricity price ranged from zero from about 10am to about 6pm, all the way to $45/MWh around 8am and 8pm, with most of the night around $30/MWh.

The operator can never perfectly predict demand and supply, and so leaves some electricity to be purchased in real-time. This is seen in the more spiky blue graph. Brief spikes on this day could be as high as $200/MWh. If you had owned a battery farm, you would have jumped in to sell power — which is also why the spikes did not last long. Real-time electricity also had a negative price from 5pm to 6pm, because the transmission grid was overloaded. If you had no way to spill power, you would have had to pay the grid to take it. At this point, if you had owned a battery farm near the hubs, you would have tried to jump in and buy it for this negative price. If you were close to the spilling generators, they would have happily paid you and not the other way around.

With conventional technologies, it seems there is not much money to be made in California on ordinary days by installing new plants and storage — the market seems to provide electricity at very competitive prices, at least on this day in spring. (In summer, there is more power demand during the afternoon for air conditioning, allowing generators to earn more. And not every day is as boring as this March 21. Moreover, California also runs an auction every three years paying utilities to install more capacity.) Some plants are still being built to earn money when demand is high and supply is low. Moreover, some plants in California are ready to be retired, opening up opportunities to build new ones. And, of course, you could make a killing if you could invent cheaper generation power to beat the prevailing price of $30/MWh at night or a cheaper storage technology to take advantage of the short spikes and variability.

Let us harp yet again on a final point. In the short term, the grid must take customer demand as given and cover it as effectively as possible. In the long term, market forces are more dynamic and responsive. Low electricity prices are not only caused by but also put a damper on the installation of more solar energy generation. Over time, however, economic forces will induce more consumers to use more power during the day if they can take advantage of cheap electricity. The prime potential consumers would be battery farms that would effectively "arbitrage" the electricity price and thereby even it out! That is, they will compete around noon for purchasing electricity and thereby drive up the price; and they will sell in the evening and thereby lower the price.

10 The Role of The Market

If the grid and related decision problems seem impossibly complex, it's because they are. The only planning device known to humankind capable of coordinating a system as complex as an electricity grid is to rely, at least in part, on market prices. Market prices induce competitive electricity suppliers to do the right thing. If there is not enough electricity supply, the electricity price rises; the price rise induces entrepreneurs to bring more power online, both short-term and long-term. Attempts to coordinate electricity provision without healthy market competition among companies – or, worse yet, attempts to fight economics – are bound to fail, as they always have. However, solutions without any regulation are also impossible. There is a need for regulators to coordinate the system and make sure that there is healthy competition among generators. Countries with good governments and the trust of their public will find it much easier to transition their energy systems.

Laissez Faire and Price Controls

Of course, no consumer likes power companies when power prices are high. Everyone wants safe, reliable, and affordable power.

Who wants to wait for new plants to be built and come online (and then bring down power costs) when their power bills have just tripled today? Thus, the public often clamors for price controls — and they do work quite remarkably at first. After power plants have been installed, they can continue to produce electricity profitably on the margin, although they cannot recoup their investment. If regulations are lax, generators can also neglect upkeep to lower prices and make consumer happier. (In the case of natural gas operator, they can provide cheaper electricity by neglecting pipeline leakage and proper capping at the end of life of the wells.)

Unfortunately, price controls also mean that not enough generators will want to build more plants thereafter in the future. It is the expectation of recouping investment at higher market prices that makes it worthwhile to build more plants. Take away high prices, and companies will become less eager to build more plants.[7] Similarly, it is the high price variability that makes it worthwhile to install more electricity storage. Take away price variability (the "arbitrage" of buying low-priced electricity and selling high-priced electricity), and companies will become less eager to build more storage.

Price controls and fear of them have ruined electricity sectors in many third-world countries, in which competitive entrepreneurs without political connections no longer want to sink the large amounts of capital required for building out power. They don't trust governments not to turn around in the future and expropriate their capital investments. Would you like to volunteer to invest your own savings on such terms?

Electricity markets require both competitive companies to build plants and sell electricity and regulators that coordinate and limit the power of these companies. Central planners influenced by political considerations or nepotism and not subject to true competitive pressure or negative consequences when they fail — whether in Washington DC, Moscow, Beijing, Kinshasa, or Brazzaville — cannot do the job. But neither can the free market alone. It's a vexing problem.

---

[7]To keep companies building new plants and stand-by power ready to provide more electricity, many U.S. grid operators hold capacity auctions, which pay generators not for energy but for power provision.

12. Electricity

Role of Regulation

However, some caveats are in order. Leaving everything to the free market is also not the best way. Regulators still need to make sure that utilities follow stringent safety and upkeep requirements; that the grid has the reliability that is more in the social than in the private interest; that competitive entrant plant builders can easily connect to the grid; and that builders do not collude with one another to keep the price artificially high or merge to reduce competition. Mergers would raise their profits by creating electricity scarcity. With fewer plants (or "maintenance" at the worst of times in a few key plants), the electricity price will be much higher than it would be in a competitive market. It is also well known that incumbents try to take advantage of the system. They will push for ever more difficult reviews and regulations for newcomers. Given the large capital requirements (and long process requirements), they will make it harder for smaller, cheaper, and better entrants to come in and compete away fat profit margins.

Thus, it is the regulatory agencies of the government that must watch the competitive market on behalf of the ultimate customers (industry and households). Without guard, the competitive system would quickly deteriorate. But guarding is itself difficult to do even for the best of governments. The regulators have no choice but to depend on information that is fed to them by power generators. If the regulators turn too adversarial, they will have to regulate in the dark (perhaps quite literally). Worse yet, power generators operate in political systems in which campaign donations and bribes often speak louder than words and lower prices. Even in the United States, the average regulator today is a former or subsequent industry lawyer or executive for a power-generation company, appointed by politicians. (Who else would know the ins and outs?) In other countries, it is often the nephews of the rulers.

Even in the United States, many economists believe that government regulation itself often becomes not a guarantor but a barrier to entry — creating a system that is abused by incumbents. Among the two evils (no regulation vs. biased regulation), it is not clear which is better in any one particular case. Maintaining impartial regulation is a never-ending and never-perfect difficult balancing act. Naïve environmentalists who instinctively prefer government regulation over capitalism may be well-meaning but often fail to appreciate the real-world dilemmas.

How To Lower Prices

Allow us some pontificating. As we are writing this, the UK is experiencing an energy shortage. A primary reason for this shortage? Price Controls! The UK has a price cap on unit electricity. Unfortunately, the price of gas has recently skyrocketed, making it unprofitable for gas generators to come online. What exactly did the U.K. expect when it instituted price controls? Utilities that would voluntarily sell for a loss?

Here is "the economists' advice" on how to lower consumer prices. This holds whether it is in the context of power or apartment rents. Price controls always backfire in the long run. Instead, the correct economic solution is to encourage as much competitive entry as possible. Prices will come down when incumbent sellers cannot prevent entrants from competing and offering lower prices. If it takes a while for entrants to arrive and solve the cost problem, subsidize the consumers. Do not punish the suppliers.

11 Current Power Plans and Forecasts

Table 12.11 is among the most important tables in our book. It summarizes the existing electricity energy generation and power capacity of the United States, China, and the world. It shows how the picture has changed in the last 5 years and the reference forecast for 2050. Except for longer-lived hydropower, 30 years is also roughly the lifetime of a power plant. Thus, most of the plants generating electricity in 2050 have not yet been built.

In the United States, coal has been on a steep decline — but it isn't done yet. Even by 2050, about 10% of U.S. electricity is still expected to come from coal. Natural gas generation is continuing to expand. Together, fossil fuels may already have peaked, but the decline will be slow. The U.S. industry has become bearish on new fossil-fuel generation power plants.[8] For now, entrepreneurs are bringing online only new wind and solar plants, whose purpose it is to cover the growth in our energy demand. In sum, the future looks rosier than the past, but it is not all that rosy. There is no "zero-carbon" electricity future on the U.S. drawing board for now.

---

[8]Much of the 6.6 GW had been in planning stages for many years and is located in fossil-fuel-friendly Texas, Oklahoma, and Pennsylvania, with 3.9 GW for base power and 2.6 GW in peak power.

12. Electricity

Table 12.11. Power and Energy Forecasts

Panel A: Generation, in TWh per year

Region	Year	Coal	NatGas	Nuclear	Hydro	Wind	Solar	(Others)	Total
USA	2015	1,410	1,317	797	249	191	39	(2.2%)	4,092
	e2020	774	1,636	785	283	343	132	(2.7%)	4,061
	e2050	593	1,953	594	294	790	1,071	(3.0%)	5,458
China	2015	3,860	148	161	1,103	186	45	(1.1%)	5,562
	e2020	4,313	267	331	1,117	574	281	(1.5%)	6,893
	e2050	3,556	803	1,002	1,448	1,001	3,379	(0.4%)	11,230
World	2015	9,621	5,585	2,440	3,843	828	263	(2.6%)	23,171
	e2020	8,244	6,458	2,630	4,034	1,741	832	(4.2%)	24,991
	e2050	8,115	7,306	3,025	5,548	6,833	10,152	(2.3%)	41,953

Panel B: Power Capacity, in GW

Region	Year	Coal	NatGas	Nuclear	Hydro	Wind	Solar	(Others)	Total
USA	2015	...758...		99	80	73	23	(2.2%)	1,074
	e2020	221	429	97	79	127	84	(2.7%)	1,155
	e2050	106	788	72	80	241	519	(5.9%)	1,919
China	2015	...990...		27	296	129	43	(2.0%)	1,516
	e2020	1,087	88	48	322	184	169	(1.2%)	1,921
	e2050	1,101	316	143	417	333	1,480	(7.8%)	4,108
World	2015	..3,919..		343	1,051	415	227	(4.4%)	6,231
	e2020	2,201	1,839	374	1,120	595	511	(7.4%)	7,172
	e2050	2,273	2,414	427	1,507	2,362	4,640	(7.6%)	14,747
Plant Changes in 2020									
USA	New	+0	+7	+0	+0	+24	+14	(4.5%)	+46
	Retire	−9	−2	−2	−0	−0	−0	(4.7%)	−13

Source: U.S. Energy Information Administration. Panel A: 2015 and e2020 and e2050 generation. Panel B: 2015 and e2020 and e2050 power. (We use estimated 2020 numbers for comparisons with estimated 2050 numbers.)

11. Current Power Plans and Forecasts

In contrast, in China, coal remains dominant. It provides the majority of electricity. Worse, it is projected to shrink only slowly. The world's biggest environmental calamity today is China's massive coal-plant building program — driven more by employment in the coal sector than by cost advantages of coal. These new Chinese coal plants will be with the world for another 30 years. If anyone has a good idea how to stop them, this is the time to speak up. In total, fossil fuels in China are forecast to grow, not shrink. Nevertheless, as in the United States, wind and solar plants are expected to cover the lion's share of China's electric energy *growth*.

The world overall is more like China than the United States. Coal will remain steady, and natural gas will grow modestly. Clean energy will grow faster to cover most of humanity's increasing electric energy demand. Thus, fossil fuels are expected to provide about one-third of humanity's energy in thirty years — down from about two-thirds. The planet does not work in percentages, though. In absolute terms, emissions in the electricity sector will no longer increase relative to where they are today, but they will also not decrease. Just holding emissions where they are today in the face of a 66% growth in electric energy consumption is a great accomplishment for clean energy, but it's not enough. This future does not look carbon-free. It would be wise to do everything possible to accelerate the transition.

12 Reliability

A critical aspect of electric power that we have so far neglected is its reliability. The electric power grid must cover both energy and power needs on demand. When consumers demand 1 TW of power (about the average power in the United States), the grid needs it *now at that very moment*. If it is not provided, the grid may "brown out" (delivering insufficient voltage, which can damage some machines) or may collapse altogether. When the grid demands 12 TWh of energy today (approximately the U.S. energy consumption per day), it is not sufficient to deliver the energy tomorrow.

Look on the bright side, Officer. I've reduced the electrical consumption of an entire neighborhood.

Customers in richer countries expect electricity to be available when they need it. Their businesses and their livelihoods depend on it. They thus greatly value reliability and are usually willing to pay for it.[9]

Yet, even in the first world, power has never been perfectly reliable, but it is so near-perfect that we take it for granted. The typical U.S. household suffers only 6 hours of outages out of about 8,800 hours per year. Most outages happen when a local power line is cut — and the utility company will almost immediately dispatch crews to fix it.

The generation itself is even more reliable, because the U.S. electric grid is designed to oversupply electric power at all times. However, it is possible for large parts of the grid to fail, and they have indeed done so recently. For example, California famously had to curtail power delivery in 2001 in order to prevent a collapse of the system. (Nowadays, California electricity utilities also regularly turn off power in certain locations when high winds threaten to topple electricity towers and start wildfires.) The Texas power outage of

---

[9] Grids tend to be more reliable in countries that are wealthier. The two reinforce one another. On the one hand, it is difficult to run an economy without a reliable grid. On the other hand, many aspects of an economy that promote economic growth and stability also promote a reliable grid.

2021 affected about 5 million people and crippled its economy for about a week. The public outrage was on the news every night. A typical headline read Despite Losing Power for Days, Texans Will Pay Higher Power Bills — Perhaps for Decades to Come.

With less base power and more intermittent power, the variability in power availability could rise. Storage can mitigate some of the daily volatility, but there need to be plans to address once-per-year or once-per-decade situations in which weather conditions are consistently bad and the standard storage will have run dry. What company would want to build a plant that is turned on only once a year or once per decade?

The economists' answer is that if customers value the presence of this once-per-year availability highly enough, they should be willing to pay for it. In the Texas 2021 outage, the retail price of electricity shot up to $9,000/MWh. Many customers saw their service cut off (the equivalent of infinite pricing) — whether intentionally by the grid (because they had to pay suppliers so much that end-use provision was unprofitable), or unintentionally because no further electricity was available.

Interestingly, Texas is unusual. It is its own island on the grid, largely unconnected to the rest of the United States. (This avoids Federal regulations). Indeed, Texas was so deregulated that it had already allowed retail customers to buy electricity at a prevailing grid-tied price, rather than at the more common fixed price. In normal times, the customers who had chosen this option paid electricity bills that were much lower than those of their neighbors who chose the guaranteed price. However, the response by these customers to the spikes in the electricity price was not one of gratitude towards the last-standing providers (at $9,000/MWh), who unlike their peers had not neglected to make their plants resistant to the cold weather.[10] Instead, these customers had a visceral reaction against "vulture providers," who took advantage of the desperate needs of their customers to charge them 100 times what electricity usually costs, and earned them billions of dollars of extra profits. In a true free-market system, such high electricity prices are precisely what is needed to induce better maintenance and more competitive entry in the future. In the real world, it doesn't work with citizens and voters.

---

[10] Ironically, the costs to weather-proof the plants would have been trivial, but incompetent or colluding electricity plant owners failed to do so. This turned out to be very profitable for the industry.

A fair characterization is that customers want reliability and are willing to pay for it — up to a point where they still want reliability but are no longer willing to pay for it. The problem is finding a good pricing solution to this inconsistency.

Just as the free market is not robust in maintaining competition, it is also not robust for protecting against and handling rare events.[11] The only solution, again, is good but difficult regulation. Regulators could charge customers in order to fund a set of fossil-fuel plants, which will lay idle almost all the time but are ready to jump in under dire circumstances. But how to ensure this? There are a lot of important questions here that lie beyond the scope of our book.

▶ Beyond Rich Countries

In contrast to the first world, electricity in the third world is often intermittent. For example, Beirut residents received only 5 to 20 hours of electricity per day in 2020. Lebanon's generation capacity simply cannot cope with demand, and no one is willing to invest to bring more generation online. Power plants make nice big stationary targets in wars, too, of which Lebanon has had plenty; and (once built), power plants make easy scapegoats that governments can force to provide power below cost. The last standing providers were barges, where operators hoped that they would not be confiscated by a desperate government and citizenry.

In Lebanon and many other poorer countries, establishments like hospitals that require reliable electric power typically run additional, but very inefficient, "mini-grids" based on their own diesel generators. Even in the United States, one of the attractions of roof-top solar with battery backup is its resiliency to wider power outages. Absent power outages or lack of transmission infrastructure, such hyper-local electricity generation is usually less economically efficient than utility-wide generation — though this may change in the future.

---

[11] The same can be said for relying on for-profit utilities that build traditional nuclear power plants. Once the probability of a blow-up is low enough (say, below 1 in 10,000 years), the day-to-day profitability concerns begin to dominate the low-probability blow-up concerns of most executives.

Conclusion

The point of our chapter was not to convert you, our reader, into an expert on electricity, but to give you a taste of the large real-world complexity of the electric energy problem and its potential solutions — and of the exciting time that we are living in today. If we were still teenagers looking for an appealing profession to pursue, electricity in all forms would be high on our list. Although we understand the outlines of the problems, there are plenty of interesting questions left. Will nuclear power become a lot safer and cheaper? Will someone offer a storage technology that will allow wind and solar to render natural gas and nuclear plants obsolete? Will residential roof solar (with car battery storage) dominate industrial solar (with a better transmission grid)? When should government take a hands-off approach, and when should it be hands-on? How can governments promote clean energy without causing voter riots? Will third-world countries leapfrog over the fossil-fuel stage in their electricity generation? How can China and other countries be induced to abandon coal-based generation of electricity in favor of cleaner technologies? What can we do to nudge decision-makers towards the better solution?

12. READINGS

Further Readings

BOOKS

- Bakke, Gretchen, 2017, The Grid: The Fraying Wires Between Americans and Our Energy, Bloomsbury USA, New York, 2017: A non-technical look at the role of the grid in America's energy future.
- Devanney, Jack, 2020, Why Nuclear Power Has Been a Flop, BookBaby, New York.
- Jacobson, Mark Z., 2021, 100% Clean, Renewable Energy and Storage for Everything, Cambridge University Press. Also, a similarly optimistic A Plan to Power 100 Percent of the Planet with Renewables, with a subsequent critique by various energy researchers.
- Macavoy, Paul W., and Jean W. Rosenthal, 2004, Corporate Profit and Nuclear Safety: Strategy at Northeast Utilities in the 1990s, Princeton University Press.
- Sivaram, Varun, 2018, Taming the Sun, Massachusetts Institute of Technology Press.

REPORTS AND ACADEMIC ARTICLES

- AMERICAN ELECTRIC POWER, AEP Transmission Facts.
- Albertus, Paul, et al., 2020, Perspective: Long-Duration Electricity, Storage Applications, Economics, and Technologies, Joule.
- Berkeley Lab, 2021, Land-Based Wind Market Report.
- Department of Energy, 2021, Wind Turbines: the Bigger, the Better
- Eash-Gates, Philip, et al., Joule, Sources of Cost Overrun in Nuclear Power Plant Construction Call for a New Approach to Engineering Design, with summary in Timmer, John, 2020, Why are nuclear plants so expensive? Safety's only part of the story.
- Eash-Gates, Philip, 2020, Sources of Cost Overrun in Nuclear Power Plant Construction Call for a New Approach to Engineering Design, Joule.
- EIA reports were the most important information in this chapter. For example:
    - Levelized Costs of New Generation Resources in the Annual Energy Outlook 2021, both generation and storage.
    - Projection of Energy Generation Out To 2050.
    - Double Share of Renewables by 2050.
    - Battery Storage in the United States: An Update on Market Trends, 2021.
    - Natural gas generators make up the largest share of overall U.S. generation capacity (Types and Amounts over Time).
    - Renewables account for most new U.S. electricity generating capacity in 2021
    - Capacity Factors.
    - [Battery Component and Chemistry Pricing]
- Environmental and Energy Study Institute, Fact Sheet, Energy Storage (2019).

- Härtel, Philipp, 2017, Review of investment model cost parameters for VSC HVDC transmission infrastructure, Electric Systems Power Research.
- Lazard, updated yearly, Levelized Cost Of Energy, Levelized Cost Of Storage, and Levelized Cost Of Hydrogen.
- Lithium-Ion Battery Cost and Performance Estimates.
- Mayr, Florian, Energy Storage News, 2021, Battery storage at US $20/MWh? Breaking down low-cost solar-plus-storage PPAs in the USA.
- Midcontinent Independent System Operator (MISO), 2019, Transmission Cost Estimates.
- Mongord, Kendall et al, PNNL, 2021, 2020 Grid Energy Storage Technology Cost and Performance Assessment .
- Morrison, Bonnie Maas Ninety Years of U.S. Household Energy History: A Quantitative Update, University of Minnesota.
- Pain, Stephanie, 2017, Power through the Ages, Nature.
- Timmer, John, https://arstechnica.com/science/2021/10/the-shifting-economics-of-solar-power-in-china/, Ars Technica, 2021/10/12.
- Projected costs of generating electricity, 2020, updated.
- Renewable Power Generation Costs in 2019.
- Ritchie, Hannah, and Max Roser, Our World in Data, 2021, World Mix, Electric and Total Energy Use.
- Schmidt, Oliver, et al., 2019, Projecting the Future Levelized Cost of Electricity Storage Technologies.
- Trabish, Herman, Utility Dive, 2014, User's Guide to Gas Generators 2014.
- Wolak, Frank, 2018 Electricity Policy Reform, Atlas, Scott W. et al, *Economic Policy Challenges Facing California's Next Governor*, Hoover Institute. Discusses policy choices to reduce the high cost of electricity in California.
- Ziegler, Micah S., et al, 2019, Storage Requirements and Costs of Shaping Renewable Energy Toward Grid Decarbonization, Joule. Covered in Vox. Estimates that storage cost would have to drop to $20/MWh to allow a 100% clean system, but $150/MWh could be enough to go to 95%.
- Grid Scale Battery Storage FAQ, NREL, ca 2017. IEA tracking report, Nov 2021.

Shorter Newspaper, Magazine Articles, and Clippings

- Clean Technica, 2021, Researchers Take A Practical Look Beyond Short-Term Energy Storage (i.e., seasonal storage).
- Corporate Finance Institute, LCOE calculations.
- Cost of electricity by source (and country).

12. READINGS

- Frazier, A. Will et al. Storage Futures Study: Economic Potential of Diurnal Storage in the U.S. Power Sector.
- Johnson, Doug, 2021, Ars Technica on Wind Power Growth.
- Johnson, Scott, Ars Technica, 2020, US grid-battery costs dropped 70% over 3 years.
- Johnson, Scott, Ars Technica, 2021, Eternally five years away? No, batteries are improving under your nose.
- Katz, Cheryl, BBC, 2020, The batteries that could make fossil fuels obsolete. (Can we reach 1,200 GWh of storage world wide within the decade?)
- Quora, What are the principle factors that affect the capacity of battery cells?.
- Roberts, Duval, 2021 Getting to 100% renewables requires cheap energy storage. But how cheap?
- Rowlatt, Justin, 2021, BBC, Why electric cars will take over sooner than you think.
- Roth, Sammy, Newsletter: The future of rooftop solar is up for grabs in California, November 4, 2021, Los Angeles Times.
- Shahan, Zachary, Solar Is Cheapest Electricity In History, U.S. DOE Aims To Cut Costs 60% By 2030, Clean Technica, 2021/03/26.
- The Solar Energy Technologies Office Solar Futures Study, planning to reach 1.6 TW of power by 2050 (from 4% to 45%), also covered in the New York Times.
- U.S. Department of Energy, Energy Storage Ecosystem Offers Lowest-Cost Path to 100% Renewable Power, Clean Technica, 2021/10/22.
- U.S. Energy Information Administration, 28% Of U.S. Coal Power Plants Plan To Retire By 2035, CleanTechnica, 2021/12/15.
- World Nuclear Organization, 2021, Nuclear Power in the World Today, 2021.
- Desjardin, Jeff, 2016, The Evolution of Battery Technology. Excellent primer. VisualCapitalist.com

WEBSITES

- `https://www.caiso.com`: California Independent Systems Operator.
- `https://www.eia.gov/`: U.S. Energy Information Administration.
- `https://energystorage.pnnl.gov/`: PNNL dedicated Storage Section.
- `https://irena.org/`: International Renewable Energy Agency.
- `https://www.nrel.gov/index.html`: National Renewable Energy Laboratory.
- `https://www.pnnl.gov/`: Pacific Northwest National Laboratory (PNNL).
- EIA Real-Time U.S. Energy Dashboard.
- Energy Storage News.

Chapter 13

Beyond Electrification

Although electrification will be the cornerstone of a transition to a clean economy, it cannot be the whole story. There are many activities that are too difficult and expensive to electrify. There are also human emissions unrelated to energy generation. We therefore now move beyond electricity.

Activities that are difficult to electrify fall largely into two types.

The first type are activities that require the high-energy density of fossil fuels. For example, commercial air travel is based on kerosene, which is both light and dense.

The second type are activities that require high temperatures, such as making steel and cement. (Some also require carbon.) Fossil fuels are very efficient at making heat, in contrast to movement or electricity, where fossil fuels are very inefficient.

Hey, you in the house! I don't run on batteries, you know!

We now look at some other sources of greenhouse gas emissions and possible ways of dealing with them beyond electricity.

1 Hydrogen

Hydrogen is the most similar clean alternative to fossil fuels. Unlike fossil fuels, though, its combustion produces no CO_2, just water.

Table 13.1. Energy Densities

	Hydrogen	Gasoline	Kerosene	NatGas	Batteries
By Weight	33	12	12	13	0.25
By Volume	0.25	0.94	1.06	0.82	nc

Source: RMI.

Table 13.1 shows that the comparative energy density of hydrogen depends on whether it is measured by weight or volume. Hydrogen is really light. Its energy density *by weight* is nearly three times that of gasoline and natural gas, and approximately 100 times that of lithium-ion batteries. Unfortunately, even when it is compressed into a liquid, its energy density *by volume* is only about one-third of natural gas.[1]

Hydrogen Production

There is no natural source of hydrogen on Earth. Hydrogen is always bound with other molecules, usually water, and, therefore, must be produced. *Brown and gray* hydrogen are created from fossil fuels. Because they come from dirty processes, they offer no advantages over fossil fuels. Their future is limited, and so we do not consider them further.

Blue hydrogen comes from splitting natural gas into hydrogen and CO_2, but its production is combined with carbon capture and sequestration. However, there is no "blue police" that checks to confirm whether a producer has really incurred the voluntary extra expense. Producers that lie and claim to be blue can always underprice their peers. Blue hydrogen is still very expensive and there is little chance that the cost will come down dramatically.[2]

---

[1] Hydrogen could also be converted into methane (i.e., natural gas), which could then be converted into alcohol fuel.

[2] Scientists have just discovered that it is possible to convert methane into hydrogen without making CO_2 in the first place. The question is whether this is commercially viable.

1. Hydrogen

Table 13.2. Hydrogen Cost in 2020-$ per MWh

Fuel	Cost Per Kg	As Primary E 2021	As Primary E 2050e	To Electricity 2021	To Electricity 2050e
Hydrogen					
Brown (From Gas)	$1	$27		$70	
Blue (With Capture)	$3	$85		$210	
Green (Electrolysis)	$5	$136		$340	
Optimistic	$1 in 2050		$27		$70
Skeptical	$2 in 2050		$56		$140
Natural Gas		$20	$20	$50	$50

Source: The Economist, 2021/10/09 and IRENA, 2012. Cost estimates are approximate. Primary energy is at the source of production or wellhead. Further conversion to electricity assumes 40% efficiency.

Green hydrogen cuts out polluting fossil fuels entirely. It creates hydrogen by splitting water molecules via electrolysis. For clean hydrogen to replace fossil fuels, green hydrogen is the way forward. However, to be widely adopted, green hydrogen needs to be not much more expensive than blue hydrogen or even natural gas.

Table 13.2 shows that green hydrogen currently costs about 7 to 8 times as much as natural gas. The reason is that 65% of its cost comes from electricity, which is still relatively expensive. Most of the remaining 35% is the cost of purchasing, operating, and maintaining the electrolysers. However, note that *even if electricity were free*, hydrogen would still cost 2 to 3 times as much as natural gas.

Both costs will drop and so the gap will narrow — the question is by how much. Green hydrogen proponents believe that the cost can be brought down to $27 per MWh. This would make green hydrogen broadly economically competitive with natural gas — *if* natural gas suffered a CO_2 tax that doubled its cost.[3]

---

[3] Airlines and shipping are highly competitive industries and fuel is a key cost. Their margins are typically under 10%. No airline could survive flying on hydrogen if its competitors could still fly on kerosene.

13. Beyond Electrification

Hydrogen Uses

▶ Transportation

Batteries cannot store large amounts of energy in small volumes and weights. Thus, the chemical nature of hydrogen energy make it a potential replacement for oil-based fuels for off-grid transportation, as in shipping[4] and airplanes.

The remaining engineering problem is compressing and storing the hydrogen. Due to hydrogen's density characteristics, hydrogen aircraft will be lighter (good) but require more fuel tank space (bad). The tanks also need to be stronger and cooler to keep the hydrogen from expanding. To be efficient, hydrogen aircraft will need to be redesigned.

Airbus is optimistic that it can solve the engineering problems. The "only" remaining problem is that green hydrogen costs many times more than kerosene — and flying is a hyper competitive industry in which every penny of fuel cost counts.

However, even if it were cheap, hydrogen would still face a large competitive hurdle when there is access to the electricity grid. As Figure 13.3 illustrates, for cars, batteries lose only 5% to charge and another 20% to move the car. In contrast, hydrogen has a net efficiency of only 30%. By the time hydrogen moves the car, nearly 80% of the original energy has already been lost. The idea of powering cars with hydrogen-combustion engines is even crazier, with a loss of over 85% of the original energy.

Thus, we believe that, regardless of the electricity generation price, hydrogen will likely never be competitive with batteries in light-vehicle transportation. The same holds true for modest-capacity electricity applications, in which the user can recharge the batteries from the grid at will. In the case of 18-wheelers, the jury is still out because weight and charging times are important, too. Hydrogen's future there will depend primarily on future innovations in battery technology. Overall, hydrogen's main role will likely be in off-grid applications — but there are plenty of them.

---

[4]There already is one hydrogen-powered pilot ship.

1. Hydrogen

Figure 13.3. Vehicle Efficiency

		Electric Batteries	Hydrogen Fuel Cell	Conventional Combustion
Well To Tank	Clean Electricity	100%	100%	100%
	Electrolysis	100%	70%	70%
	CO_2 Capture	100%	70%	44%
	Fuel Production	95%	52%	44%
Tank To Wheel	AC/DC Inversion	90%	52%	44%
	Battery Charge	86%	52%	44%
	H2 to Electricity	86%	26%	44%
	DC/AC Inversion	81%	25%	44%
	Engine Efficiency	73%	22%	13%
	Final Efficiency	73%	22%	13%

Source: WTT (LBST, IEA, Worldbank, TTW, T&E calculations), via Zachary Shahan, Clean Technica. Conventional Combustion is a "power to liquid sustainable" gasoline-like way of fueling standard vehicles.

---

anecdote

> **Saudi Arabia** is placing a big bet on the future of hydrogen. The Saudis are building a $5 billion electrolysis plant powered entirely by sun and wind that will be among the world's biggest green hydrogen makers when it opens in the planned megacity of Neom in 2025. The Saudis have plans to become a global hydrogen exporter — which may make sense given how expensive it would be to build transmission cables to send their cheap solar power to major markets.

▶ Electricity Storage

Can hydrogen be used for electricity storage? Compared to batteries, the cost of hydrogen storage is low. A cool big sturdy tank will do. Thus, any excess solar and wind electricity during peak generation could be diverted for electrolysis into storage tanks.

However, the problem for hydrogen as short-term modest-capacity electricity storage is still the same: the alternative of batteries. The electron-based technology of batteries is more efficient than the chemical-conversion-based technology of hydrogen. Utility-scale batteries can return about 85% of the energy used to charge them. Hydrogen can only reach about 35% efficiency, because 30% of the electrical energy is lost in the production (electrolysis) and another 35% is lost in the turbine or fuel cell used to generate electricity.

Batteries are already much cheaper for low- to medium-capacity storage needs. Yet if future utility-scale batteries cannot solve the capacity problem — i.e., with giant-tub batteries full of electrolytes and a couple of anode/cathode sticks, rather than the volatile small lithium battery packs — then hydrogen could play a "last resort" and large-capacity storage role that covers grid needs after the (daily or weekly) batteries have been exhausted. Hydrogen might also be used for long-term seasonal storage.

➤ Industrial Heat

Hydrogen has also been proposed as an alternative to fossil fuels for producing industrial high heat. The production of steel and cement alone accounts for about 15% of total CO_2e emissions. The question is whether it will ever make economic sense to start with electricity and make hydrogen to burn for heat, rather than using the electricity directly and avoiding the conversion loss. Further research should resolve this issue in the near future.

➤ Fertilizer

Hydrogen is already used today for the production of industrial ammonia, the main ingredient in artificial fertilizers. Without them, agricultural productivity would plummet, and the planet could not sustain 8 billion people. This demand for hydrogen is almost certain to increase as population grows.

2 Industrial Heat

Fossil fuels are not particularly efficient when they are used to create electricity or kinetic energy, because most of their primary energy is lost in the conversion. However, they are supremely efficient when they are burned for heat. This feature makes them economically difficult to replace in heating applications.

Almost Done

Steel and Cement

The two most prolific consumers of high heat from fossil fuels are steel production and cement production, accounting for about 7% and 8% of the world's 51 $GtCO_2e$ emissions, respectively.[5] Cement is primarily a building material. Steel is also used in building, but plays a role in many industrial activities, too. Bill Gates believes that by 2060 the world's building stock will double — mostly in India, China, and Nigeria. More cement and steel will be needed for infrastructure.

Again, we believe that the dream of reducing steel and cement emission through linked CO_2 sequestration offsets is naïve. Sequestration would add about 20% to the cost of steel and about 100% to the cost of cement.[6] There is no reason why any steel producer would want to incur the expense; little reason why India, China, and Nigeria would want to hamstring their producers for the greater global good; and no police that could easily confirm that a producer has really paid the cost. Worldwide competition in steel would probably bankrupt the most compliant steel producers. Domestic competition in cement would probably do the same for compliant cement factories if competition could get away without sequestration.

This situation leaves the world with essentially two ways to reduce the emissions of these two key processes.

The first way is using cleaner energy to create the necessary high-intensity production heat. Although electricity can create high heat via <u>electric arc</u>

---

[5] Glass and Aluminum also use a lot of energy in their creation. Aluminum already uses primarily electricity, though.

[6] Besides, if it is worthwhile to sequester CO_2, why should other firms not do so, too, regardless of what they produce? The only reason to link production with sequestration is if it is a lot cheaper to sequester CO_2 at the plant than elsewhere.

13. BEYOND ELECTRIFICATION

technology, the economic case for clean energy will be tougher, because fossil fuels do not suffer the conversion handicap that they do when it comes to electricity generation. Thus, heat will be among the last applications to be taken over by electricity. Nevertheless, with solar power falling below $35/MWh and to $15/MWh by mid-century, and marginal costs for DC power (rather than grid-synchronized AC power) perhaps then as low as $5-10/MWh, heat reservoirs for steel production could become cost-competitive *while* the sun shines.

Nuclear reactors would seem almost ideal for high heat generation, but this would raise a host of other proliferation and safety problems. Right now, nuclear power is not even on the drawing board for industrial high-heat applications.

Finally, as mentioned in the last section, there is green hydrogen. However, when competing with direct use of electricity, hydrogen has to overcome the energy loss involved in its production. Electric arc technology seems more viable.

The second approach is to invent cleaner processes or materials. Many promising technologies have been suggested for producing cement in a more efficient way. Making cement not only pollutes because it needs high heat, but also the required "clinker" production itself releases CO_2. (The two parts are about equally responsible for CO_2 emissions.) Fortunately, some cement innovators have been heavily funded by Venture Capital (VC). Fly ash cement mixes have a good chance to be both stronger *and* better than traditional cement. It's not clear yet whether VC technology can revolutionize cement production for the global benefit, but the capitalists are already betting on it.

There are some possibilities that almost seem too good to be true. Forms of hemp (not the stuff you smoke) produce materials that are stronger than steel. They can be used as direct substitutes or mixed in with cement. We first had to do a double-take to make sure that the source was not the Hemp Council of America, but reputable outlets like the New York Times. It is indeed the case that, like wood, hemp could be a clean next-generation composite building material.

3 Agriculture

The world's 8 billion people need to be fed. And farming emits plenty of GHGs, not just CO_2, but also methane and nitrous oxide. In Chapter 3, we mentioned that methane is about 33 times more potent and nitrous oxide is about 265 times more potent than C02. Because of these large multiples, agriculture was responsible for about 7 $GtCO_2$ of our 51 $GtCO_2$ in 2019, which was about 14% of total human GHG emissions.

Worse, the human population is expected to reach approximately 10 billion by 2100. Worse again, it is getting richer — and rich people not only consume more food but also like to consume more meat and dairy whose production emits more GHGs than crops. Some scientific forecasters suggest that agriculture could contribute 0.7°C to global warming by 2060. Agricultural emissions are a serious problem.

Interestingly, economists of the past (Thomas Malthus in the 1840s, Paul R. Ehrlich in the 1970s) thought that agricultural production on Earth would limit human population. This turned out to be vastly pessimistic.[7] Simply put, it looks as if humanity will be able to produce more food than it will consume for the indefinite future.

Crops and Nitrogen

Plants need nitrogen in order to grow. There is plenty of nitrogen in the air, but gaseous nitrogen is useless to most plants. Instead, plants mostly rely on bacterial processes that "fixate" the air nitrogen into compounds that plants can then use.

Lack of nitrogen is a primary constraint to plant growth. This is where the aforementioned fertilizers come in — concentrated nitrogen in plant-consumable form. Traditional natural fertilizers are shit (i.e., manure) — from almost any kind of animal living on Earth. Synthetic fertilizers are cheaper even than manure. Without them, the world could probably feed only half as many people as it does today.

We have already discussed the CO_2 released in the production of synthetic fertilizer, but there is a bigger problem. Only about half of the nitrogen in

---

[7]We can thank the chemist Fritz Haber, whose first claim to fame was the artificial fertilizers mentioned above. Unfortunately, his second claim to fame was his work on then already-banned poison-gas warfare.

13. BEYOND ELECTRIFICATION

agricultural fertilizers is taken up by the targeted plants. The rest runs off, eventually causing unwanted local pollution elsewhere (e.g., algae blooms in lakes and oceans) or escapes into the air in the form of nitrous oxide (NOx) created by microbes. As we noted in Chapter 3 — nitrous oxide warming potential is 200-300 times higher than CO_2.

From a social perspective, fertilizers are overused by farmers — and again not just in the OECD. What can be done to discourage it?

We can dismiss the most obvious one — an appeal to the social conscience of farmers. Appeals will indeed resonate with a few, but not with the many. If appeals could solve the problem, they would have already done so.

The first realistic solution is the next obvious one: make it in farmers' interest to reduce fertilizer use, preferably by taxation. Equally obvious is why this approach is difficult to implement. Farmers are among the most powerful voting groups in every country on earth. The only hope is to offer them more carrot than stick: paying farmers to compensate them for higher taxes on fertilizers. (Another complication is that few farmers are progressive global climate activists. Instead, most tend to instinctively dislike government intervention.)

The second solution is less obvious: technology.

All farmed crops and animals today have been genetically engineered by humans for thousands of years. Scientists can just do it faster now in the lab than through selective breeding. They can now engineer <u>plants that require less nitrogen or that can fixate some nitrogen themselves (mostly with the aid of symbiotic bacteria)</u>. Moreover, such plants cannot only reduce the need for fertilizers but also for herbicides and insecticides.[8] It is also impossible to distinguish between a plant that has been engineered with the latest <u>CRISPR/Cas9</u> technique (which only cuts out a part of existing genomes and does not insert any new genes) and a plant that has naturally undergone the same kind of mutation and lost a gene. This is because there is no difference.[9]

Unfortunately, this is where science and environmentalism often collide. The "natural foods" movement — especially in Europe — seems to detest all

---

[8]Fertilizers, herbicides, and even insecticides as were in use 150 years ago before chemistry improved, are <u>allowed in organic farming</u>. Many of these substances are highly toxic.

[9]Disclosure: One of us has invested in a startup that works with CRISPR/Cas9, e.g., to reengineer coffee in order to avoid the toxic decaffeinating process.

genetically modified organisms (GMOs). Indiscriminate resistance to progress in food technology is — apologies for the strong phrase — stupid and harmful for crops and farm animals, for producers and consumers, for wild fauna and flora, and for the world in general. Genetic engineering is a powerful tool. It could even allow growing plants in the desert again and without the need for fertilizers. Like all powerful tools, genetic engineering can be used for good or evil. Of course, it requires good regulation in the common interest.[10] Properly shepherded, better plants and animals for humans could be *much* better for farmers, consumers, and the environment.

It is not an important distinction whether a food is engineered, conventionally farmed, or organic. (In many cases, this is not ascertainable.) It is sad that we have to point out that natural foods does not mean what many consumers think it means. First, natural foods can be treated with "natural" but permitted fertilizers, herbicides, and pesticides. These can be *more* toxic than those used in conventionally grown foods. It is not possible to grow appealing fruits and vegetables without them. Second, "natural" does not mean healthy. Cherries, apricots, plums, and peaches contain cyanide. If we warned naïve consumers with mandatory signs that they contain cyanide, they would probably no longer sell.

A cyanide label on cherries would be counter-productive. Of course, other food warnings are not. A little natural mushroom named Amanita phalloides, which looks like many edible mushrooms, when mixed into your food practically guarantees the need for a liver transplant. A food prohibition — and not a warning that says "all natural" — is obviously warranted. Experts need to set smart default food standards, not knee-jerk ones, to guide the public at large. There is even a simple standard here: if the overwhelming majority of experts are willing to eat the food themselves and feed it to their children, then it is probably safe for the general public. It should not carry a general health warning or non-organic sign.

---

[10] It is a problem that very few companies have now monopolized the seed market. The Top-10 now control 67% of the world seed market! This has raised prices and made farmers badly suspicious of their products. It is a good question whether seeds should be patentable at all.

13. Beyond Electrification

The third solution is reducing demand by reducing the human population. But as we have said before, the means by which population might be managed, other than through economic development, is beyond the scope of this book.

> sidenote
>
> Scientists are beginning to figure out how to make starch from CO_2 and hydrogen on an industrial scale — essentially mimicking the same process that plants use to make starch. Our grandchildren may no longer need 40% of the world's land for farming.

Rice and Methane

The world's largest food crop by calories consumed is rice. More than 3.5 billion people count on rice to provide 20% of their daily calories. Rice is typically grown in flooded fields, not because rice needs it, but because rice is indifferent to it. Weeds, on the other hand, do not survive standing water. The problem is that flooded fields emit huge amounts of methane. Rice fields contribute about 1.5% to global warming.

The problems and solutions largely mimic those of other crops.

The first solution is to charge farmers for flooding their fields. For the same reason that farmers are unlikely to be taxed for fertilizer use, this will likely not happen. The second solution is to genetically reengineer rice to require less water and compete better against weeds even in unflooded fields. The third solution is to reduce the human demand for rice. The most realistic and best choice for the planet is technology — if only environmentalists could be convinced not to poison the minds of consumers by insisting on "all natural" rice varieties grown the old-fashioned way.

Cows, Meat, and Methane

History suggests that as the rest of the world becomes wealthier, it will gravitate towards consuming more meat and dairy. It would be great both for the world and consumers if this trend were to reverse, but it's just naïve to believe that this is likely to happen out of concern for climate change.

Globally, approximately one billion cattle are raised for meat and dairy products. These cows produce annually about 2 $GtCO_2e$, or about 4% of all global emissions, through a process called "enteric fermentation." Bacteria inside cows' stomachs break down the cellulose in grass and plants and ferment it. Methane is a byproduct, which the cow then belches out. Meat may account for as much as 60% of all GHGs from food production. If the entire supply chain is counted, meat and dairy are responsible for as much as 7 $GtCO_2e$.

The problems and solutions for animals largely mimic those for crops. The first solution is to tax cows, meat, and dairy. However, ranchers are not going to take this easily.

The second solution is again technology — finding a way to produce better foods with fewer cows and cow emissions. One approach, taken by Beyond Meat and the Impossible Burger (as well as some laboratory R&D startups), is to produce plant-based meat substitutes. It's still too expensive and not fully equivalent in texture and taste, but U.S. consumers have started to take to it. The next step will be a lab-grown steak.

Before I say yes to a date...tell me, do you contribute to global warming very often?

Rather than replacing cows, it is also possible to make them more efficient — that is, to reduce their methane production. When treated with the (reasonably cheap) compound 3-nitrooxypropanol, cows' methane emissions fall by about 30%. Unfortunately, the drug must be given daily — an expensive and impractical procedure for free-grazing cows. More preliminary research suggests that Asparagopsis, an edible warm-water seaweed grown in Australia, contains a compound called Bromoform. In trials as a food additive, Asparagopsis reduced cows' methane emissions by 80-98% when comprising as little as 2-3% of the diet (though cattle growth also slows a little and the long-term effects are still uncertain.). Unfortunately, cows don't like it. Thus, it does not work for free-grazing cows with a choice, which accounts for most cows' lifetimes. Nevertheless, it is worth investigating Asparagopsis. With some government subsidies for use, it could reduce humanity's cow-methane problem. The final and possibly best solution may however be to breed cows that belch out less methane. Estimates are that this could reduce emissions by about one quarter — and it would work for both free-grazing and feed-lot cows.

13. Beyond Electrification

Need we state the third solution — reducing the size of the human population — again? Of course, climate-conscious consumers could just decide to eat less meat and dairy, which would have the same effect. Unfortunately, we have little faith in individualized solutions. If they worked, obesity would not be a problem in America.

> anecdote
>
> And now for an entirely innovative solution to climate-change: potty training cows! Really not an April's Fool joke, either. Here is a Video.

> anecdote
>
> The New York Times reports that some French green politicians have finally found the problem and it is French men. "If you want to resolve the climate crisis, you have to reduce meat consumption, and that's not going to happen so long as masculinity is constructed around meat."

Food Waste

You could help save the environment by eating it all here

Ironically, another meaningful source of methane emissions is food that is thrown away. Rotting food produces methane with a warming impact equal to 3.3 $GtCO_2e$ (out of 51 $GtCO_2e$ total worldwide GHG emissions). Bill Gates reports that in Europe, industrialized parts of Asia, and sub-Saharan Africa, more than 20 percent of food is simply thrown away, allowed to rot, or otherwise wasted. In the United States, it's 40 percent.

The ideal voluntary solution is behavior change — better logistics, more careful shopping, and preparation of food. But to sound like a broken record, we doubt that this will happen. It hasn't yet.

Raising food prices with taxes would help, but such a suggestion would probably be rejected by just about everyone, not just by the poor who would be hardest-hit.

Once again, technology looks like the best hope. The challenge is always figuring out how to do it economically. Not enough landfills are outfitted with methane capture equipment. Better government incentives could improve adoption. Smart bins could help people track how much food is wasted. Saving food can help the poor and the climate at the same time. Technology can also alter foods so that they spoil more slowly or release less methane when decomposing. Small-scale solutions will not change global warming but can nevertheless be cost-effective for local adopters.

Soil Tilling Practices

Some estimates are that soils remove about 10 $GtCO_2e$ per year, mostly in dry and cold environments. Estimates are that better cropland management could sequester an additional 1-1.5 $GtCO_2$ per year for about 20-40 years. The main improvement would have to come from tilling. Tilling involves turning over the first 6 to 10 inches of soil before planting new crops. This practice blends surface crop residues, animal manure, and weeds deep into the soil. It also aerates and warms the soil.

However, tilling exposes carbon buried in the soil to oxygen, allowing microbes to convert it to Methane and CO_2. The alternative is no-till farming. By keeping excess oxygen out of the soil and away from microbes, no-till farming keeps the carbon that builds up when plants die and decompose below ground.[11] The process becomes planting in the spring, spraying with *more* herbicide than in ordinary farming, and applying fertilizer dropped on the surface or injected in a slot.

With regard to no-till farming, the same three alternatives that we have routinely discussed apply. One could hope that a voluntary approach would work, but don't hold your breath. Conceptually, it might be possible to tax tilling, but it makes little practical sense. If it hurt their bottom lines, farmers would oppose it and other people would not care, paying it little mind.

Once again the best possible alternative would be to research methods of making fields equally productive and profitable but without widespread tilling. There is active sustainable agriculture research in early stages.

---

[11]Tilling also does some long-term harm for farming. Tillage also loosens and removes any plant matter covering the soil, leaving soil bare. Bare soil, especially soil that is deficient in organic matter, is more likely to be eroded by wind and water. Untilled soil resembles a sponge, held together by different soil particles and channels created by roots and soil organisms. When the soil is tilled, its structure becomes less able to absorb and infiltrate water and nutrients.

Ethanol

And while we are discussing agriculture, the United States should stop subsidizing Ethanol fuels immediately. These subsidies amount to as much as $100 billion per year. They live on largely because Iowa is an early primary electoral state. They were never green. It takes more than 1 gallon of fossil fuel to make 1 gallon of Ethanol fuel. When the Ethanol is burned, it releases CO2. (And Ethanol is actually harmful as an additive in combustion engines, too.)

4 More Methane Problems

Agriculture is not the only important source of potent global-warming methane. Methane comes primarily from four sources: (1) natural ones (decay of organic materials in the absence of oxygen); (2) agriculture (especially cows and rice fields); (3) landfills; and (4) oil & gas operations.

We have already discussed (1) and (2). In this section, we discuss (3) and (4).

Figure 13.4 shows the location of the four different types of emitters in the lower 48 states. Agriculture has low emissions per acre but covers almost the entire U.S. territory. Still, there are a few hot spots: California, the Texas panhandle, the corn belt, the Mississippi Valley (Oklahoma), and North Carolina. Oil & gas operations are more concentrated. There is one belt of emitters running a diagonal line from Texas to Wyoming, another smaller one from West Virginia to Pennsylvania, and smaller ones on the east coast and California. Landfills are near population centers and less important than agriculture and oil & gas.

Landfills

Most landfill content is organic matter: food scraps, yard trimmings, junk wood, wastepaper. Their decomposition produces biogas, a roughly equal blend of carbon dioxide and methane accompanied by a smattering of other gases. As a result, landfills are a significant source of emissions, releasing 12% of the world's methane total — about 1 $GtCO_2e$.

The good news is that these emissions are localized, and we know where they are. Even better, the methane can be captured. The technology is relatively simple. Dispersed, perforated tubes are sent down into a landfill's depths to collect gas, which is piped to a central collection area where it can be vented or flared. It can also be compressed and purified for use as fuel in generators and garbage trucks, or mixed into the natural gas supply. (Although burning methane produces CO_2, the residual CO_2 is negligibly small.)

Oil and Gas Operations

Figure 13.4. 2012 US Methane Emissions

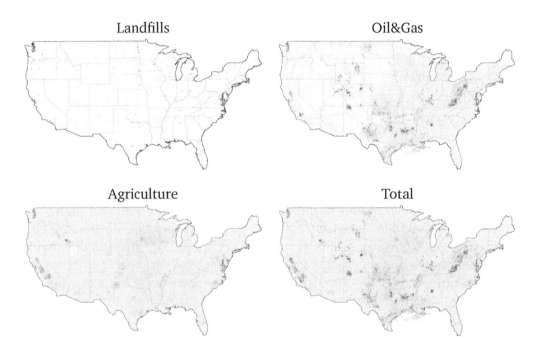

Note: Darker colors mean more methane emissions. Landfills contribute only near some cities. Oil & gas operations are prominent in California, and along a belt from Texas to Wyoming, and around Pennsylvania. (1) Landfills are mostly near cities; (2) Oil & gas is located near reservoirs; (3) Agriculture is spread across the entire continent; (4) the net total is "everywhere plus hotspots."

Source: Maasakkers et al., 2012, EPA. The scale is the same in all four plots.

13. BEYOND ELECTRIFICATION

The second most important emitter of methane are oil & gas wells. Some emitters are operating wells that simply allow too much leakage for no good reason, often to their own economic disadvantage. Others are more deliberate. The EPA estimates that there are 3 million abandoned leaking wells in the United States alone, and a further three million in Canada. (Furthermore, abandoned coal mines can also leak large amounts of methane.)

It costs about 2% of a fossil-fuel well's revenues (not profit) to plug it at the end of its life. Flaring is often cheaper, but many abandoned wells are not even flared. Walking away is always cheaper, and this is what many drillers have done for over a century. For many of these wells, even the owners themselves no longer exist. If the company is too big to walk away, it can always sell the well near the end of its life and "focus on more lucrative projects." Indeed, the five worst methane emitters in the United States today are not the oil & gas giants, but fairly small drillers now owned by private equity firms and legally insulated from liability by clever organizational structures.

Some wells can leak for decades or even centuries and pollute the groundwater. Others stop by themselves after a while. Remarkably, some estimates attribute 65% of all U.S. leak emissions to just 10 "super-emitters"![12] Fortunately, super-emitters can be easily identified. This is good news. We can plug or flare them immediately. The cost of doing so is insignificant, especially in light of the co-pollution that these leaks generate. For the broader set, there is large variability in how much it costs to plug or flare a well. For many, it is socially worthwhile (even from a local co-pollution perspective without global climate concerns); for others, it is not. Clearly, the U. S. government should have collected and set aside the cost of plugging wells at the end of their lives, but it did not do so — and it is still not doing so. This is the worst kind of subsidy imaginable to the fossil-fuel industry. Fortunately, it is a relatively easy fix. With proper policies and proper incentives, it should be possible to quickly reduce the amount of methane produced from these sources.

Understanding of methane emissions has improved greatly in the last decade. The European Copernicus Programme has launched a series of "sentinel" satellites that can measure concentrations in real-time. Even cheaper micro-satellites (Claire and Iris, see Figure 13.5) are now coming online, allowing private companies to monitor local or worldwide GHG emissions.

---

[12]There is some disagreement here. A study by JPL and Arizona Universities suggested much smaller numbers for superemitters.

4. More Methane Problems

Figure 13.5. Turkmenistan Methane Emitter Spotted by Iris

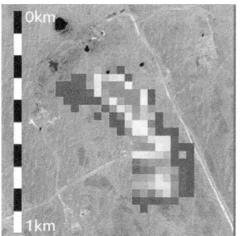

Source: GHGSat / Bing Aerial, via BBC

Before the introduction of these satellites, the world had no idea how much methane was leaked outside the United States and Canada. And many of these sources are short-term temporary flares that are vented (without burning) by facilities in the US and Russia. Unless there is a satellite working 24/7, these flares would never have been noticed. Moreover, having verifiable satellite data that can detect emissions within a radius as small as a square mile will also make it possible to locate polluters quickly and to institute an effective stick-and-carrot program to reduce their emissions.

Figure 13.6 graphs the most important human point emitters of Methane. Do you remember the analogy about a "peeing section in the swimming pool" (from Chapter 6)? Even if the United States mitigates all its major methane leaks (and it probably should do so immediately if only to reduce local copollution), it is not enough to solve the problem, much less make a global difference. There are still many other methane leaks elsewhere in the world. Canadian abandoned wells alone are large enough to overwhelm anything the United States might do.[13] So are Saudi Arabia's, Iran's, Australia's, and Russia's, etc.

---

[13] These particularly problematic wells are ironically located in the area that would have been fed by the Keystone pipeline project.

Figure 13.6. Copernicus Methane Sentinel Sat Images of Large Emitters

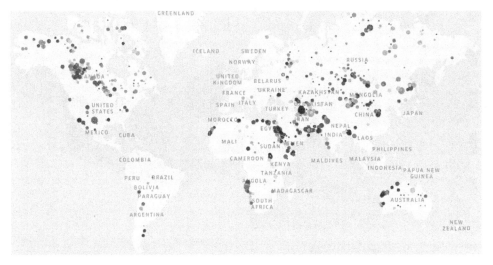

Source: ESA Copernicus Sentinel, Methane Emissions, Global Map 2019.

Unlike the Paris CO_2 climate treaty, which we believe has little chance to curb CO_2 (Chapter 9), methane super-emitter mitigation could be negotiable. The costs of methane mitigation are reasonably well-dispersed and low relative to the mitigation benefits (including local benefits), plus compliance is much easier to measure. It will not require dramatic sacrifices, and the potential benefits are large.

5 Construction and Efficiency

Building construction emits GHGs, but so does living in buildings. There are some obvious changes — and remarkably, many of them pay for themselves over time. However, despite the long-run benefits, they typically require upfront payments. This can be a hindrance for poorer people and poorer countries. It may not be an option if it prevents the residents from spending money on the food and health care necessary to stay alive.

Better Insulation

Better insulation often seems like a "no brainer." Insulation is cheap and can reduce average home heating and cooling costs by around 30% even in warm climates such as Australia and New Zealand. The cost of installing insulation there pays for itself in around 3-5 years through reduced energy bills. Upfront investment in insulation is more effective than other green investments. It is a once-only cost that lasts for the life of the building (typically 50-70 years) and requires no further maintenance.

This leads to the obvious question as to why not all people don't automatically buy more insulation. Even middle-class people in developed countries often fail to properly insulate their homes. One reason may be that they fail to appreciate the benefits. Here a nudge from the government could help. Just as labeling the nutritional value of foods helps people manage their diets, providing consistent information regarding the benefits of insulation could help residents make better economic decisions for themselves and for the environment.

Thermal Energy Storage

A storage heater or heat bank is an electrical heater that stores thermal energy when electricity is available at lower cost, and releases it when heat is required but electricity is expensive. Alternatively, solar storage heaters are designed to store solar energy as heat, to be released during the night or other periods when it is required, often making it more cost-effective than selling surplus electricity to the grid and buying it back at night. Storage heaters are usually used in conjunction with a two-tariff electricity meter, which records separately the electricity used during the off-peak period so that it can be billed at a lower rate. In most countries, storage heaters are

13. Beyond Electrification

only economical (compared to other forms of heating) when used with such special tariffs. Once such tariffs become standard, thermal energy storage and clean intermittent sources will complement one another.

Furthermore, the same process can be used in reverse to providing cooling. Thermal storage cooling solutions are a cost-effective and reliable option for cooling offices, schools, hospitals, malls, and other buildings. By producing low process fluid temperature during off-peak times, this environmentally friendly cooling solution reduces energy consumption and greenhouse gas emissions. Once again, the future of this technology depends on the rational pricing of electricity from intermittent sources of generation.

Heat Pumps

Because furnaces convert 95% of the energy in natural gas to useful heat, it may seem as if there is no better way to make heat — but that deduction is surprisingly incorrect. Heat pumps do not have to create heat — they just move it around (often into the ground). (The principle is similar to that of a refrigerator.)

Thus, for every kilowatt hour of power drawn from the electric grid, a heat pump can transfer three kilowatt hours of heat energy from the outside of the house to the inside for a total heat output of four kilowatt hours. The effective efficiency of the heat pump is 400% compared to 95% for the gas furnace.

Heat pumps do have limitations. They only work in moderate climates and have modestly shorter lifespans than furnaces. If it gets too cold outside it becomes difficult for the pump to provide energy to the interior at comfortable temperatures. For this reason, leading heating companies are now selling dual fuel systems that switch between heat sources depending on the outdoor temperature and home heating needs. These systems are not only environmentally beneficial, but they are also economical in that they are the lowest-cost source of heating. It is a perfect example of how technology improvement can play a key role in reducing emissions.

Conclusion

This chapter has broadened the perspective beyond electricity as a solution to greenhouse gas emissions. It has described some of the more important sources of GHG emissions and opportunities for reducing them. There is no great common theme here, because the activities are so diverse — except perhaps that the key will have to be more research, technology, and deployment.

13. Readings

Further Readings

Books

- A summary of various opportunities and payoffs related to the reduction of methane emissions.
- Gates, Bill, 2021, How to Avoid a Climate Disaster: The Solutions We Have and the Breakthroughs We Need, Penguin Books, New York.
- International Renewable Energy Agency Positive Hitpiece on Hydrogen.

Reports and Academic Articles

- Gustavsson, Jenny, et al., 2011, Global food losses and food waste, Food and Agriculture Organization of the United Nations.
- IRENA, 2020, Green Hydrogen Cost Reduction: Scaling Up Electrolyzers to meet the 1.5°C climate goal.

Shorter Newspaper, Magazine Articles, and Clippings

- Garrison, Cassandra, 2021, Concrete makers face heavy lift on climate pledges, Reuters, on cement/concrete efforts to reduce CO_2.
- Hanley, Steve, There's more to lowering atmospheric carbon dioxide levels than driving electric cars. Decarbonizing the cement and steel industries will be equally as important. Clean Technica, 2020/09/14.
- Hasler, Arthur Frederick (Fritz), ₂ Clean Technica, 2021/10/15.
- Kahn, Brian, 2021, Heat Pumps Are Ready to Have a Moment
- Mufson, Steven, et al., October 19, 2021, Russia allows methane leaks at planet's peril, Washington Post.
- Puko, Timothy, October 1, 2021, Who Are the World's Biggest Climate Polluters? Satellites Sweep for Culprits, Wall Street Journal.
- Shahan. Zachary, This Stunning Chart Shows Why Battery Electric Vehicles Win, Clean Technica, 2020/06/10.
- USDA, Some basics from the USDA on agriculture and climate change.
- Tabuchi, Hiroko, June 2, 2021, Here are America's Top Methane Emitters. Some Will Surprise You, New York Times.
- The tab on Agriculture has a good list of problems and solutions.
- The Foods That Reverse Climate Change by the BBC.

Chapter 14

Remediation and Geoengineering

At this point, you may be wondering whether humanity could pull greenhouse gases out of the atmosphere. The answer is yes! In fact, pulling out CO_2 is what the planet itself already does every year in great quantities for free. In some cases, humanity could just pay for speeding up existing natural processes.

This is called remediation. It refers to actions taken to counter the climate effects of past GHG emissions. (Climate mitigation refers to actions taken to reduce future GHG emissions.) You should not think of GHG remediation as undoing the chemical reactions, such as splitting CO_2 back into its carbon and oxygen constituents. This would be far too expensive. Instead, it usually means some sort of geoengineering whose purpose it is to push GHGs from the atmosphere into the ground, materials, or ocean. This goal could be accomplished, e.g., by speeding up natural rock weathering processes. Or by increasing forests. Or by compressing CO_2 at industrial plants into liquid CO_2 and storing it underground in exhausted gas wells.

An altogether different geoengineering approach would be not to tinker with the atmosphere but to reduce incoming sunlight. This goal can be accomplished (perhaps) by cloud seeding into the troposphere (lower atmosphere) or sulfur-dioxide particle seeding into the stratosphere (middle atmosphere). Nature herself is running such geoengineering experiments all the time. For example, volcanic eruptions emit reflective sulfur-dioxide particles, which have always cooled the planet – more in some years, less in others. If you are wondering whether this could ever be effective enough, the answer is "yes, easily." Just one volcano on the other side of the globe was powerful enough to cause a year without summer in Europe in 1816.

Environmentalists often worry about the unintended consequences of solar radiation management, and for good reason. It's dangerous. But keep perspective. Humanity has always been geoengineering. For that matter, so has the biosphere. Even the oxygen in the atmosphere is the result of indiscriminate tinkering by photosynthesizing plants. And so has the universe, blessing Earth with supervolcano eruptions and blasting it with asteroids every few ten thousand years.

A related question is *what is the optimal temperature?* — a question we first raised in Chapter 4. If the bad effects of global warming are really terrible, and if humanity for all its shortcomings fails to stop them, then maybe the bad alternative is simply less bad.

In our view, humanity should research to be prepared to intervene, regardless of whether the climate problem is natural or man-made, and whether it is cooling or warming. Interventions with fast response rates could be important to stop bad temperature feedback loops or push the planet back under a tipping point before everything gets out of hand. Meanwhile, let's hope we will never need it.

1 The Social Cost of CO_2 (Yet Again)

In Chapter 7, we explained that climate scientists and economists universally lament the fact that today's global CO_2 tax policies are perverse. On average, the world is subsidizing fossil fuels. Instead, the proper tax should be somewhere between $50/tCO_2$ and $100/tCO_2$, rising over time. The main disagreement among scientists and economists is about the question of whether the tax should be high or higher, and how steeply it should rise. (Unlike most scientists and environmentalists, we think this debate is mostly irrelevant.)

Chapter 7 also explained that if the world were governed by one benevolent dictator and it cost $100 to remove one ton of CO_2 from the atmosphere, then $100 would be an upper limit to any CO_2 tax. That is, if everyone who emits one ton of CO_2 were charged $100, the Utopian government could then spend this money to take it out again. The social problem would be solved. Our dictator's best solution would be pairing fossil fuel use with aggressive removal of CO_2 from the atmosphere — called <u>carbon sequestration</u> — *if* only taking out CO_2 from the atmosphere is cheap enough.

So what is humanity's cost of removing CO_2? It depends. Different CO_2 removal solutions have different costs in different quantities. We will explain below that the lowest-cost solution — tree farming — would already pay for itself today. Yes, in effect, the price tag to remove the first $GtCO_2$ could be $0/tCO_2$! Naturally, the opportunities for such lowest-cost solutions are limited. If CO_2 entrepreneurs were to build more tree farms, eventually the cost of forestable land would go up and the price of lumber would go down. This creates soft limits on the capacity of tree-farming as a removal solution — but the world is a long way from bumping up against limits. Removing, say, the first 1 $GtCO_2$ from the atmosphere (or equivalently, not emitting it in the first place) may cost under $10/tCO_2$. From our *moving the needle* perspective, this is *the* number environmentalists should care most about *today*. The world should implement these least costly processes to remove the first few $GtCO_2$ as soon as possible and not argue so much about of how much all 50 $GtCO_2$ will eventually cost.

More speculatively, it is in the realm of the possible that, with more R&D, the cost could remain under $50/tCO_2$ even for removing all 51 Gigatonnes of annual human CO_2 emissions! This $50/tCO_2$ is also an amount that humanity could reasonably afford in order to rid itself of its climate change

14. Remediation and Geoengineering

problem. Over time, the costs of CO_2 removal processes will change. They could not only go up (as in the case of tree farms), but they could also go down (if research discovers better techniques). As with battery technology, it may be better to spend billions of dollars on R&D today than trillions of dollars on giant CO_2-sucking facilities that could become outdated within a few years.

Before you get too enthusiastic, the real problem is again neither technology nor cost. Instead, it is the same free-rider problem from Chapter 7 that makes assessments of the collective social cost of carbon-dioxide largely irrelevant in the real world. No country finds it in its own interest to pay for removing emissions at large scale on behalf of the world's other 200 countries. If it seems tough trying to convince countries to take responsibility for cleaning up their own emissions, wait until someone explains to voters that CO_2 in the air does not have country labels attached to it. How do you think American and European voters would feel about paying for sucking out CO_2 that India and China are emitting? Nobody wants to pay for sucking nation-less CO_2 out of the atmosphere. And when India and China realize that other countries will pay for removal of their emissions, would they not happily emit even more?

Attempts to set up a global system have predictably failed. Global rules under some cap-and-trade systems designed to curtail CO_2 (in place of an emissions tax) have created lots of funny money and shenanigans. You would not believe how inventive people can become when free money is involved!

In the end, most removal solutions stand in stark contrast to cleaner energy generation technologies in two important ways. First, CO_2 removal would not remove the nasty local copollution of fossil fuels to which most people truly object. The local benefits that induce local reduction in the use of fossil fuels are just not present for local removal of CO_2. (Removing CO_2 is effectively paid for domestically but mostly enjoyed abroad.) Second, if one country were to invent a clean and super-cheap electricity plant, it could sell it all over the world. Both the researching country and the deploying country would want it *in their own interests*. No country will want to pay others to suck out large amounts of CO_2.

2 CO_2 Removal

The cost of removing CO_2 is logically an upper limit to the dictatorial social cost of carbon-dioxide, whenever CO_2 policies, the IPCC, or the Nordhaus and Stern models are discussed. But our view has always been that the environmental focus has to shift towards promoting solutions that are in the self-interest of individual countries. If you doubt our hypothesis that the social cost of carbon-dioxide is conceptually useful but irrelevant for practical purposes, we hope that reading this chapter will convince you.

Our evidence against the cost-of-carbon view is that countries cannot even agree to solutions today that have CO_2 removal costs as low as $5/t$CO_2$. They will never agree to solutions that cost as much as $100 or $200/t$CO_2$ — and especially while some other countries are still emitting CO_2.

Nevertheless, this section describes methods of removing CO_2. In our opinion, only the forestation-based methods are currently viable for non-trivially large amounts of removal. The others are pipe dreams, because they are and will likely remain much more expensive for a long time to come. At most, we can recommend further research. Implementation would be economically and likely also ecologically wasteful.

Forestation

The world has been cutting forests at record rates. The global loss of tropical forests contributes about 4.8 GtCO_2 per year (about 10% of annual human emissions). The most common reason to cut trees has been to make way for more agriculture to feed growing populations. The most common way of making way has been burning, often done by and for the poorest of the poor and without governmental permits. This is doubly bad: it burns carbon and it removes the trees which previously were sequestering more CO_2 every year.

From 2011-2015, humanity cut down about 20 million hectares of forest every year — about 50 million acres, the size of the state of New Jersey or the country of Israel. Since 2016, this amount has increased to an average of 28 million hectares per year. Brazil alone is responsible for about one-third of world deforestation (and growing). In 2021, it is at its worst level in 15 years. Nigeria, which has one of the world's highest population growth rates, has lost more than 60% of its forest cover since 1990. In Indonesia, forests are being cut for palm oil plantations. Americans and Europeans no longer cut much of their forests — not out of virtue, but because they have already cut down most of them in the past.

► Technology

The average hardwood tree absorbs about 20 kg of CO_2 per year, 1 tonne of CO_2 over 40 years of life. Fifty trees can absorb 1 tonne of CO_2 every year. Thus, 1 trillion trees could sequester about 20 $GtCO_2$ per year out of thin air. Even better, harping on the low cost again, estimates are that the net cost of doing so are on the order of less than $15/t$CO_2$ for about 15 $GtCO_2$ of removal, and this includes the cost of the land! (Pushing tree planting into more expensive and harder-to-reach locales to effect the last 5 $GtCO_2$ of removal would become much more expensive.)

The way to make this work at such a low cost is to plant trees and harvest them for lumber (or higher-tech composites that can even substitute for steel) — wood keeps the carbon nicely inside. On the margin, tree-farms are already profitable today even without any subsidies, although this depends upon the price of lumber. The business case is solid.

However, forestation can only work if environmentalists understand that they must not go on the barricades to stop the harvesting of trees. The whole reason why tree planting is so efficient is that trees convert CO_2 into wood that can then be sold — and, after they are harvested, new trees can be planted to transform yet more CO_2 into yet more wood. Preventing the harvesting of trees defeats the whole purpose. (When old trees burn or die, the CO_2 returns to the air, which is the worst of all worlds.) Sure enough, cutting down trees does not appeal to environmentalists, but it is the right thing to do to stem climate change.

This may sound too good to be true, and indeed it may be. Trees can also emit methane and other greenhouse gases. Some evidence suggests that tree farms are not as efficient as natural forests. There is also a serious concern that if one plants trees in many deserts (which are light and thus good at reflecting sunlight), the darker color of trees can reduce the relevant albedo and actually make global warming worse! However, for now, it looks as if there are plenty of opportunities for viable tree-planting solution that are cheap and for real — and it may become better if scientists can learn how to fine-tune it. For example, it may be better to plant trees that can grow faster with a mix of other trees (and different climate zones require different kinds

2. CO_2 Removal

of trees). Scientists could also genetically engineer trees that are better at removing CO_2 even in inhospitable areas. It's not for sure, but it looks very promising. Trees could viably move the needle now at extremely low cost for the first few gigatonnes of CO_2!

➤ Opportunity

Remarkably, scientists are only just beginning to understand forests. Until just a few years ago, they thought Earth had about 0.4 trillion trees. Yet we now know from satellite imaging that a better estimate is 3 trillion trees — only about eight times as many!

More importantly, how many trees could humanity potentially add to our planet? A reasonable estimate is about another 1 trillion trees. Figure 14.1 shows where more trees could viably be planted. In order, Russia (151 million hectares), the United States (103), Canada (79), Australia (58), Brazil (50), and China (40) have the most potential. Forests in the American West could fight desertification and protect biodiversity. And there is more good news: if there ever were to be a global treaty that pays some countries to plant trees, satellite imaging can cheaply confirm whether countries that have agreed to reforest are actually holding up their end of any bargains.

Is tree planting viable in large scale? The short answer is yes. In one state in India, Uttar Pradesh, 800,000 volunteers planted 50 million tree saplings in one day. If all the saplings grow into trees, and this process could be replicated 800 times, it would cancel out all the CO_2 humans pump into the atmosphere.

➤ Failure

The cheapest way to pull CO_2 out of the air, including the cost of land, is thus also the least objectionable. Better yet, trees have many local benefits, too — water filtration, flood buffering, soil health, biodiversity habitat, enhanced climate resilience, anti-desertification. This mitigates the cross-country free-rider problem. Countries benefit economically and environmentally from their own forests.

So why has humanity not managed to pursue even this low-hanging fruit? In fact, in large areas of the globe, the opposite of reforesting is happening. Brazil, Indonesia, and Africa are actively destroying their rain forests, while the rest of the world is largely watching passively.

14. Remediation and Geoengineering

Figure 14.1. Potentially Reforestable Areas

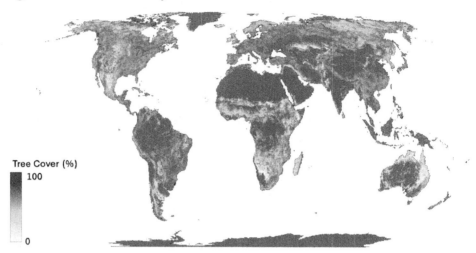

Note: Land can be unsuitable for reforestation for different reasons. The Amazon and the Congo are already forested, while the Sahara is unsuitable sand without enough rainfall.

Source: Crowther Lab ETH Zuerich.

The problem is that it is in the interest of individual farmers to cut down trees on public land to gain what is effectively free land. Why don't the controlling countries stop them? It depends on the locale. Farmers are powerful political constituents in many countries, where they often control large states in an upper house of Congress (the equivalent of our U.S. Senate). Moreover, even when countries decide to protect their forests, small subsistence farmers are difficult to keep in check. When their children are starving, who can blame them for cutting down trees to clear fields for grazing or planting? This is the case even if the loss to their country (and the planet) greatly dwarfs their personal gains. Drowning people often don't care whether they drag down others in their desperate attempts to survive. (And how do the police prove who started the forest fire?)

Although capitalism and industry are often claimed to be the culprits for deforestation, it is actually the opposite. In an ideal capitalist market, the owner of the forest would protect it. But when it comes to government-owned land, the enforcement costs are high and the property owner is often disorganized and conflicted. (At one point, a majority of Brazilian Congressmen were under indictment for corruption. And Indonesia and Africa typically rank below Brazil on most corruption indexes.)

We have argued that the world has no effective government that could enforce global solutions, and that country solutions are our best hopes, because countries represent the largest organizational levels with true power. Yet even this may have been too optimistic. In many circumstances, countries may not be sufficiently able and willing, either. Ironically, India's urban air quality is devastated by <u>farmers</u> burning off residual plant matter. One would think that India could solve this simple problem, but interstate conflicts and governmental incompetence have stymied common-sense solutions.

Nevertheless, if there is a solution, it will have to involve better governments putting restrained and managed capitalism back to work. Capitalism cannot survive without benevolent governments, just as governments cannot survive without benevolent capitalism. One reason for hope that solutions can be found is that reforestation often makes economic sense, too. Within a few years of the initial burns, former forest soil often loses its potential for productive agriculture. More efficient farming and subsidies for poor farmers in designated areas could render forest burning uneconomical, as could general reductions of poverty. A solution will require a balance between the needs of tree farmers to earn good livelihoods and the ecology of the forests. By involving local villages and providing Western environmental subsidies, incentives could be put in place to curb overly aggressive harvesting.

In sum, the current state of affairs is that reforesting solutions are both surprisingly cheap and effective, yet depressingly difficult to implement in the real world. If humanity cannot even manage reforesting, it is difficult to see how any other non-technical collective solutions will ever work.

▶ Biology Beyond Forestation

We discussed agricultural changes in the previous chapter. We covered trees above. However, ocean plants may be similarly important.

Trees are not the only plant capable of capturing CO_2. In Chapter 13, we mentioned the great potential of biological engineering. In their evolution, plants went down some clever and some not-so-clever pathways. It may be possible to engineer <u>plant roots that can capture more CO_2</u> and fix nitrogen out of the air (thereby reducing the even more potent nitrous oxide GHGs).

Of course, if successful, we have to be careful that plants won't capture too much CO_2, or we may end up with another Snowball Earth. It's all about balance. Earth's oxygen is an example of how plants mutated once before to change the global environment.

As we noted before, civilization should prepare not only for temperature increases but also for decreases. As we are writing this, the 350m asteroid Apophis passed inside our geosynchronous satellites. If it had hit Earth, it would have carried more energy than all nuclear weapons combined and caused a nuclear winter-type catastrophe.

Enhanced Rock Weathering

The next cheapest solution seems to be accelerating the process of rock chemical weathering on the surface of the earth. The principal stone that can bind CO_2 is Olivine. If you have never heard of it, you are not alone. However, olivine comprises about 60-80% of Earth's upper crust, mostly found (but not confined to) basalt. When olivine is weathered, it turns into magnesite and quartz. Weathering olivine is also an efficient process, not requiring extra energy, and binding the equivalent of the emissions from one liter of gasoline in one liter of stone.

If we just waited a few thousand years, olivine and some related minerals would take care of all the excess CO_2 in the atmosphere by themselves. But we do not have a few thousand years. If olivine is crushed into small grains, exposed, and doused with water, the weathering process can be reduced to a few years. Unfortunately, crushing and exposing are expensive. Nevertheless, in some areas (the Deccan traps in India, the Columbia river basalts in the United States, and the Siberian traps in Russia), the cost to do so may be as low as $50/tCO_2$. In other areas, the cost can be as high as $150/tCO_2$. Chances are that the costs could go down further if humanity resolved to weather olivine in large quantities and learned when, where, and how it would be least expensive.

Realistically, this is not likely to happen. Unlike tree planting, there are no local benefits to olivine conversion, so it seems unlikely that any countries will volunteer to pay for implementation at a scale large enough to remove even their own country's CO_2 emissions, much less those of others.

Enhanced Ocean CO_2 Capture

About 70% of the planet is covered by oceans, and it is the primary destination of most atmospheric CO_2. Unfortunately, however bad humanity's treatment of ecosystems on land is, humanity's treatment of ecosystems in the ocean can only be described as abysmal. Out of sight, out of mind. With no country owning the oceans of the world, it is a race by fisheries to exploit them as aggressively as possible before others do so before them. It's not capitalism, but failure of shepherded capitalism that has gotten us here.

Moreover, as modest as our knowledge is about large-scale atmospheric CO_2 removal onto land, it is even more modest when it comes to large-scale CO_2 removal into the oceans.

A 2012 study in Nature suggests that seagrasses store more than twice as much CO_2 as forests per square mile. And like trees, seagrasses have plenty of local benefits, from cleaning water to providing habitats. But, as is the case with forests, humanity is depressingly not raising seagrasses but destroying them. Moreover, scientists don't know enough about whether seagrass planting could be made to work.

Another approach is to coopt algae. They can be up to 400 times more efficient than a tree, albeit in a bioreactor. Instead of making wood out of CO_2, they make more algae. It might even be possible to use algae to make plastic-type polymers or fuel out of thin air. Ironically, one of the challenges is that algae grow too fast to be easily manageable — surely a problem that can be overcome with more research.

For a while, many scientists were bullish about seeding oceans with fine particles of iron to stimulate plankton growth. Recent evidence has suggested more skepticism. Seeding may not work on a large scale, because it could deplete nutrients that are needed elsewhere. Scientists just do not know.

Ocean liming — adding a form of calcium to seawater — could also allow the oceans to absorb more CO_2. Cost estimates range from $70/tCO_2$ to $150/tCO_2$, with $100/tCO_2$ a good middle.

Other ocean research is just beginning. We are keeping our fingers crossed. Project Vesta is researching whether it is possible to use ocean wave energy to accelerate natural stone weathering. One of our UCLA colleagues wants to produce hydrogen while pushing CO_2 into the seawater. His cost estimates are $100/tCO_2$, with possible long-term reductions down to $50/tCO_2$. Unlike many other schemes, these could be nearly limitless processes.

Industrial CO_2 Capture and Sequestration

I can tell you the meaning of life, but I can't explain sequestration...

Finally, there are industrial solutions to CO_2 removal. Humanity has become good at industrial solutions over the last century. Civilizations have built entire industries before. There are even some commercial uses for captured CO_2. For instance, CO_2 can be injected into gas wells as solvent to improve extraction. However, in this case, nature already offers so many good biological self-replicating solutions that industrial solutions will have a tough time competing—unless breakthroughs reduce costs by two orders of magnitude. This seems unlikely.

For example, Climeworks is already aggressively pursuing industrial CO_2 removal, though in minute amounts. They quote removal costs of about $300/tCO_2$, with the potential to bring the cost down to $200/tCO_2$. The U.S. Department of Energy has announced "earth shot" research to bring the cost down to $100/tCO_2$, which may or may not be possible.

Scientists are also researching different technologies, from capture of CO_2 out of the thin air (which can be done anywhere, including near volcanoes to tap cheap heat) to capture at industrial plant exhaust stacks. For now, exhaust capture can only be installed in a limited set of locations (soon we hope even more limited as coal plants will hopefully be disappearing). But capturing highly concentrated CO_2 is three times cheaper than direct air capture. Howard Herzog from MIT estimates capture from flue gases on chimneys to be around $100-$300/tCO_2$ by 2030, and direct air capture at $600-$1,000/tCO_2$. Other estimates are more optimistic, going as low as $50/tCO_2$ by 2050. None are likely to reach the $10-$30/tCO_2$ that lumber could offer.

Right now, Climeworks and other industrial operations are funded by Stripe (a payments processor), where rich people can buy carbon offsets.

Some do so out of ethical considerations, others for public relations purposes. While laudable, Climeworks is a model that will never scale to the world as a whole.

We view industrial carbon capture as an interesting research venue, although we believe it is unlikely ever to become commercially viable. Even if the costs were eventually to come down to $100/tCO_2$, it would still not be the lowest cost of carbon sequestration for a long time. We think that the market problems are so serious that industrial carbon capture will never be deployed on a large scale.

3 Solar Radiation Management

The second class of geoengineering solutions does not attack CO_2. Instead, solar radiation management (SRM) involves reflecting some solar energy back into space, ideally before it reaches the troposphere (Earth's lower atmosphere). Any such shield must also not block outgoing infrared radiation or it would worsen the greenhouse effect.

Not climate geoengineering again!

Space-Based Parasol Shields

As lovely as the concept of an umbrella in space is to us Star Trek fans, this is non-sense. Launching large structures into space is too expensive. It takes a huge amount of energy to put a mass into orbit. There are even more expensive schemes — Texas Congressman Gohmert pondered whether we could move Earth to the great amusement of the press and the educated public. (Even Jean-Luc Picard could not do this sort of geoengineering.)

Sulfur-Dioxide Particles in the Stratosphere

The most promising SRM technique is based on jetliners dispersing small reflective sulfur-dioxide particles into the stratosphere. These particles would then reflect incoming sunlight and cool the climate (as they are intended to). They circulate for a few years before falling back to Earth.

Scientists know how this type of SRM works, because volcanoes have demonstrated its efficacy many times before. However, SRM is a patch, not a solution. It does not address other harmful effects of greenhouse gases, such as ocean acidification. (In fact, sulfur-dioxide particles will make it worse.)

The big advantage of SRM is cost. The estimates are best described as ridiculously cheap, ranging from about $2 billion to $10 billion per year. This is 100-1,000 times cheaper than CO_2 elimination.

Furthermore, cooling the world with sulfur-dioxide may be ill-advised as long as the harm from climate change remains limited. Yet if global warming were to ever become world-threatening (although we do not see how this could happen) or just extremely harmful, sulfur-dioxide would work almost immediately. Contrast this to the many decades that it would take to start cooling the planet with CO_2 reductions.

Artificial Clouds

Another interesting alternative is seeding artificial clouds. It is fairly cheap to send a few boats out onto the ocean and spray water from atomizers into the air that then form clouds. If we can do it for snow, why not for reflecting sunlight? Maybe 200-300 boats could do the job. Unlike sulfur-dioxide particles, cloud seeding can be started and stopped almost instantaneously — plus it has no negative ocean acidification consequences. Unfortunately, science is not even sure whether clouds help or hurt global warming. There could also be regions where seeding helps and other regions where it hurts.

3. Solar Radiation Management

Be Ready For Plan C

It's important for scientists to learn more about SRM, even though it seems premature to deploy it. Hopefully, humanity will never have to. No one should be excited about solar radiation management. It shouldn't even be Plan B, but rather Plan C. Many environmentalists and scientists legitimately warn about unintended dangerous side-effects. Even the most ardent proponents only advocate for SRM as a stopgap bridge solution for a few decades, a complement to emission reductions. The second danger is that some countries may use it as a substitute. They may simply prefer to spend a little on solar radiation management rather than a lot on the underlying problem of greenhouse gas emissions.

There is an important asterisk here. Countries could disperse particles without asking other countries for permission. Countries in the Sahel may be thrilled to receive more rainfall, but what if it negatively affects rainfall in Europe? What if India suffered a heatwave killing 20 million people and decided unilaterally to deploy sulfur-dioxide particles? Would this be a good reason to start a war?

For now, we agree with many other academics interested in climate science (including Profs Dressler, Parson, Pindyck, and the National Academies) that humanity should learn more about the costs and benefits of SRM. It would be wise to understand these technologies. We advocate testing them in small-scale temporary pilot projects, where the negative effects are almost surely outweighed by the positive ones. For example, we should run a one-year experiment to find out whether a little bit of cloud seeding could increase rain in a small part of the Sahara (reducing rain over the Indian Ocean). It would be worth the learning cost. Yes, there is the possibility of unintended consequences — but they need to be weighed against the benefits of the intended consequences.

It would be foolish for humanity to count on solar radiation management as a long-term solution, if only because it may not work or the side effects may be too bad. There are good reasons to fear unintended consequences. However, it would be even more foolish not to be able to deploy a rescue if a catastrophic temperature-feedback effect were to occur.

Humanity may need to take unprecedented steps to moderate its climate. We hope use of the power to alter climate will come with far greater wisdom and forethought than what civilization has shown so far. Developing this wisdom may well be a taller order than the geoengineering itself.

Conclusion

Large-scale expensive industrial CO_2 removal schemes are currently not feasible. To be economically viable, the cost of CO_2 removal has to be under $20/tCO2, perhaps less. Forestation in many places already has even lower costs *and* provides positive local benefits — and yet humanity is still passing it by! Environmentalists should work on moving this needle now — incentivizing entrepreneurs to plant more trees for profit. Hoping to fight climate change by simultaneously bringing down the cost of industrial CO_2 removal processes to $100/tCO2 and overcoming the free-rider problem seems quixotic.

We view solar radiation management as important for two reasons. First, it is cheap — so cheap perhaps that one rich philanthropist like Bill Gates could potentially stop world-wide warming. Second, its effects could be quick — unlike CO_2 removal, which has effects only on time spans of decades. Environmentalists should push not for deployment of solar radiation management, but for small-scale scientific research experiments — if not using sulfur-dioxide particles, then at least using ocean cloud seeding.

Further Readings

BOOKS

- Busch, Jonah, 2019, Earth Innovation on deforestation and reforestation choices.
- Dressler, Andrew and Edward Parson, 2019, The Science and Politics of Global Climate Change: A Guide to the Debate, 3rd ed. Cambridge University Press.
- Robinson, Kim Stanley, 2020, The Ministry of the Future. A science fiction novel about India undertaking geoengineering to stop a deadly heat-wave.
- Robert Pindyck, 2022, Climate Future: Averting and Adapting to Climate Change covers adaptation (such as SRM) in better detail than we do.
- Unnamed, 2018, The Big Bad Fix: The Case Against Geoengineering, Heinrich Böll Stiftung.

REPORTS AND ACADEMIC ARTICLES

- Aldy, Joseph E., and Richard Zeckhauser, April 2020, Three Prongs for Prudent Climate Policy, Resources for the Future.
- Herzog, Howard, 2022, *Direct Air Capture*, Royal Society of Chemistry, Chapter 6.
- Tollefson, Jeff, 2018, First sun-dimming experiment will test a way to cool Earth, Nature News Feature.
- Trisos, Christopher et al., 2018, Potentially dangerous consequences for biodiversity of solar geoengineering implementation and termination, Nature Ecology & Evolution.
- Unnamed, 2021, Reflecting Sunlight: Recommendations for Solar Geoengineering Research and Research Governance, The National Academies of Sciences, Engineering and Medicine.
- Zarnetske, Phoebe L., et al., 2021, Potential ecological impacts of climate intervention by reflecting sunlight to cool Earth, PNAS. Also covered in Ars Technica by Doug Johnson, Ecological impacts of solar geoengineering are highly uncertain.

SHORTER NEWSPAPER, MAGAZINE ARTICLES, AND CLIPPINGS

- Almost everything written by James Temple, senior editor for energy at the MIT Technology Review, focused on clean energy and the use of technology to combat climate change.
- Carrington, Damien, July 4, 2019, Tree planting 'has mind-blowing potential' to tackle climate crisis.
- Crowther Laboratory, April 7, 2019, How trees could help to save the climate.
- Johnson, Scott K., There's a lot we don't know about ocean CO_2 removal. Ars Technica, 2021/12/14.

14. Readings

- Let Your Trees Grow For Profit, July 13, 2016, Wood Magazine.
- Rotman, David, February 8, 2013, A Cheap and Easy Plan to Stop Global Warming, Technology Review.
- Simon, Matt, November 30, 2021, Think Climate Change Is Messy? Wait Until Geoengineering, Wired.
- Wood, Charlie, 2021, Is blocking out the sun a good solution to the climate crisis?, Popular Science.

Websites

- Harvard Solar Geoengineering Research Program.
- Joanne Chory, leader in reengineering plants to capture more CO_2 into plant roots.

Video

- NOVA: Can We Cool The Planet?

Part

The Transitioning Problem

Chapter 15

Making It Happen

The world is indisputably warming — probably by another 1–2°C from where we are now. This 1–2°C rise will be painful, especially for the poor, but it does not endanger humanity as a whole. (It is already 6°C warmer than it was 15,000 years ago.) A more dangerous problem is a less likely but not impossible scenario: This more modest 1–2°C rise could set into motion dormant forces that have not been awoken for millions of years and that could in turn cause far larger further temperature increases. In an even more exceedingly unlikely scenario, the short-term domino effects could even be so apocalyptic that they could extinguish our species. (Then again, so could many other unlikely events, too.)

Fossil fuels are almost surely a primary cause of global warming, even if a minority of scientists are still wondering whether it is the only one. Moreover, the harmful health effects of fossil fuels beyond their global warming effects are also highly significant. It is now in the collective interest of humanity to replace fossil fuels with cleaner alternatives more aggressively than ever before. So we end our book with observations about what countries and individuals can realistically do to speed up the worldwide transition to a cleaner planet.

Although we view ourselves as environmentalists, our recommendations are not quite the same as those promulgated by many other better-known environmentalists, such as Greenpeace or the School Strike For Climate (Greta Thunberg's Organization). We have our reasons.

First, we don't have to play to a home audience. We can thus say out loud that environmentalism needs to be not just against but also in favor of

15. Making It Happen

big policies, occasionally even painful ones. Too many environmentalists are simultaneously against everything — fossil fuels, nuclear power, hydroelectric dams, geothermal plants, lithium mining, solar cells, windmills, new electric transmission lines, and tree felling. The requirement of zero environmental harm on every dimension cannot change the world for the better. It only empowers the status quo.

Second, we are more concerned about what policies can maintain large-scale popular support for a long time — not just in rich countries but all over the world. If greener policies turn into expensive vanity projects or greatly hamper the economic development of poor countries, the backlash could delay or even stop them altogether. Plans that allow for regular energy outages or that "go back to nature" are bad plans. Earth cannot sustain 8 billion people without modern industry and agriculture. Even if some environmentalists are prepared to accept the misery consequences of radical energy-use reductions, most people are not. They would revolt and the environmentalists would lose.

Third, we are more concerned about what is realistic and cost-effective and less concerned about what is Utopian and high-minded. The latter makes for good salon conversations but it will not *move the needle*, not now and maybe never. Yes, we too would love countries to spend their military budgets on humanitarian causes instead — but it won't happen.

Fourth, we are focused on initiatives that environmentalists can start today. (There will be more problems in 50 years, but we leave their contemplation to others.) Quoting an old Chinese proverb, we need to begin the journey of a thousand miles with the first steps today. We cannot sit down at the starting line and lament that humanity has not yet gotten far enough or will have to travel a thousand miles (eventually), while billions of people have not even arrived at the stadium. Ideally, we would stop talking and develop (clean) trucks that can drive all of us a thousand miles. Walking may be inspirational, but it won't get us there as fast and as safe as the trucks will.

In this, our final chapter, we describe what governments, organizations, businesses, and individuals can do here and now. Note the order. Global pollution is a big problem and requires big solutions. Countries are as big as effective human organizational structures get.[1] Business, organizations,

---

[1] However, it is also true that governments, because they are so big, are also often more dysfunctional than many smaller organizations — and sometimes governments are outright

and individuals can play useful roles, but their means are more limited than those of governments. Environmentalists are in a war that can only be fought by large collectives, not by individual heroes — and not just a short but a very long war that will have to be fought for many decades and over many generations

Realistically, all viable prescriptions for *moving the needle* now involve low-hanging fruit. High-hanging fruit is effectively fruitless, because it is unreachable. For the highest-hanging fruit, this is also for the better, because they are indigestible — they cost way too much for the good they do. Unfortunately, indiscriminate activism has endorsed some of them. Fortunately, there are so many examples of better low-hanging fruit that we don't have the space to list them all. This chapter focuses on our favorite ones. Many of them involve the government acting as a catalyst — overcoming bumps rather than climbing mountains. Environmentalism should push for these catalytic actions as soon as possible — ideally, yesterday.

But let's start with where we are today.

1 What Can Countries Do?

We have emphasized the appropriate scales of both problems and solutions throughout our book. The problem is big and the solutions are slow. We explained why even the entire OECD could not stop the growth in world emissions, much less push it down to net zero. The arithmetic does not add up. The majority of emissions are from the 80% of people living in poor countries and their emissions are growing faster than the OECD could ever reduce their own. If you need a reminder, go back to Section 6.2.

Furthermore, we have already discussed remedies that are likely to fail (in Chapter 9) and remedies that are likely to succeed (especially in Chapter 10). There are a lot of steps we could take *now*, whatever your broader views and political philosophies are. In the current chapter, we are summarizing our favorite approaches in the hope of leaving an impression. They are not listed strictly in order of importance, effectiveness, or cost. But they are limited to those we consider highly important, highly effective, and relatively cheap.

---

malicious. Co-opting government for a purpose is still usually more effective than co-opting just your friends and their friends or even your neighborhood.

(1) Increase Innovation

Recall the truck example from the start of this chapter — the one supposed to carry us all to the finish line. The truck is technology. We need to requisition it. Technology does not appear out of the blue. Countries should do everything they can to help develop new clean technologies.

We covered relevant technologies in the third part of our book (Chapters 12-14). Ultimately, technologies that make clean energy *cheaper* are the *only* way to reduce world-wide emissions to the point where atmospheric CO_2 concentration will decline again. Every other environmental measure will, at best, slow the accelerating increase.

▶ Research, Development, Deployment

New technologies require research, development, and deployment (RDD). Research is the part where scientists are poking around and really do not know whether the results will ever be useful. Development is the part where the scientists know that the concept works, but they do not know whether it can work in the real world. Deployment is the part where companies begin commercialization, with a few first pilot tests in the field. (Engineers call this kind of deployment FOAK, First Of A Kind. The FOAK cost is typically much higher than the NOAK, Next of A Kind, cost.) The borders between the three areas are fluid. For example, the first deployment often uncovers new problems that lead back to new research.

We are not alone. There is near-universal agreement among experts that the best path to green energy adoption worldwide is through innovation (even though these experts may disagree what else should be done). Period. Investment in green innovation is likely to provide the most bang for the buck, much better than green deployment. Lomborg estimates that every dollar spent on fundamental green energy research expects to pay off ten dollars. Although this estimate is on the high end, it is not out of line with other estimates. Even if the expected payoff were only $5, governments should still do a lot more to foster RDD than they do today. And it would be in their

1. What Can Countries Do?

own competitive interests, as well. This is the easiest lowest-hanging fruit — increasing the funding for clean-energy RDD.

This is not a socialist proposal. Government research support in the OECD has driven technological development and growth for many decades. Even companies that work without government funding still draw heavily on government-funded basic research (often started at research universities) and on talent educated and trained by schools and universities. The Internet and batteries are just two of many such examples. It's just that our governments do too little of it. Granted, real-world governments have many inherent conflicts and inefficiencies and can easily be corrupted, but even given their shortcomings they should still increase their relevant support and work on doing it a lot better. (And government research support has arguably been focused too much on military than commercial potential. The future of the OECD vs. more autocratic regimes may now depend more on its global competitiveness than its military muscle.)

With such payoffs, recommending more funding for clean-energy RDD is the easy part. Doing it well is the hard part. We need a conceptual framework. How should governments allocate funding? What role should the market play? What are the problems?

➤ Externalities

There are at least two economic externality problems with private RDD, already explained in Chapter 6. The first is that the private rewards for inventions are only a fraction of the social benefits. If RDD is left to private entrepreneurs, there won't be enough of it from a social perspective. The second is the desire of inventors, once successful, to patent their inventions and prevent others from using them. Of course, this is understandable: no one would invest $1 billion in a new battery technology, with its great risk of failure or of the arrival of a superior competitor, if there were not high profits in case of success. But after the technology works, the world would be better off if it were available to everyone at the lowest cost.

Economists do not have universal solutions to these two problems. They do know that the best solutions in the real world are often imperfect mixes of government subsidies and private markets. (Just because markets are not perfect does not make the government perfect. Governments often respond to short-term political pressures and do not seek out the best long-run alternatives.)

Despite the just-mentioned problems, the benefits of clean-energy RDD are so high that the government should fund more of it. The best choices of which technologies to support should be left to panels of expert scientists and engineers. Some far-out alternatives could pay off big, too, and deserve funding. (However, sometimes it does not take an expert to understand that some research is so relatively cheap and could have such positive effects on humanity that it should be subsidized in any event. A few of our favorites are listed in Appendix Section App. A.)

➤ Public-Private Collaboration

Worldwide research funding already largely operates in a mixed public-private partnership way. Governments fund research, often in universities, and the results, if any, are later commercialized by companies. In exchange for generous government subsidies, governments could demand more knowledge-sharing of inventions with other companies.

University researchers and experts tend to be better in assessing basic research than companies. Companies are better at the technology development and deployment stages. At these later stages, the potential profits and losses align much better with corporate incentives. The government may still want to help with a first "FOAK" plant deployment, but thereafter it becomes high time for governments, researchers, and experts to leave the field to as competitive a market of firms as possible.

➤ Risk and Failure

An important difficulty in funding research is that government employees (and bureaucracies) often have no stomach for embarrassing public failures. Yet the whole purpose of R&D is to venture into the unknown. If we knew a concept worked, we wouldn't need research.

In 2010, the U.S. company Solyndra received hundreds of millions of dollars to manufacture and deploy novel solar cells. It ultimately failed because the prices of raw materials for competing Chinese polysilicon cells dropped too quickly. The resulting fallout made great headlines for the GOP opposition, because Solyndra was funded by the Obama administration. Whether right or wrong, it was Monday-morning quarterbacking. A quarterback who has never thrown an interception on a long pass should probably throw more marginal passes — he has not taken enough risks. If every R&D investment worked out, we would not have made enough of them. We should have aimed higher. And herein lies the problem. Risk-taking is not what politicians and bureaucrats excel in. Ideally, innovation research would be funded by a bipartisan panel.

Failure seems to be more (but also not sufficiently) tolerated in academic research grants, if only because the research results often remain more obscure. This is perhaps one of the reasons why government scientific grants to universities are a good solution.

▶ Prizes and Funding

Traditional research grants to universities are important, but we would recommend some more daring supplemental funding mechanisms, too.

The X-Prize Foundation has offered highly visible prizes that have proved to be great catalysts for creative research. Winners of X-prizes gain not only funding, but also instant publicity and credibility. Elon Musk has now offered a $100 million as a prize for carbon capture. Governments should supplement such prizes. Why can't prizes be $1 billion or $10 billion instead of $100 million?

The government could also offer prizes that consist of guarantees to buy the first product, such as the first 100 GWh of fusion electricity provided for a price point of $300/Mwh or the first 1 $GtCO_2$ removed via accelerated weathering for a price point of $20/MWh. Would it work? We won't find out unless we try. (The same "fail sometimes" approach is needed here.)

▶ Engineering and Public Education

Finally, a more long-run aspect of R&D is education. This ranges from training more engineers to educating voters and consumers. We need science and engineering to become "sexy" again.

15. MAKING IT HAPPEN

► Funding

Our immediate and most important recommendations are also the easiest:

1. Expand the budgets earmarked for green technologies at the National Science Foundation and the Department of Energy — perhaps double, perhaps quadruple. Keep politics out of it and keep it science- , research- , and development-based.
2. Establish large prizes and guarantees for milestone achievements for FOAK plants.

► A CO_2 Tax

We are in favor of appropriately limited CO_2 taxes that will help direct research and development towards innovations designed to reduce CO_2 emissions and with it the required taxes. Innovation is a complement not a substitute for a CO_2 tax.

► What?

Although we want to restrain in general from too many engineering opinions (except for a few in the appendix), we can't restrain ourselves with respect to energy storage.

Efficient pool-sized energy storage grid battery solutions are the biggest prize there is. If you work for a company that only has a 1-in-a-1000 chance of success, but whose technology could bring down storage cost to half of what it is today, your expected contribution to humanity may be much higher than what a million others could contribute by painful GHG reductions. It is the opposite of the free-riding problem (of curtailing energy use and CO_2 emissions), where you can only be a minor contributor — at best.

If you either want to change the world or become rich (or both), then take your chances here. It's an exciting time.

1. What Can Countries Do?

(2) Share Technology Globally

Most of the RDD funding, and as a result most new discoveries, will probably come from developed countries. Although breakthroughs are more likely to happen when there are many brilliant scientists working together, this is not all good. U.S. universities have been responsible for the largest brain drain in human history, from poor to rich countries — though immigration antipathy in the United States has recently been slowing it down.

Morally, we owe it to poorer countries to help them with their transitions on so many levels. Pragmatically, CO_2 emissions are a global problem. They are as bad when they occur in India as when they occur in Indiana. It is in the world's interest to share technology and expertise. Is it also in countries' self-interests? This is less clear. If we could wave a magic wand or direct global negotiations, our emphasis would be on ways for the world to collaborate more on clean-energy RDD.

How can clean energy be made more accessible to poorer countries? Despite a lot of general waffling and political lip service, most governments have been defending their own industries and not been advocating for the interest of the world, much less the interests of the poor of the world. Public relations talk of *equity* is one thing. Actual sacrifice and sharing are another. There is also a second complication. The best solutions may not be the same in poor countries. In many politically less stable countries, energy technologies need to be different. It makes little sense to build a nuclear plant or a dam in a war zone. Countries without an electric grid may be better off with roof solar cells. And so on.

Our immediate recommendation is therefore for the West, East Asia, and China to establish a joint program that discounts technology license fees or waives patent fees for countries that meet certain poverty criteria and that want to install clean technology domestically. Barge-based near-shore nuclear power plants could provide subsidized electricity to many countries that want clean power but cannot be trusted with nuclear technology.

(3) Tax Local Fossil-Fuel Pollution

We economists love taxes on negative externalities such as pollution. Prices provide incentives to reduce harmful pollution and to develop and deploy alternative sources of energy. Therefore, we would advocate that countries should impose fossil-fuel taxes (instead of today's fossil-fuel subsidies) to reduce local harm; and in reasonable amounts it will also make them better off themselves. Fossil-fuel taxes reduce co-pollution and adverse health consequences and help develop competitive clean-energy export sectors.

Why are we not advocating a major effort for a global CO_2 tax to combat climate change? Forcing 200 other countries to institute global-targeted taxes would be like Sisyphus rolling the proverbial stone up a hill. Even if CO_2 taxes can be passed on behalf of global rather than local interests, they would likely take decades to come into force and not survive some next electoral cycle, recession, or energy crisis in many countries.

In contrast, local CO_2 taxes that provide local benefits are more like putting a wedge under the rolling stone. Such taxes can be catalysts. The government only needs to run the trick one time and get people used to it. Once established, going back to allowing high emissions that make one's own population worse off is going to be more difficult for the fossil-fuel industry — especially, once the public in places like New Delhi and Beijing realizes how much better life can be without asthma and visibility limited to 30 feet.

Nevertheless, let's not kid ourselves: even local CO_2 taxes will be difficult to institute. The biggest hindrances are powerful mining and fossil fuel lobbies, both on behalf of companies and employees. And CO_2 taxes often hurt poorer people more. To institute local CO_2 taxes will require excellent politicians, carrots, and sticks. But locally justified CO_2 taxes and controls stand at least a fighting chance for long-term public support. Globally justified CO_2 taxes do not.

You may disapprove of the modesty of our goal, but we wouldn't be surprised if local CO_2 taxes alone could make a big difference, halving global CO_2 emissions. However, we admit that we have no evidence to back up our assessment. Whereever possible, let's try it out!

1. What Can Countries Do?

> *anecdote*
>
> Most governments have been deficient in basic tasks. Gratuitous methane leaks from oil&gas wells are low-hanging fruit. Burning off leaking methane would be cheap. However, it remains even cheaper for producers to abandon wells than to appropriately plug them at the end of their lives, and few governments have had the attention bandwidth to do much about it. They should impose harsh penalties. Governments worldwide — especially those in the Middle-East and North America — have been dysfunctional in failing to institute such.

(4) Forestation

In Chapter 14, we explained that the world is still actively <u>deforesting</u>. Nevertheless, there is also a lot of space to plant more trees elsewhere, including in the United States. A lot of today's forests are simply in the wrong locations, places where poor people need their space for planting subsistence crops.

Forestation has a lot of local environmental benefits, it is widely popular on a bipartisan basis, and it is cheap (to the tune of \$10 per tCO_2e). There is no reason for countries not to go ahead immediately with spending more money incentivicing the planting of more trees.

It is important that the plan be to harvest the timber. Wood sequesters the CO_2. It must not be allowed to burn and decay. And harvesting is what makes the enterprise economically viable in the first place.

There are commercial timber companies that could be subsidized if they lay out a sustainable model, in which they receive reduced-cost access to public lands, grow timber, cut them down, but allow wildlife to move to adjacent parcels. Not every environmentalists will be thrilled about such plans, but many will be. It's a sacrifice well worth making.

We can do more yet. There are probably even better ways to plant CO_2-consuming forests. (It may also make sense to work on other CO_2-absorbing plants, but this is not mutually exclusive.) The forestation enterprise could be made more effective and economical with additional research. What trees should be planted where for most effect? As we have stressed global warming is a problem that needs to co-opt economics, not fight it.

15. Making It Happen

(5) Price Electricity By Demand and Supply

We only cut our CO_2 emissions to piss off the utilities company.

Wind and solar are already the cheapest forms of power in history. The problem is electricity storage.

But why do we need so much storage in the first place? Part of the reason is because few of us are used to dirt-cheap electricity from 9am to 5pm and expensive electricity from 5pm to 10pm and 5am to 9am. How many of us would be willing to buy smarter appliances and adjust if we could count on saving half of our electricity bills *and* at the same time do good for the environment? And we could make it twice as expensive for those of us who refuse to adjust. We suspect most of us would learn pretty quickly — and saving money on expensive electricity is something that should appeal even more to countries and people that are poorer.

Even in my own city, Los Angeles, electricity pricing is difficult to understand for me, someone with an interest in it. Large providers now usually offer customers both flat rates and time-of-day plans. However, most customers stick with whatever the default is. (The price differences are not that big, either.) One vendor has a time-of-day plan that perversely considers day time to be a peak period, with higher electricity pricing while the sun is shining. This is a perverse situation, probably a legacy of the old days before solar power was cheap. The other vendor has a time-of-day plan that makes sense, with lower daytime pricing. It's certainly not the kind of situation where most customers are privy and aware of dirt-cheap electricity during day-time hours and super-expensive electricity when the sun is going down.

Governments can help facilitate the switch to demand-sensitive pricing on many levels. They should help electricity companies sign up as many customers as possible to demand-sensitive pricing plans. They should tax fossil-fuel plants to increase their after-hour prices further. They should help make electricity price information ubiquitous. They should open a frequency in the RF spectrum on which providers could broadcast the current and anticipated prices of electricity (the same way we broadcast atomic clock signals and hurricane warnings). They should standardize power-embedded signals and encourage standardized Internet-based two-way signaling of prices and impending customer electricity demands.

Again, our suggestion is for governments not to fight market forces in order to reflect global CO_2 externalities over decades or centuries. Instead, it is for governments to act as one-time catalysts to bring about the changes that switch consumer habits to consuming when energy is both dirt-cheap and no-dirt-clean — and then to get out of the way as soon as possible.

(6) Uproot Environmentally Bad Habits

Many emissions can be cut <u>without spending a penny</u>. But, as Benjamin Franklin noted, old habits die hard. Many people are not only too busy to worry about changing (a <u>status quo</u> bias) but intrinsically distrustful of anyone trying to alter what has worked for them for a long time — even if they barely <u>remember</u> the reasons why they are doing what they are doing in the first place. The good news is that habits can be altered by governments acting as catalysts. Governments can jump-start changes and then get out of the way once habits have changed. Governments won't have to fight this battle forever.

Here are a few important bad habits in the environmental context.

First, most people don't know or don't care about electricity pricing. They have busy lives. They still think of electricity as being *more* expensive during the day rather than the night, which is how it was when coal plants supplied factories with electricity mostly during the day-time. How can we get people to notice and change? If we can get consumers onto time-of-day plans *and* make them aware that electricity is cheap during the day and expensive at night, then we, the people, will probably do all the rest voluntarily.

Second, habits influence agricultural tilling and farming practices (Chapter 13). Turning over the soil reduces weeds and increases yields, but releases more CO_2 than <u>No-Till</u> farming. Worse, farmers tend to be intrinsically even more conservative and distrustful of government than the average person. How can we nudge farmers to adopt practices that will not cost them much and help improve the environment? We could offer them direct subsidies for better tilling practices and tax them for harmful tilling practices. But the big deal will be to change farming habits. Getting farmers to try out an alternative at least once would be winning more than half the battle.

Third, habits can be based on and reinforce mistrust and myopic consumption patterns. The *Energy Efficiency Paradox* is that most people decline to spend more upfront even when the lifetime energy and cost savings are far greater. For example, a majority of people decline to spend more on energy-efficient washing machines, even though they would come out financially better.

Changing habitual behavior patterns is not easy, but it can be done. Here are three possible approaches.

➤ Coercive Mandates

A forceful way to overcome inertia is to mandate the purchase of more efficient devices. The government effectively forced inferior incandescent bulbs, inferior housing insulation (with building codes), and inferior gasoline engines (with MPG standards) off the market. The flip side of this approach would have been to subsidize LED bulbs, insulation material, and more efficient gasoline engines to make them cheaper. (The two could go together, too.)

An important aspect of coercive mandates is that they help the ultimately better solution steal the appropriate economies of scale from the prevailing worse solution. Economists tend to be skeptical of mandates for good reasons. There is a lot of potential for abuse and unintended consequences. Thus, they are probably best used only if the social disadvantages of the current solution are so large that there is little chance that the government could get it wrong.

➤ Nudges

There is often a better, less coercive, and brilliant alternative: Nudges, i.e., gentle prods, courtesy of Richard Thaler (a Nobel-Prize winning economist) and Cass Sunstein.

Nudges are at their most powerful when they can put people's intrinsic inertia to good use by selecting good defaults. In an example, 42,000 households in Germany were asked to choose between a green-energy provider and a fossil-fuel provider for their electricity. The green choice was slightly more expensive. For those households for which the traditional fossil fuel provider was the default, only 7.2% switched to the green alternative. When the green alternative was the default, 69.1% of households chose to stick with it. This was the case even though everyone could choose whatever they wanted.

1. What Can Countries Do?

Thaler and Sunstein also advocate taking advantage of social norms. Many utility companies now include Home Energy Reports in their bills, which tell consumers how their usage compares to that of their neighbors. This tactic has led to remarkably large reductions in energy usage. It can be pushed further in many ways. If someone has switched and saved a lot of money doing so, the government could tell neighbors and friends, or even reward switchers when they themselves tell neighbors and friends. (Some advertisers post rewards for bringing in other customers.)

Nudges can also help overcome an information problem. Why would the government need to tell people that they can save money with better lighting, insulation, and cars? Are they not smart enough to ask themselves? They may be, but it would be difficult for buyers to compare different products when every seller can measure and claim benefits in their own way. Vendor claims would degenerate into a race to the bottom. In such cases, government can promote better products not by forcing everyone to abandon inferior products, but by disclosing standardized cost estimates. The classic example is the standardized Monroney sticker on cars, which informs shoppers of fuel efficiency. In many cases, information disclosure is cheap for sellers and salient to consumers. Of course, it does not work everywhere. In some cases, too much good intent can lead to uselessness — as everyone who has ever had to sign 120 disclosure documents when obtaining a mortgage can attest.

Nudges can be brilliant and cost next to nothing. It's just that someone more clever than us has to think of them in the first place and then implement them.

➤ Product Introduction

Governments can also help implement beneficial social changes in the same ways that companies try to increase their sales:

Advertising is the most common way to make the public aware of changes.

Price discounts for early adopters of better climate practices make it easier to try out different practices. For a time, governments should subsidize electricity consumption during the day and tax it at night, *beyond* what is otherwise optimal, making cheap daytime electricity even cheaper. It's the same strategy by which Uber weaned us from taxis and alerted us to its presence. Uber rides started out cheap but they no longer are.

15. MAKING IT HAPPEN

Guarantees for those who are willing to adopt new practices can reduce the fear of the unknown unknowns. For instance, there should be strong one-time "no-worse-off" guarantees for all farmers who are willing to try out more environmentally friendly tilling practices.

(7) Reverse Bad Technological Lock-In

There are many cases in which economies suffer from Technological Lock-in. These are situations in which the economy is too committed to an existing technology to allow it to change to a better one. At some point, when the social benefits of a switch become much higher than the one-time transition costs, then governments should step in to make us better off. Here are two examples: lighting and cars.

Just one decade ago, LED light bulbs cost about three to five times as much as incandescent light bulbs. However, their lifetime cost was only one-fifth as high. LED bulbs last longer and consume less energy. Nonetheless, the aforementioned energy efficiency paradox kept most consumers using incandescent bulbs. In turn, this buying pattern maintained economies of scale in incandescent production and reduced economies of scale in LED production.

In 2007, the Bush government banned the manufacture of particularly inefficient household light bulbs.[2] After some detours,[3] the end result is that today's LED bulbs have become cheaper to purchase than incandescent bulbs ever were! Competitive incentives and mass production have worked wonders for per-unit production costs. When incandescent bulbs ruled the market, they had the existing economies of scale on their side. Now LED bulbs have them. Who could argue with the result? Cheaper, better, less polluting!

Take another example. For the longest time, gasoline cars were thought to be irreplaceable. Their production had economies of scale. Yet it has become increasingly clear that electric cars are superior. Combustion cars have served

---

[2] We would not have advocated for an outright ban, analogous to an infinite tax, but merely for a much higher tax instead.

[3] In hindsight, the mandate may have been three years too early, because LED needed a few more years to overcome the advantages of fluorescent bulbs, another technology. However, without the mandate, the development incentives would have been lower and it could have taken a decade.

1. What Can Countries Do?

civilization well for a century, but their time has passed. The problem is: How can electric cars overcome the advantages of gasoline cars in terms of mass production, available infrastructure (especially gas stations), and consumer familiarity?

Governments helped. They did not invent electric cars, but they did support relevant fundamental R&D for many years. They also offered generous subsidies to car makers for early zero-emission cars. We all know what happened next. Tesla showed the world that electric cars are not only more efficient and pollution-free but also no more expensive than gasoline cars, courtesy of the economics of mass production and falling battery costs. Soon, electric cars will be cheaper. Every car maker on the planet is now planning to phase out combustion engine cars by the end of this decade.

With the exception of the fossil-fuel industry and combustion-engine makers, everyone has won. Perhaps best of all, having acted as the catalyst that drove the switch from a worse equilibrium to a better one, government is now no longer needed. Its job is done. It can now get out of the business of deciding winners and losers and let market forces take over.

(8) Coordinate Transitions

Technological lockin is especially severe when it comes to problems that require many simultaneous changes. Selling electric cars requires public charging stations. Tesla not only had to invent practical electric cars, but it also had to

Strength and speed are useful, son, but coordination is *crucial*!

install a charging network, because there were no electric gas stations. Other car makers are still working on the problem. To build profitable charging stations requires widespread adoption of electric vehicles. A classic chicken-and-egg problem!

Let's zoom out to a wider perspective. There are two coordination problems that are so paramount that they could make or break the transition to a clean energy economy, and only government is in a position to move them along.

The first is the capacity of the electric grid. It has served us well in the past, but it has grown into a messy tangle of poor interconnections dominated

15. Making It Happen

by local regulations and interests. Without the ability to connect to a grid that can make good use of clean electricity, it makes little sense to generate more clean electricity. Making it cheap for clean-electricity providers to sell electricity into the grid is of first-order importance and requires national involvement.

The second is the coordination of electricity supply and demand. We need a universal open bidirectional communications protocol for generators, end-consumers, and storage on the electric grid. Establishing a protocol is harder than it appears. It requires addressing issues such as geography (which price matters to what house?), time (what is the price now and what will it likely be?), and cyber-security (how can the grid mitigate wrong and/or malicious signals?).

One example we mentioned earlier that requires coordination is the build out of charging stations in conjunction with the growth of electric vehicles. In a comprehensive article, the Economist states, "Look beyond the glamorous, high-tech filled automobiles and a merciless bottleneck appears. Governments are only waking up to the problem. Put simply: how will all the electric cars get charged? The current number of public chargers — 1.3 million — cannot begin to satisfy the demands of the world's rapidly expanding electric fleet."

Our governments need to tackle both problems.

(9) Reduce Green Red Tape

Real-world governments enact not only useful regulations but also many bad ones. Many start out good but turn bad over time. Economists consider this examples of unintended consequences.

Most of us (especially our lawyers) like the ability to sue parties that harm us, but the law can also become our own worst enemy. Most environmentalists want wind power, just not in their own backyards. The typical wind project in the United States already takes over a decade (!) to get lease approvals and permits. This is one reason why American offshore wind capacity is less than 5% of Europe's.

Yet even in Europe, neighbors don't like rumbling from windmills. Farmers and fishermen often don't like windmills, either. In America, offshore wind opponents are often the wealthy and powerful who live the near the shorelines. But what is the alternative? Yes, someone may deserve consideration even in

1. What Can Countries Do?

the case of clean wind power, but year-long lawsuits are not the ideal way to handle the problem.

Most of us environmentalists like clean cars and clean grids. As we explained in Chapter 12, the necessary batteries are made out of cobalt, nickel, and lithium today. Without a lot more of these elements, there will be no clean-energy transition. Yet, few of us environmentalists like mining — but civilization's choice now is between more mining and no energy transition. We can't have our cake and eat it, too. Of course, we should not want unregulated mining—that would also be a terrible idea. However, it now takes 16(!) years to approve the average global mining project (*before* it can start breaking ground). We must make green-related mining decisions better and faster.

I kind of regret objecting so strongly to the wind farm they originally had planned

anecdote

> Nevada has some of the richest Lithium sources in the world. The environmental harm of lithium mines are modest (unlike, say, for lead, gold, or coal mines). Alas, standing in mid 2021, new Lithium mines in Nevada have hit some "minor" snags:
>
> Ioneer Corp wants to build a mine halfway between Reno and Las Vegas. Unfortunately, three years into the process, the U.S. Fish and Wildlife Services discovered a rare plant named Tiehm's buckwheat, and later named it an endangered species.
>
> Lithium Nevada Corp wants to build a 20,000 acre mine in the Thacker Pass. Unfortunately, the Reno-Sparks Indian colony, about 300 miles southeast of the project, has filed a lawsuit based on the National Historic Preservation Act (plus some process violations about environmental approvals). It is where Native Americans in the late 1800s hid and were slaughtered by U.S. soldiers. The relevant area is only half an acre.
>
> The environmental harms may be modest and plausibly resolvable. The economic harms of the law suits are not. They could take years to resolve.

We like safety, fairness, and competitive regulations, but these regulations also prevent new competitors from entering. For example, today only utilities are allowed to buy energy from the grid. You cannot build a wind farm to power a data center if it also needs occasional backup from the grid. Don't ask how difficult it would be for a data center to obtain utility status. (And don't ask how many different agencies and bodies have to approve anything that wants to be connected to the grid or wants to extend the grid.)

We like *extremely* stringent safety regulations for nuclear power, but it seems as if the NRC considers reactors safest when they are not built. It has become impossible in many countries to design and build better and safer plants. The time, effort, and uncertainty to get regulatory approvals have killed off most of the nuclear construction industry over the last five decades. We are keeping our fingers crossed that Terrapower's new Wyoming plant will be able to overcome the hurdles and build the safest nuclear power plant in the world.

Good regulations are not easy. They require constant struggling. Unfortunately, once in place, even bad rules are difficult to overturn. How many unnecessary stop signs have ever been taken down when traffic patterns changed later? Business as usual has become too slow to deal with the world's climate crisis. The rate of regulatory evaluation and change has to be accelerated when it comes to environmentally better solutions.

➤ Concierge Services

Many regulations make sense by themselves but not when considered in conjunction with hundreds of others. Figuring out how to start a new clean-energy project is only slightly less painful than a root canal. If we want clean energy, we have to try to entice more competition by making it easier. Governments should:

- Guarantee regulatory "concierge" service, an assigned shepherd with expertise in the regulatory and permission processes and good connections to the relevant agencies, who can facilitate much faster reviews by agencies ideally with firm short deadlines and without undue compromises on safety and environmental standards.
- Guarantee and pay for the interconnection of a new entrants' first plant into the electricity grid.
- Guarantee a stated price for a fixed amount of *clean* dispatchable electric energy over the first decade. The specific terms (e.g., time-related

pricing) can be revised every five years. They could even be auctioned off.

This model contrasts with the current *modus operandi*, in which the government funds plants, mostly by incumbents who know how to navigate the process and who have placed past and future employees into the key government posts.

(10) Lease Out Land for Solar and Wind

The Federal government owns over a quarter of all the land in the United States, much in sparsely populated states in the West. In Nevada, the government owns 80% of all the land — prime locations for wind and solar farms without great alternative uses. This is also the case in many other countries: governments typically own their countries' deserts and mountains.

An immediate step would be to make it easy for clean-energy developers to lease such land cheaply for 30-50 years if it is for the purpose of building wind and solar farms, with penalties for non-use.

Good News Update in August 2021: More Federal land leasing for clean energy projects has just been made policy in the U.S.!

(11) Kill the Worst Emitters

Among the lowest hanging fruit is shutting down the worst polluters. They are also remarkably easy to identify.

➤ Methane

Natural gas is worse than suggested by its plant emissions. Indeed, it may be no cleaner than coal! This is because too many wells and some pipelines are leaking methane. It is a world-wide problem. Fortunately, the majority of emissions comes from a minority of locations. Even better, satellites make large emitters easy to detect. Governments should immediately send crews to flame off or close the leaks. It can be decided later who has to pay for the cost. If need be, allow private-party lawyers to sue for recovery and retain some of the settlement.[4]

---

[4] A good solution would be to make any large fossil-fuel company that has ever received income from a well jointly and severally liable for plugging the well at the end of its life and specified damages if this has not happened *even* if it has long since sold the well.

15. MAKING IT HAPPEN

➤ Coal

New coal plants are bad, because they release a lot of CO_2. However, old coal plants are worse. They release not only CO_2 but also many other harmful pollutants. Many of these old coal plants are barely economical to run even without fossil fuel taxes. Thus, governments should push them over the edge and immediately close them. With appropriate one-time subsidies and special waivers for many regulatory delays, cleaner plants could substitute for the lost power relatively quickly in many countries. If the subsidies are based on, say, 2015 emissions, they would also not create perverse incentives to build more polluting plants. Yes, such programs cost money — but they are worth it even for local citizens (and, of course, for the world at large).

(We wish we had good ideas how to stop the imminent construction of coal plants in China, India, and beyond. We do not. What a missed opportunity for the world.)

(12) Negotiate *Some* International Agreements

Most international treaties are not low-hanging fruit. Negotiating over CO_2 emissions seems largely futile to us.

However, there are situations in which international negotiations could work. Our favorite one is methane emissions control that would make it in the interest of countries to eliminate super-emitters. The cost of plugging or flaming off leaks for the worst emitters is low (relative to the worldwide harm) and their actions can be easily verified by satellites. An international treaty, in which rich countries could share some of the cost, could speed this along.

(13) Adapt?

Adaptation to climate change will greatly reduce its harmful effects. It is why hurricanes (Chapter 4) kill far fewer people today than they did 100 years ago. It is why most earthquakes in California have become nuisances rather than catastrophes. It is why Venice and the Netherlands are still above water. The Global Commission on Adaptation 2019 Report estimates that investing $1.8 trillion globally from 2020 to 2030 could generate $3.5 trillion in total benefits – a hefty return on investment when considered from a social perspective!

1. What Can Countries Do?

Yet our book has barely touched on adaptation. There is a reason to this madness. Most of the time, it is in the interest of the involved parties to adapt. It is (or at least should be[5]) in the interest of people not to build houses on the ocean shore at zero elevation or next to dry forests that will burn sooner or later. It is in the interest of countries to build warning systems, dykes, and fire control systems. Adaptation is not really a global problem plagued by a global externality, like climate change, which is the subject of our book.

Could adaptation be dangerous by substituting for the necessary global fossil fuel detox? Maybe. But we cannot steer the boat (i.e., Earth) back so quickly that we won't need adaptation. And we wouldn't want people not to protect themselves.

With a nod to the Buddha, as far as global adaptation goes, "it is what it is" (or "it will be what it will be"). And as far as our book goes, it is already too long, so we have to punt on this important subject.

(14) Crises Beyond Climate Change?

Bjorn Lomborg has a whole list of global problems that are worth tackling. For the most part, he concludes that fighting climate change through ordinary means today gives too little bang for the buck. (At least, it used to be too expensive. With improvements in technology, which he also predicted, the tradeoffs are shifting.) Other environmental issues brought about largely but not only by our global population explosion — like eradicating global hunger or malaria — are comparably much cheaper. They could be accomplished for a tiny fraction of the cost of premature decarbonization..

It is difficult to choose among worthy causes for humanity and beyond. habitat destruction, species extinction, and overfishing are examples of impacts that extend beyond the human species. How can we weigh the misery of 3 to 6 million children starving to death or half a million children dying from malaria every year today against the misery of a potentially looming climate catastrophe?

Ideally, "we" would tackle all these problems. As economists, we are schooled in the science of scarcity and tradeoffs. As humans, we find it difficult to judge which miseries are more important than others. These are

---

[5]Some well-meant government insurance schemes have created a moral hazard that will make the matter worse in the future.

questions of ethics, and the moral dilemmas posed by these questions weigh heavily on us.

Yet, if our book's main thesis is correct, there really are no tradeoffs between the world spending resources on fighting climate-change vs. fighting, say, extreme poverty. This is because there is no world decision-maker who would trade them off. These very real problems are not going to be solved by a social world planner. They will have to be solved by a collection of about 200 self-interested nation states, thousands of governments, about 1 billion richer people in OECD countries, and about 7 billion poorer people in non-OECD countries. The only way to change the world is to influence the tradeoffs that these individual parties face.

2 What Can Individuals Do?

We now shift to considering voluntary choices made by individuals — good choices and bad choices; choices that could make a disproportionate difference and choices that will not.

(1) Change Your and Others' Behavior?!

There have been many bestsellers that have held forth about how to reduce your carbon footprint. They sell many copies to the faithful, but they are misguided. They would be amusing distractions if only the issues were not so serious, if only the beliefs were not so widely held, and if only the diversions would not delay what really needs to be done.

Why haven't most people voluntarily changed their behavior? Is the problem that they just don't realize how they can reduce their personal carbon footprints or how much it would help the environment?

We would love people to change their ways selflessly, but it's unrealistic. Economics suggests that not enough people will do so if it is not in their self-interest. This implies also that clean energy must not be much more expensive than dirty energy to achieve widespread adoption. It is a fallacy to think that voluntary changes against personal self-interests could transform the world. It won't happen. Don't shoot the messenger. It's not our fault. We did not design the world this way.

2. What Can Individuals Do?

Part of our skepticism stems from the fact that behavioral changes significant enough to affect climate would not only have to be widespread but also long-lived. Otherwise, changes have little impact — in fact, almost surely only an immeasurably small impact. Even large changes in response to a crisis lasting only a few years would barely move the needle.

Recall our discussion of David MacKay's brilliant book <u>Sustainable Energy Without the Hot Air</u> from 2009 (Chapter 9, Page 259). It advised to dress appropriately and moderate your thermostat; to read your meters; to stop flying and drive less; to use old gadgets and avoid clutter; to replace incandescent light bulbs; and to eat vegetarian. (*Duh!*) The appendix shows a few similar lists from the EPA and other climate-conscious websites. Unfortunately, as we noted, good intent neither tends to last long nor is it contagious on a worldwide scale.

What about setting an example? Of course, if you are the Pope, the Archbishop of Canterbury, or the Orthodox Ecumenical Patriarch, your example and <u>joint appeal</u> may matter at least a little — though even their influence has been too small to be empirically detectable.

If you are like the rest of us, don't overestimate your importance. Frankly, the world does not care what your thermostat reads, whether you fly across continents, or whether you eat vegetarian or not. Of course, we encourage you to follow the recommendations and to exercise more. Just don't think that your actions and those of your friends will ever make a meaningful difference to the CO_2 concentration in the atmosphere.

The statement "if everyone did it" is a fallacy. If everyone does it, it will have had nothing to do with you. Your eating or not eating steak won't change even a few million people — and even if you add the people that you indirectly influenced through your own imitators. (If you still think you have more meaningful influence, you should see a shrink for a <u>Napoleon Complex</u>.)

What about carbon-shaming? Fat chance. It may make you feel morally superior (and ask yourself why you need this), but it is more a sign of ignorance about what really matters than it is an effective climate-change strategy.

Even if you are the world's greatest carbon-shamer, convincing everyone you will ever meet (and who they in turn will ever meet), *it doesn't matter*. The world only cares what hundreds of millions of people do. The CO_2 concentration changes only when hundreds of millions of people change their behavior. Realistically, you can't shame even a significant fraction of so many people. Even Greta Thunberg, the world's most visible activist, has not managed to shame people into eating less meat and flying less.

Don't blame the messenger. We did not make the world. Sometimes defeatism is just realism. Our strong preference and advice is to focus the world's attention on what can make a real difference in changing atmospheric CO_2 levels instead.

(2) So What Can You Do?

We are all little cogs in a big machine of eight-thousand million people. By far, the best contribution you can make is to play a small part in moving the large collective — not playing a large part in moving a small collective.

Thus, you can do more to combat climate change by working on inventing and deploying new, clean technologies than you could ever possibly do by decarbonizing your local neighborhood. Only better technology can move the needle. Your neighborhood deployment cannot.

The best advice we can give to climate activists is to bicycle to work not for the sake of the planet but for the sake of fun and exercise. The best advice for the sake of the planet is to get involved in clean-energy research, development, and implementation.

Business and economics are not the enemy. They are part of the solution. Often the most difficult aspect of new technologies is cost-effective implementation. Elon Musk is not an inventor. He is, however, the greatest technological innovator of our time. His entrepreneurial talents have almost single-handedly pushed the United States back into a leadership position in both cars and rockets. Inventors are a dime-a-dozen. Visit the laboratories of your local university, and you will find hundreds of fascinating inventions. Elon Musk's brilliance has been his ability to jump start the deployment of important large-scale generation-leaping inventions.

There are also other ways to help. Politicians can help convince the public to make better choices and occasionally even sacrifices. Journalists can help capture the attention of readers. Academics and authors can help educate the next generation about what is important. Climate activists and environmentalists can help prick our conscience and help maintain public support. Religious leaders can help foster the common good and appeal to our less-selfish instincts. Philanthropists can help, at least a little, where

2. What Can Individuals Do?

government has failed — this includes Bill Gates, whose initiatives in the third world and in the energy sector are a blessing for humanity.

In our minds, the best way to help humanity now remains researching, developing, and implementing scalable clean-energy technologies. Getting rich in the process is merely a nice bonus for those who end up making a contribution.

What would we recommend to our own children? What can they do? What can you do?

- Build your career in science and technology related to climate change. One path would be studying climate change directly. Another would be doing research related to green energy provision. Yet another would be research on improving agricultural processes.

- Become an entrepreneur or work in the clean energy space. Elon Musk has moved the needle far more than the entire United Nations — with all its Rio, Kyoto, Copenhagen, Cancun, and Paris conferences. So have Lewis Urry and Sony with their Lithium battery work. So has our colleague, Lesley Marincola, who is trying to bring small-scale solar energy to the poorest of the poor. There is room for thousands of start-up firms exploring new ways to accelerate the transition for millions. Don't feel guilty if you get rich off it, too.

- Lead others. As much as you may dislike politics, government is the only institution capable of significantly accelerating the transition to a clean economy. For those who have the stomach and the talent, government service is a route to consider. If you do so and when you get there, don't fall for futile showy green policies that won't accomplish much in the end. Instead, expand your country's research, development, and initial deployments of green technology.

- Pursue a relevant teaching career. There is so much misinformation out there — some intentional, some ignorant — that helping educate the public about the science and economics behind climate change is an important undertaking. Educate people about what really matters. Inspire them.

- Write a book. Remain honest even when it is uncomfortable. Give the people the red pill. Tell them what the problems are, what can be done about them, and what truly matters. Don't argue for the sake of winning

15. Making It Happen

or dogma. The point is not to win an argument. It is to search for better solutions — including solutions that we did not propose.

That is what we have tried to do here. It is why we wrote this book.

Conclusion

Why did we think the world needed another book about climate change?

First, we thought we could explain the issues better and do so in a way that had no partisan or home agenda. Second, we wanted to contribute to pushing our readers towards more realistic approaches and away from unrealistic ones. Little of what we have written has not already been stated somewhere else. If you already knew it all, we apologize for having wasted your time. If you did not, we hope we helped focus your thinking.

Our book claimed that there is really only one red ace when it comes to reducing CO_2 in the atmosphere: technological advancement. There are two more picture cards: locally justifiable taxes on fossil fuels (for the sake of reducing co-pollution) and tree planting. There are also many non-face cards that can and will help. But nothing else will come close in importance to technology improvements.

Other commonly proposed solutions are too limited, too unrealistic, too expensive, suited only to rich countries, or all of the above. If the 1-2 billion people living in the richer economies cannot find solutions that will induce the poorer 6-8 billion people to leapfrog over fossil fuels, the world's CO_2 emissions will not decrease but increase for decades or centuries.

Humanity has been luckier than it could have hoped. Technology has been improving at a rapid rate despite far less government support than collectively optimal. (Humanity must increase it!) As economists, we believe that real-world governments' best role is to facilitate the transition. They cannot execute it by decree. With intelligent government support, individual self-interest and competition can work wonders. We are optimistic that human creativity can then quickly reduce greenhouse gas emissions. But we must not be complacent. It is in the self-interest of virtually all countries and especially rich countries to work on accelerating progress.

From our small selves to the governments of the world: Please subsidize green research, development, and first deployments far more than you have in the past.

2. What Can Individuals Do?

This is the best way to "move the needle now."

15. Making It Happen

Further Reading

Books

- John Doerr, 2021, Speed and Scale, Random House.
- Gates, Bill, 2021, How to Avoid a Climate Disaster, Knopf, New York. A guide to reducing emissions from the leading philanthropist of our times.
- Global Commission on Adaptation, 2019, Adapt Now: A Global Call for Leadership on Climate Resilience. A comprehensive analysis of the costs and benefits of adapting to climate change.
- Lomborg,Bjorn, 2020, False Alarm, Hachette Book Group, New York, 2021. An alternative view of the costs and benefits of climate change choices — well-intended and posing profound dilemmas — but misguided in one sense: there is not a global entity that could make such choices.
- Nordhaus, William 2018, Climate Change: The Ultimate Challenge for Economics, Nobel Prize Lecture.
- Thaler, Richard H. and Cass R. Sunstein, 2021, Nudge: The Final Edition, Penguin Books, New York. Steps that government can take to overcome human inertia.
- We share many views with Ken Caldera, who now works as an advisor to Bill Gates — from geoengineering merely being a mask, to exploring many different technologies, to keeping a skeptical but hopeful perspective on nuclear power.

Reports and Academic Articles

- Aldy, Joseph E. and Richard Zeckhauser, 2020, Three Prongs for Prudent Climate Policy, Resources for the Future Working Paper.
- Borenstein, Severin, 2005, The long-run efficiency of real-time electricity pricing, The Energy Journal.
- Matthews, H. Damon, et al., 2021, An integrated approach to quantifying uncertainties in the remaining carbon budget, Communications Earth & Environment. (About 0.44°C per 1,000 $GtCO_2$.)

2. What Can Individuals Do?

SHORTER NEWSPAPER, MAGAZINE ARTICLES, AND CLIPPINGS

- Marshall, Aarian and Matt Simon, <u>Adapting our rich cities</u>21st-century storms are overwhelming 20th-century cities Ars Technica, 2021/09/06.
- <u>Climate Appeal by Medical Journal Editors</u>, September 6, 2021.
- <u>DOE Signs Up 125+ Local Governments to Fast-Track Solar Permits</u>, September 28, 2021.
- Taylor, Adam, et al., Nov 10, 2021, <u>2C or 1.5C? How global climate targets are set and what they mean</u>, Washington Post.
- <u>Timmer, John</u>, <u>Most of the power sector's emissions come from a small minority of plants</u>. Ars Technica, 2021/08/11.

WEBSITES

- <u>National Academies of Sciences, Engineering, and Medicine, Division on Earth and Life Studies</u>. Has, e.g., suggestions for foundational research for ocean carbon sequestration, fusion, and reflecting sunlight.

App. A Some Exciting Green Tech

We are not engineers, but we want to share our own interest and excitement about some technologies that could potentially change the world.

- Any new technologies that are grid-scale electricity storage related and promising are of great interest to us. Lack of adequate scalable electricity storage is the only remaining hindrance to wind and solar taking over the world.

  For instance, where can exhausted gas wells and other underground caverns serve as compressed-air storage at large scale?

- Safe nuclear fission power plants with minimal waste. Once built, their power could be so cheap that even natural gas would be more expensive. There are no scientific reasons why it should not be possible to design fission reactors that can intrinsically no longer explode and that can reuse their fuel a thousand times more often than they do today. However, for decades, the world has not deployed many new reactors, and nuclear technology learning has crawled along way too slowly.

  Regulation (with good intent but perhaps not good reason) has made any plant changes almost impossible, leading companies to prefer to work with known but ancient, intrinsically dangerous technologies (pressurized water reactors, where cooling failure can lead to meltdowns) rather than with unknown but potentially safer technologies (where cooling is passive and not dependent on a backup power source, so that even if the operators make stupid mistakes, as they did in the Chernobyl disaster, the plant cannot blow up). The goal should be to build a reactor in which even the most malicious black-hat operators and hackers could no longer make the plant release radioactivity.

  Such a reactor should also not produce weapons-grade material, be small in size (to be shipped and assembled on-site), and mass-producible. Pebble bed reactors seem like excellent candidates. Offering leasing / financing to countries that are willing to replace coal plants with these small reactors would further broaden their reach. Finally, there should be a concerted plan to reprocess the fuel in breeder reactors instead of governments guaranteeing non-existent long-term storage of spent high-level waste.

  This is *not* the nuclear industry of today. However, it could become the nuclear industry of tomorrow.

- Nuclear fusion: The National Academies of Sciences' target for a prototype plant should be 2040. There are still many technical challenges to overcome, but fusion promises virtually limitless safe energy without waste products *if* it could be made to work. It may not work, but it's worth a shot.

  Note that from an economic perspective fusion is often misunderstood. Fusion plants will be more akin to super-safe nuclear fission plants with no waste fuel than something entirely different. Both fission and fusion plants have extremely high fixed costs and negligible fuel costs. The fact that fusion uses a different, inexhaustible fuel is unimportant. There is more than enough dirt-cheap uranium and thorium to run traditional nuclear power plants at almost zero fuel cost for a thousand years.

- Industrial high heat. Could small nuclear reactors be used not only for electricity production but for industrial heat?[6]
- Geothermal power could potentially tap more heat from our planet. All it really requires seems to be a very deep hole (and some water). Could the cost be reduced by an order of magnitude?
- Though we are generally skeptical of carbon sequestration because there are few private incentives here, tree planting and advanced olivine weathering for accelerated carbon removal are potentially cheap and deserve further RDD.
- Solar radiation management (SRM). This could involve injecting reflective sulfur particles into the upper atmosphere to reduce the amount of radiation that is absorbed by the Earth. The cost is remarkably low. Can we try this in very small scale and learn what it does?

App. B Recommendations By Others

Many individuals and organizations have offered suggestions (often in list form) for how greenhouse gas emissions could be reduced. This appendix describes some prominent such lists. **Please bring prominent important lists to our attention. We would be happy to reference more of them.**

Note that we describe the suggestions here even when we do not believe that they will move the needle on climate change. We have no qualms doing so: many of these suggestions are laudable, wholesome, and commendable — as is exercising more. We would love populations all over the world to follow them, but we suspect that their uptake will be limited. However, some of these recommendations are also cost-efficient even for the parties involved, and nudges could further improve their uptakes. Without further ado, here they are.

---

[6]Nuclear reactors may not be what you think they are. Even a 14-year-old managed to build a basic reactor by himself.

15. Making It Happen

(1) Environmental Protection Agency

The Environmental Protection Agency (EPA), founded in 1970 by a Republican president, is the most prominent and powerful environmental agency in the world.

The EPA is pursuing a number of <u>initiatives</u> intended to reduce GHG emissions:

- Measuring and reporting GHG emissions and sinks.
- Works with industry to reduce emissions.
- Cost-benefit analysis for policy.
- Science support (incl. wildfire research).
- International partnerships.
- Community aid.
- Information and education.

The EPA also posts recommendations for individuals and businesses. Following the EPA's own order, they are:

Energy:
- Look for Energy Star certified energy-efficient products.
- Live in an Energy Star certified dwelling.
- Heat and cool more efficiently (insulation, etc.).
- Adjust thermostats.
- Switch to green power (e.g., rooftop solar, rooftop garden); buy green power.

Waste:
- Buy less new stuff, if so with more durable, sustainable, and recyclable components.
- Reduce food waste, compost.
- Reuse old clothing, bags, etc.
- Buy used.
- Recycle.

Transport:
- Bike, walk, carpool, public transport.
- Drive less aggressively.
- Switch to energy-efficient and/or electric vehicle.
- Make fewer trips.

Water:
- Turn off running faucets, run clothes washer with full loads and cold water.
- Use energy-star water-sense certified dishwasher — better than by hand.
- Plant water-smart landscape.
- Low-flow showerheads, shorter showers.

Environmental Justice:
- Get to know your community and neighborhoods.
- Plant trees, especially in urban environments.
- Learn where large industrial GHG facilities are.
- Learn about local powerplants.

- Communities can find lowest-price energy-star certified products.
- Learn about and empower near-port communities.
- Work with neighbors and community to integrate smart growth and environmental justice.
- Join local advisory boards.

More:
- Educate children and young adults.
- Environmental stewardship to reduce GHG emissions.
- Estimate annual GHG emissions.
- Participate in citizen science by sharing data.
- Tell others.

(2) Project Drawdown

Project Drawdown is a prominent nonprofit organization that collects and disseminates suggestions for how to reduce global warming. Following their recommendations in order:

Electricity:
- Enhance efficiency to reduce demand.
- Shift production to avoid fossil-fuel use.
- Improve transmission grid and energy storage.

Food, Agriculture, Land use:
- Shift diets lower on the food chain and address food waste.
- Protect land and ecosystems. Improve farming productivity.
- Change farming practices (e.g., rice and cows).

Industry:
- Improve materials, esp. plastic, metals, and cement.
- Reuse waste.
- Improve refrigerants (CFCs).
- Enhance process efficiency.

Transportation:
- Public transport and ride-sharing.
- Improve fuel efficiency in combustion engines.
- Electrify vehicles.

Buildings:
- Enhance efficiency (insulation).
- Use alternatives for heating.
- Improve refrigerants (CFCs).

Land Sinks:
- Reduce food waste, eat vegetarian. Reduces deforestation.
- Protect and restore ecosystems.
- Change agricultural practices.
- Improve degraded land.

Coastal Sinks:
- Protect and restore ecosystems, esp. mangroves, salt marshes, seagrass meadows.

15. Making It Happen

- Investigate seeweed and kelp farming.

Other: Project Drawdown also suggests investigating engineering sinks and improving health and education.

(3) Clean Technica

Clean Technica is an prominent environmentally focused website. It's list is somewhat unusual, in that it starts not with what we should be doing, but what we should *stop* doing that is outright stupid. It is hard to argue with Clean Technica's analysis.

- Stop ethanol subsidies.
- Stop fuel cells for light vehicles (cars and trucks), rather than batteries.
- Stop carbon capture and sequestration. Don't use fossil fuels.
- Continue existing nuclear reactors, but don't build new ones.
- Stop fossil fuel subsidies.
- Stop blue hydrogen (requires carbon-capture and sequestration).
- Stop biofuels. Wind, water, solar is much more efficient.
- Stop using natural gas (methane).

Reading the above suggestions is depressing in that we also wish they were addressed sooner rather than later. Clean Technica also does offer "smart" positive suggestions, though:

- Wind, water, solar.
- More wind, water, solar — overbuild.
- R&D, especially grid-based storage technologies.
- R&D for long-distance high-voltage DC transmission lines.
- Move to battery-electric vehicles.
- R&D for electric ocean propulsion, long-distance trucking, short-haul aircraft.
- R&D for hydrogen propulsion for long-distance aircraft.
- R&D on industrial heat-related processes.
- Follow Marc Jacobson's suggestions. Note: there is controversy among scientists whether his ideas are enough to decarbonize the US. However, there is little controversy that many of his suggestions would make for good policy.

For personal use, Clean Technica recommends

- House solar panels.
- Don't buy another gas car.
- Don't eat meat.
- Use only LED lighting.
- Use heat-pump based heaters.

App. B. Recommendations By Others

- Check insulation.
- Buy energy-star rated appliances.
- Set thermostat appropriately.
- Turn off devices when not in use.
- Avoid refrigerators and freezers, unless absolutely necessary.

(4) Global Methane Initiative (GMI)

The Global Methane Emissions and Mitigation Opportunities Factsheet describes not only sources of methane emissions, but also opportunities to reduce them. Because Methane is such a powerful GHG with shorter-term effects than CO_2, mitigation can work more quickly and efficiently than CO_2 mitigation.

The recommendations are always to capture and channel methane gas instead of venting it. At the end, the methane can be used to power devices (such as engines) or if this is not possible, flared off. This can be done in

- Agriculture, mostly manure-management.
- Mining, active or abandoned mines.
- Municipal landfills.
- Oil & gas systems.
- Wastewater and sewage.

In addition, the GMI recommends measurement and detection. (Methane leakage shows nicely on infrared cameras.)

(5) A Few More Suggestions For a List

We want to emphasize a number of further "listable" suggestions:

- Establish strong penalties for methane leaks that are not immediately fixed. Allow government to intervene and later charge for fixing leaks that are not immediately fixed.
- Offer financing to poorer people (and countries) for the higher upfront cost of clean energy.
- Buy out and close the worst coal plants, not just domestically but internationally, *and* help retrain their workers, possibly in clean-energy installation jobs — a position now endorsed even by the Coal Miners' Unions). Have the OECD subsidize some of the cost in third-world countries, especially if these countries agree not to build more coal plants.
- Investigate seasonal roof color choices to capture solar heat in winter and reject it in summer.

15. Making It Happen

- Improve awareness of heat pumps and heat storage. Heat pumps are almost magical devices. They produce heat at far lower cost than all other alternatives.
- Stop installing natural gas lines into new buildings (as in California).

And, of course: clean R&D, more R&D, and more R&D.

Appendix

Crib Sheet — Summary of Facts

We may use slightly different sources and numbers in this crib sheet than those we have used in the book. The numbers are reasonably consistent, yet convey typical estimation differences that show up when different sources or years are used.

Unless otherwise noted, the principal data source is US EIA International Energy Outlook, Oct 2021.

Please bring any errors to our attention asap.

CRIB SHEET — SUMMARY OF FACTS

Chapter 1: Population

See Fig 1.3: Population (in billion)					
	1960	2020	2100e	Δ	%Δ
OECD	815	1,369 ~	1,400	+31	**2%**
USA	187	331 ↗	434	+103	**31%**
EU27	356	445 ↘	364	−81	*−18%*
not OECD	2,220	6,426 ↑	9,475	+3,049	**47%**
Asia	1,705	4,641 ~	4,720	+79	**2%**
China	660	1,439 ↓	1,065	−374	*−26%*
Othr Fr East	963	2,206 ↘	1,826	−380	*−17%*
South Asia	517	1,605 ~	1,689	+84	**5%**
India	451	1,380 ~	1,447	+67	**5%**
Africa	203	1,341 ↑↑	4,280	+2,939	**219%**
Sub-Sahara	220	1,094 ↑↑	3,776	+2,682	**245%**
Nigeria	45	206 ↑↑	733	+527	**256%**
World Total	3,035	7,795 ↑↑	10,875	+3,080	**40%**

Primary Data Source: <u>Worldbank</u> and United Nations.

Chapter 2: Primary Energy (PE)

See Fig 2.9 and 2.8: PE By Region (ca 2022)			
	Popln	Total	pPpD
OECD	1.38	71 PWh	141 KWh
USA	0.33	28 PWh	232 KWh
Europe	0.60	24 PWh	109 KWh
Not OECD	6.50	116 PWh	49 KWh
China	1.45	48 PWh	90 KWh
India	1.41	12 PWh	23 KWh
Other Asia	1.18	14 PWh	32 KWh
Africa	1.37	7 PWh	14 KWh
Sub-Sahara	1.04	2 PWh	5 KWh
USSR (CIS)	0.25	11 PWh	121 KWh
Mid-East	0.26	10 PWh	111 KWh
Latin America	0.52	8 PWh	41 KWh
World	7.88	187 PWh	65 KWh

Primary data source is US EIA. Population is in billions. pPpD is *per Person per Day*. USSR, Mid-East and Latin America are from British Petroleum (BP).

See Tbl 2.5: Energy Purpose

	USA	World
Home/Work	40%	30%
Transport	30%	20%
Industry	30%	50%

Primary Source: NAS.

See Tbl 2.10: Primary Energy Growth

1965	2019	2050e
49 PWh	185 PWh	260 PWh

The 1965 estimate is inferred from Our World in Data and the EIA 2019 number.

See Tbl 2.10: Primary Energy Use Growth

	2022	2050e	Δ	
OECD	71	82	+11	PWh
USA	28	32	+3	PWh
EU	24	28	+4	PWh
Non-OECD	116	177	+62	PWh
China	48	58	+10	PWh
India	12	35	+23	PWh
Other Asia	14	25	+11	PWh
Africa	7	13	+6	PWh
World	187	260	+73	PWh

Over the next 30 years, the world is expected to increase its energy consumption by about 40%.

See Pg 47: Data Disagreements

BP	EIA[1] HDB	EIA[2] IEO
162 PWh	186 PWh	173 PWh

All three estimates are for total primary energy consumption for the world in 2019. Our World in Data also uses the BP data. The first EIA number is from the Historical Data Browser, the second is from the International Energy Outlook.

See Fig 2.11: Energy Sources (2019)

Biomass	11 PWh	6%
Coal	44 PWh	25%
Oil	54 PWh	31%
Natgas	39 PWh	22%
Nuclear	7 PWh	4%
Hydro	10 PWh	6%
Wind	4 PWh	2%
Solar	2 PWh	1%
Total	173 PWh	100%

Non-fossil fuels are grossed up as if they had similar efficiency losses as fossil-fuels.

Crib Sheet — Summary of Facts

Chapter 3: Emissions

See Fig 3.1: CO$_2$ Equiv, By GHG

CO$_2$	38 GtCO$_2$	86%
Methane	9 GtCO$_2$e	20%
NOx, CFC, +	5 GtCO$_2$e	11%
Land Charge	4 GtCO$_2$e	9%
Total	55 GtCO$_2$e	100%

See Fig 3.3: By Emitting Use

Energy	37 GtCO$_2$e	73%
Agriculture	10 GtCO$_2$e	20%*
Other	4 GtCO$_2$e	8%

If the land charge accrues to agriculture, then agriculture's share increases from 20% to 25%.

§3.2: Annual Atmosphere Change:

- Human Emissions: +38 GtCO$_2$.
- First-Year Natural Atmospheric CO$_2$ Removal: ≈20 GtCO$_2$.
  (Total removal: 100s-1000s of years.)
- Extra Human-Caused Atmospheric:
  +18 GtCO$_2$/year ≈ +2.5 ppm/year. (Relating Emissions in GtCO$_2$ to PPM.)
- 1870: 2,200 GtCO$_2$ ≈ 280ppm.
- 2021: 3,200 GtCO$_2$ ≈ 420ppm.

See Tbl 3.14: CO$_2$ Emissions, 2022→2050e

	2022	2050e	Δ	
OECD	12.1	12.1	−0.0	GtCO$_2$
USA	4.8	4.8	−0.0	GtCO$_2$
EU	3.8	3.7	−0.1	GtCO$_2$
Non-OECD	24.2	30.8	+6.6	GtCO$_2$
China	11.0	10.5	−0.5	GtCO$_2$
India	2.7	5.8	+3.1	GtCO$_2$
Other Asia	2.8	4.9	+2.0	GtCO$_2$
Africa	1.3	2.0	+0.7	GtCO$_2$
World	36.8	42.8	+6.6	GtCO$_2$

The table in the text quotes log-growths. The table here shows GtCO$_2$ instead.

See Tbl 3.17: CO$_2$ Emissions (ca 2022)

	Total	pPpY
OECD	12.1 GtCO$_2$	8.8 tCO$_2$
USA	4.8 GtCO$_2$	14.4 tCO$_2$
Europe	3.8 GtCO$_2$	6.4 tCO$_2$
Not OECD	24.2 GtCO$_2$	3.7 tCO$_2$
China	11.0 GtCO$_2$	7.6 tCO$_2$
India	2.7 GtCO$_2$	1.9 tCO$_2$
Other Asia	2.8 GtCO$_2$	2.4 tCO$_2$
Africa	1.3 GtCO$_2$	1.0 tCO$_2$
Sub-Sahara	0.4 GtCO$_2$	0.6 tCO$_2$
World	36.3 GtCO$_2$	4.6 tCO$_2$

Fossil-fuel based CO$_2$ emissions. pPpY = per Person per Year.

Chapter 4 and 5: Temperature

§4.4.4. Atmosphere State:

- Long-Run: $2 \times CO_2$ (ppm) $\Rightarrow$ +1.0°C. Includes water vapor.
- $\Rightarrow$ 50% increase from 280-420 ppm (+50%): $\approx$ 0.5°C.

See Fig 4.5 and 4.9: Estimated Planetary Conditions

	Year	CO_2 in ppm	Temp in °C	SeaLvl in m
Vostok	−100,000	236	−2.1	
	−30,000	206	−6.8	−80
	−20,000	200	−8.1	−133
	−10,000	240	−2.5	−62
	0	280	−0.4	−0.1
	1400	280	−0.3	0.0
Mann	1700	276	−0.8	0.0
	1800	281	−0.5	0.0
	1980	**339**	**0.0**	**0.0**
NASA	2000	370	+0.3	+0.2
	2020	415	+1.0	+0.2
NASA and IPCC 2021 Report, Page SPM-29				
RCP 4.5	2050e	500	+1.5	+0.3
	2100e	560	+2.5	+0.3
RCP 6.0	2050e	500	+1.6	+0.3
	2100e	720	+3.0	+0.4
RCP 7.0	2050e	600[?]	+1.7	+0.3
	2100e	850[?]	+3.6	+0.5
Clark	10,000e	630	+3.0	+37

The base year is 1980. Clark et al's estimate is based on RCP 6.0 extrapolated.

Data Basis: mostly IPCC 2021 6th Report for RCP 4.5 and 7.0. RCP 6.0 is now interpolated. Sometimes IPCC 5th.

CRIB SHEET — SUMMARY OF FACTS

See Fig 5.5.1: RCP Emissions

	2050e	2100e
RCP 4.5	45 $GtCO_2$	15 $GtCO_2$
RCP 6.0	55 $GtCO_2$	50 $GtCO_2$
RCP 7.0	60 $GtCO_2$	80 $GtCO_2$

Equivalent 2020 emissions: 39 $GtCO_2$. RCP 6.0 was interpolated from RCP 4.5 and RCP 7.0.

§5.2: Expected Economic Damages:

- Terrestrial effects are difficult to assess: hotter but wetter. Uneven.
- More energetic weather phenomena.
- Sea level effects: 400 million displaced, primarily in Bangladesh and Indonesia.

§5.6: Dangers:

- Fast speed of increase.
- Dormant feedback loops.
- Tipping points.
- (Very rare asteroids, supervolcanos)

Chapter 6: Economics

§6.3: Social Cost of CO_2:

- Also optimal tax on CO_2:
  - more ⇒ curtail too much.
  - less ⇒ pollute too much.
- Sequestration cost is one ceiling to SCC.
- Many problems: judging harm, inefficient administration, corrupt administration, differential harm, escape.

See §6.2: GDP by Region (2020)			
	Total (t$)		pPpY
OECD	$52.3	62%	$38,000
USA	$20.9	25%	$63,000
Europe	$15.3	18%	$34,200
not OECD	$32.4	38%	$5,000
China	$14.7	17%	$10,400
India	$2.7	3%	$1,900
Sub-S Africa	$2.1	2%	$1,500
World	$84.7	100%	$10,900

See GDP Forecasts, PwC			
	In US$	in PPP	
	2020	2020	2050e
OECD	62%		< 50%
USA	25%	16%	12%
Europe	18%	15%	9%
Not OECD	38%		> 50%
China	17%	18%	20%
India	3%	7%	15%
World	100%	100%	100%

Estimates can vary. The IMF estimate of world GDP for 2021 is $94 trillion. The population estimate for 2021 is 7.9b (8.0b for 2022).

§6.5.5: Marginal Thinking and Cost/Benefit:

- COP are not about eliminating global warming but about reducing it by "10–20%."
    - Consider RCP 6 to RCP 4.
    - Reduction of global warming by 2050 by 5% (from about 1.7°C to about 1.6°C).
    - Reduction of global warming by 2100 by 20% (from about 3°C to about 2.6°C).
- Est. required reduction: $\approx$ 15 $GtCO_2$/year.
- 4–5 $GtCO_2$ for each 0.1°C reduction by 2100.
- All US CO_2 emissions: 4.7 $GtCO_2$.
- 15 $GtCO_2$ at $50/$tCO_2$ about $750 billion:
    - About 1% of World GDP. About $100 per person per year.
    - About 1.5% of OECD GDP. About $500 per OECD inhabitant.
    - About 3.5% of US GDP. About $2,000 per US resident.
    - About size of US military spending.
    - About size of US Public School education spending.
- $50 SCC is reducible through (a) smart ramping up of CO_2 tax; (b) smart delay (better tech).

(Warning: All above numbers are immensely huge.)

Crib Sheet — Summary of Facts

§6.3: Cost Concepts:

- Diminishing Returns;
- Sunk Costs;
- Learning Curves (FOAK);
- Returns to Scale;
- Optimal Delay.

Chapter 7: IAMs

See Tbl 7.5: $50/tCO$_2$ Tax

Product	Cost Change
Oil & Gasoline	+50%
Coal	+400%
Natgas	+100%
Tree	−$3/tree

§7.5: Key IAMs Issues:

- sensitivity to discount rate.
- estimating future parameters.
- uncertainty and risk.
- omitted choices: population, income inequality, opportunity costs of other philanthropic activities.

See Fig 7.2 and 7.3: Important Scenarios

			Nordhaus		RCP	
	Year	Base	Prefers	"2°C"	4.5	7.0
CO$_2$ Tax	2020	$0	$45	$60		
	2050e	$0	$110	$150		
	2100e	$0	$300	$500		
Welfare	2020	0	−0.15%	−0.14%		
	2050e	0	−0.23%	−0.53%		
	2100e	0	+0.42%	−0.71%		
Emissions (CO$_2$)	2020	39 Gt	33 Gt	32 Gt	39 Gt	39 Gt
	2050e	60 Gt	40 Gt	34 Gt	45 Gt	60 Gt
	2100e	71 Gt	16 Gt	−10 Gt	10 Gt	80 Gt
Temp	pre-ind	.... ≈ −0.45°C				
	1980	.... 0.0°C				
	2020	.... 1.0°C				
	2050e	2.1°C	2.0°C	2.0°C	1.5°C	1.7°C
	2100e	4.1°C	3.5°C	3.3°C	2.5°C	3.6°C

Chapter 8: The Wrong Question

Irrelevant:

- Problem is understanding choices by decision-makers.
- World outcome is *not* the engineered solution to a world problem.
- OECD countries are no longer big enough to solve the problem.
- Non-OECD countries are too poor to fight it.

Chapter 9: Fantasy

Key Problems:

- §9.2: A global (SCC) carbon tax is impossible without a global government.
- §9.3: Treaties not in self-interest. Excludability and free-riding incentives. No similar treaty ever effective.
- §6: Carbon footprints have been known for decades. (Carbon-shaming or setting an example?) What will change?
- What will change? Need to convince 8–11 billion people, not just 25% of the (more climate-conscious) population in the 25% that the OECD represents.

Chapter 10: Reality

Best *Viable* Choices:

- §10.1: Adaptation.
- §10.2: Locally justifiable fossil-fuel taxes (PM Health costs: $10/tCO_2$ to $100/tCO_2$).
- §10.3: Clean Technology.
- §14.2: Reforestation with lumber harvesting.

Chapter 11: Fossil Fuels Vs.

§11.2: Fossil Fuels:

- Achilles Heel: High mining and transport costs;
- 75% of fossil-fuel primary energy ends up as waste heat.
- Primary energy vs. Nameplate Power.
- PM Health costs: $10/tCO_2$ to $100/tCO_2$.

Fossil Fuel Alternatives:

- §11.3: **Hydrogen**: similar to NatGas, but likely far too expensive for many decades.
- §11.4: **Nuclear Power**: • 1 Meltdown / 3,704 reactor years; • 500 (old) nuclear power plants worldwide; • waste disposal solution; • need safer reactors (pebble-bed?).
- §11.5: **(Li) Batteries** • <1/10 energy density of fossil fuels, but reusable; • High power, Low capacity; • Almost perfectly in/out-efficient; • Tiny capacity on grid ($\approx$ 10 min total); • Expensive.

Crib Sheet — Summary of Facts

§11.7: Propaganda Clarifications:

- Most clean-tech in lab will fail (true), but there are dozens of exciting techs in lab.
- All numbers are immensely large — think 1/10 of all agriculture.
- Space and materials needed for clean tech, but plenty are available *long-run*.
- Clean-tech enjoys some subsidies, though small compared to fossil fuels.
- Expect bumps on the road.

Chapter 12: Electricity

§12.3: Fundamentals:

- High-quality energy. Jack of all trades. High conversion efficiency to kinetic energy.
- Typical daily electricity demand pattern today: Low at noon; Peaks at 7am and 8pm.
- Typical clean-energy supply: High at noon, low at 7am and 8pm.

See Tbl 12.5: LCOE per MWh

	2020	2050e
Solar	$35	$15
Wind	$35	$20
Nuclear	$70	$60
Natgas, 24/7	$40	$45
Natgas, Peaker	$200	$200
Coal	$75	$65
Hydro	$55	

Costs are in 2020-$ and representative utility-scale but vary by location.

See Tbl 12.8: Storage Cost, 2030e

	Cost per MWh
Batteries	$120 or $200-$250
Natgas Peaker	$100-$200
Pumped Hydro	$130
Compressed Air	$100

See Fig 12.9: Needed Grid E-Storage

% Clean Elec	Needed Hours
50%	1 hour
80%	10 hours
90%	100 hours
100%	1,000 hours
Currently	minutes

See Tbl 12.6: Coal Plant Status 2022, in GW

	Oprtg	Cnstrct	Prmt	Anncd
OECD	501.0	16.0	5.0	3.9
USA	232.8	-	-	-
Europe	117.8	12.2	-	-
China	1,046.9	96.7	43.0	72.1
India	233.1	34.4	11.7	11.7
All others	≈280	≈37	≈20	≈24
World	2,067.7	184.5	78.9	111.8

See Tbl 12.11: E-Generation in TWh				
Region	Year	USA	China	World
Coal	2020	774	4,313	8,244
	2050e	593	3,556	8,115
NatGas	2020	1,636	267	6,458
	2050e	1,953	803	7,306
Nuclear	2020	785	331	2,630
	2050e	594	1,002	3,025
Hydro	2020	283	1,117	4,034
	2050e	294	1,448	5,548

See left				
Region	Year	USA	China	World
Wind	2020	343	574	1,741
	2050e	790	1,001	6,833
Solar	2020	132	281	832
	2050e	1,072	3,379	10,152
Total	2020	4,061	6,893	24,991
	2050e	5,458	11,230	41,953

These are secondary energy estimates, EIA base scenario.

§12.7: Transmission:

- About $2 million per GW per mile.
- Cheap now only because generation is near use. Will become more expensive as generation has to be farther away.
- Giant regulatory mess.

Chapter 13 Beyond Electricity

See Fig 13.3: Vehicle Use Efficiency			
	Fuel×	Mvng	≈Total
Battery	95%	75%	70%
Hydrogen Fuel Cell	50%	40%	20%
Hydrogen Combustion	45%	30%	13%

Chapter 14: Remediation

Key Points:

- §14.1: Removal cost is one upper ceiling to the Social Cost of Carbon Dioxide.
- §14.2: Reforestation with lumber harvesting is cheapest method, perhaps as low as $10/tCO_2$ for first marginal GtCO$_2$ (that world is not taking).
- Industrial CO_2 removal projects seem hopelessly expensive for decades to come. Economics work only to arbitrage government subsidies.

- §14.3: Solar radiation management is worth investigating, but not (yet) deploying. Danger of unintended consequences.

Chapter 15: Transition

Favorites:

- Increase innovation.
- Share technology better.
- Tax fossil fuels for local health.
- Forestation.
- Price by supply cost (time).
- Uproot bad habits / nudges.
- Reverse tech lock-in.
- Coordinate transition.
- Reduce green red tape.
- Targeted Federal land leases.
- Kill worst emitters.
- Minor international agreements.

Made in United States
Orlando, FL
17 November 2022